Climate

Into the 21st Century

Australian Bureau of Meteorology

Meteorological Service of Canada

Météo-France

Deutscher Wetterdienst

Japan Meteorological Agency

Norwegian Meteorological Institute

Swedish Meteorological and
Hydrological Institute

Federal Office of Meteorology
and Climatology

Met Office, UK

U.S. National Oceanic and
Atmospheric Administration

Climate
Into the 21st Century

Edited by
William Burroughs

PUBLISHED BY THE PRESS SYNDICATE OF THE UNIVERSITY OF CAMBRIDGE
The Pitt Building, Trumpington Street, Cambridge, United Kingdom

CAMBRIDGE UNIVERSITY PRESS
The Edinburgh Building, Cambridge CB2 2RU, UK
40 West 20th Street, New York, NY 10011-4211, USA
477 Williamstown Road, Port Melbourne, VIC 3207, Australia
Ruiz de Alarcón 13, 28014 Madrid, Spain
Dock House, The Waterfront, Cape Town 8001, South Africa

http://www.cambridge.org

First published 2003

Printed in the United Kingdom at the University Press, Cambridge

A catalogue record for this book is available from the British Library

ISBN 0 521 79202 9 hardback

Contents

Foreword

G.O.P. Obasi
Secretary-General
World Meteorological Organization

Climate is one of humankind's greatest natural resources. Life itself and the existence of human beings on planet Earth depend on a favourable climate regime. People are very much affected by climate factors, including in terms of availability of food and water as well as choice of shelter and clothing. It plays an important role in our culture, health, leisure and general well-being.

More recently, we have also become more aware of the increasing interaction between society and climate, particularly the human effect on the changing climate. This interaction is among the significant considerations pertinent to the quest of nations for sustainable development.

In this regard, there has been increasing interest to have a better appreciation of what climate is, what constitutes the climate system, what progress has been made in this area and what lies ahead. It is in this light that this book, *Climate: Into the 21st Century*, was prepared. This project was undertaken by the World Meteorological Organization (WMO) in line with its long history of interest and involvement in climate issues. This goes back to its predecessor institution, the International Meteorological Organization (IMO), which in 1929 already had a technical commission for climatology. Since its establishment in 1950, WMO has been instrumental in drawing attention to climate-related issues, particularly through its World Climate Programme.

This book highlights the enormous progress in our understanding of the climate system and also seeks to promote a better community awareness of the importance of climate in our lives. It may be recalled that substantial loss of life and property destruction in natural disasters relate to dangerous extremes of climate variability. With the recognition that the climate system is changing as a result of human activities, there is also the realization that the impacts on society may prove increasingly significant and challenging. Hence, knowledge about climate needs to be applied in an enhanced way for a safer and more productive world, for the benefit of humankind.

This book provides a perspective of the global climate system across the twentieth century; it identifies some of the major climate events and their impacts on societies. In addition, it traces the development of our capabilities to observe and monitor the climate system and outlines our understanding of the predictability of climate on timescales of months and longer. The book culminates with insights into issues for the twenty-first century, drawing heavily on the work of the WMO co-sponsored Intergovernmental Panel on Climate Change (IPCC) and its recently released Third Assessment Report.

This publication provides a readable and selective account of some of the important climatic processes, examples of the enormous impact of some climate events and the challenges that climate poses at the start of the twenty-first century. It is written for a wide audience and covers a vast spectrum of interests in climate and climate events.

The book also demonstrates the role of, and developments in, international cooperation in meteorology, especially in data collection and exchange, in research as well as in applications. The National Meteorological and Hydrological Services (NMHSs) of each country have, for a century and more, carried out the important task of observing and recording the atmospheric parameters and its associated weather. These observations were initially for immediate applications in weather forecasting and warning; now, they have resulted in an accumulated climate record, forming the backbone of our understanding of the climate system and its variability. As our understanding of the sensitivity of the climate system to human interference has grown, it is even more important that the climate observing networks are retained, and even enhanced, and that international cooperation is strengthened to better meet the challenges of climate variability and change.

In this connection, it is fitting to acknowledge the contributions of millions of dedicated professionals and volunteers who have in the past, and will continue in the twenty-first century, to record their observations of daily weather that help to build a full picture of the fascinating climate of this planet.

I take this occasion to also thank the countries, groups and individuals who have contributed to the realization of this work.

The comprehensive global perspective of this book was made possible by the contributions from the NMHSs in all regions of the world. Financial and in-kind contributions were received from several countries. Overall guidance on the structure and content has been through a Task Team of the WMO Commission for Climatology led by Ms Mary Voice of Australia. In addition, the Task Team has had the assistance of lead authors for the different sections in the formulation and review of the content. Furthermore, several eminent climatologists provided reviews of the book at the draft layout stage, and these provided valuable guidance for its further development. This has been an undertaking fully supported by staff from the WMO Secretariat.

WMO is particularly grateful to Dr William J. Burroughs for his dedication to the challenge of compiling, selecting and editing the considerable body of scientific information, and structuring it with the accompanying graphics. His style and balance are reflected in the presentation.

It is my hope that the readers of this book will find in it a wealth of information and knowledge that will lead to a better appreciation of climate, the role it plays in our lives and the need to protect and preserve it for future generations.

(G.O.P. Obasi)
Secretary-General

Preface

Task Team Members

Mary Voice
(Chairperson)

Phil Jones,
Chet Ropelewski,
Brad Garanganga,
David Phillips

The structures of everyday life are fashioned to a surprising extent by the local climate. The climate, its seasonal patterns and the chances of extremes influence what we wear, our housing and transport, sport and recreation, what we eat and how we work. Weather and climate can even influence our emotions and moods. When we observe a clear blue morning sky, feel the gentle warmth of a rising sun, a zephyr of breeze with perhaps the scent of flowers wafting on the air – we can't help but feel a sense of peace and tranquillity. In contrast, an approaching storm with a rising wind, a swirling of dust, sand or snow and a leaden sky all evoke very different emotions. More crucially, weather and climate have the potential to enhance community wealth creation and sustainable development on a global scale, or conversely to cause severe disruption to the rhythm of life and even its destruction. It is essential then for us to understand and learn how to deal with the wide-ranging impacts that our Earth's climate is capable of delivering.

Clearly, the vigorous greenhouse debate of the past two decades about the possible extent and likely consequences of global warming indicates that in spite of modern technology and know-how, we still have a long way to go in climate-proofing our societies and ourselves. Even more significantly, we are being told that it is our own actions which are introducing new uncertainties into how the global climate machine is working. At the same time, a fear of future climate change from the greenhouse effect or other causes, accompanied by a lack of knowledge and understanding of the climate variations that our parents and grandparents have lived through, is not the best foundation for good community planning.

Thus, we on the Task Team believed there was an important story to tell. In putting this story together, we remain convinced that to cope better with the climate of the future a greater awareness is needed of what has happened to our climate during the 20th century and even further back through time. Armed with this knowledge and understanding, communities that already know how to manage season-to-season and year-to-year climate variability will more easily learn how to cope with climate change.

From this background grew the seed of the idea for this book. The seed germinated during discussions among colleagues of the Commission for Climatology of the World Meteorological Organization (WMO). It was the international perspective of these discussions and the need for international cooperation that brought a proposal to produce this book to the attention of the Twelfth Congress of WMO. Congress enthusiastically endorsed the plan and an international Task Team was formed to guide the project through to completion.

The challenge was to cover a century of climate with both geographic and temporal balance, recognizing at the same time that even one hundred years of climate is but a snapshot of what has been experienced by the planet over thousands and millions of years. During the 20th century the population of the planet approximately quadrupled with much of the expansion being increasingly concentrated in the larger cities of the world. These facts alone distort our perspective on what have been the most significant weather and climate events of the past one hundred years or so. They also help to explain why an extreme weather event may have a greater impact today than an event of similar severity occurring decades ago. Another factor distorting our view of the past is that there is progressively less information available about climate events as we go back in time. Other factors inhibiting a comprehensive coverage include the varying degree of record keeping that often depended on local needs, as well as historically limited communications between peoples speaking different languages. We are therefore especially grateful to those contributors who have supplied materials translated into the working language of the project. As a result of these and many other factors, there was the risk that a book of this nature could give a deceptive impression of trends in the occurrence and severity of extreme events. Cognizant of this challenge, the Task Team has done its best to delve into those archives that extend back to the beginnings of meteorology as a modern science along with its sister disciplines of hydrology and oceanography, and to early studies of weather and climate phenomena. In doing so, we hope that we can bring to the reader a clearer view of how the Earth's climate has varied in the past and a better perspective on the extreme events that continue to dominate our perceptions of the turbulent natural environment in which we live. Notwithstanding our best efforts, it is inevitable that the distant past and some geographic areas are under-represented within this book.

The book has five sections beginning with a broad accessible synopsis of what has been happening in the climate system over a century and more. Section 2 on the climate system includes a survey of a number of major weather and climate events of the past interspersed with somewhat more challenging topics on the sciences that underpin our knowledge of climate. Section 3 describes the many influences and impacts of a varying climate and is followed by a section analysing in more detail how they affect the Earth and its peoples. The final two sections review the contribution of science and technology to understanding the climate of the 20th century, and the prospects for applying that knowledge and the accompanying experiences for the benefit of society in the 21st century. Each section comprises a number of two-page spreads that cover particular topics relevant to the section, and it is the intention that each spread or in some cases a group of spreads is complete in itself, without the need to cross-reference to other parts of the book. Nonetheless, since the Earth is for the most part a closed system, with the energy coming in from the Sun balanced by that radiated back to space, everything is ultimately connected. Hence we see topics such as El Niño, monsoons, floods, droughts and so on cropping up in many contexts. We trust that the reader will make good use of the index in order to gain a fuller appreciation of the complexities behind both the causes and the effects of these weather and climatic phenomena.

The Task Team would like to thank the many contributors and reviewers who have helped to shape this book. We especially acknowledge the assistance and guidance of the section leaders who formulated the intent of each section and scrutinized the integrity of each two-page spread as it emerged from the efforts of many expert views and opinions: Neville Nicholls, Section 1; Ann Henderson-Sellers, Section 2; Michael Glantz, Section 3; Reid Basher and Hiroki Kondo, Section 4; and Richard Moss and Stephen Schneider, Section 5.

We, the people who made this project happen, have had one eye on surveying the past century and one eye on the future in offering you this book. It contains, we believe, a collective and faithful perspective of the Member countries of the World Meteorological Organization on many important climate issues.

1.1 A century of discovery

Climate affects our lives in many ways. We are comfortable with the regularity of the seasons but dangerous events are profoundly worrying. Here we explore the huge advancement of climate-related science and technology, together with the expansion of information and capability for planning and early warning, during the 20th century. Now, in the 21st century, new climate challenges confront us.

The troposphere is the relatively thin and dense layer of the atmosphere extending 10 to 14 km above the Earth's surface and within which weather systems are found. Naturally occurring ozone is generated in the overlying stratosphere and protects life from harmful UV radiation. The air density decreases rapidly with altitude through the troposphere and stratosphere.

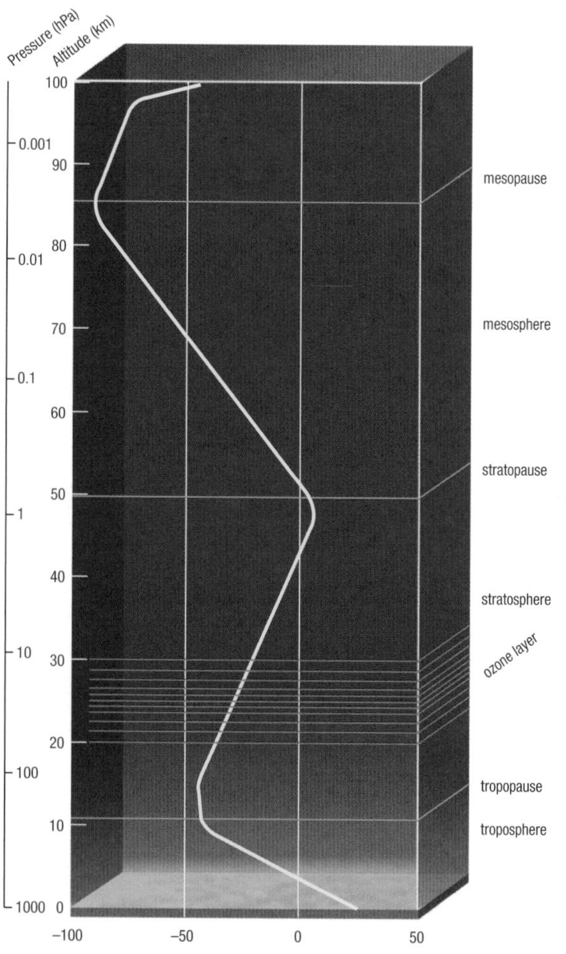

Scientists were already discovering and unravelling the complexities of the atmosphere and its motions as the 20th century was dawning. Systematic meteorological observations in many places, particularly the scientific observatories of Europe and the Americas, already stretched back more than 100 years. These records provided information about local climates and enabled some comparison of climates between different places.

Many of the fundamental physical and chemical properties of the atmosphere were also known. How certain gases, such as water vapour and carbon dioxide, made the atmosphere like a 'greenhouse' was recognized, as was the fact that the greenhouse effect prevented the Earth from being a frozen wasteland. The presence of ozone in the high atmosphere had also been identified. Then, the discovery of the stratosphere in 1902 led to understanding of the formation of the ozone layer, and an appreciation of how it shields us from harmful ultraviolet (UV) radiation and hence is vital for life on Earth.

Climate and society

The impacts of the agricultural revolution and the industrial revolution were in turn transforming society. The seeds were also being sown for a profound impact on climate. Farming practices had changed landscapes. Mechanization was to further clear native forests for extensive agriculture and livestock grazing around the world. The delicate balance of heat and moisture exchange between the land surface and atmosphere was changed forever as forest clearing altered how much sunlight was reflected back to space, increased rainfall runoff and reduced the amount of moisture returning to the atmosphere.

Coal and wood were the main fuel sources at the beginning of the 20th century. As a consequence, smoke was choking the growing urban environments and within the expanding cities the air quality further deteriorated. It also became evident that transport of pollutants was harming life far from the original sources. Although regulation and changing technologies reduced much of the local smoke pollution, motor vehicles brought a new threat in the form of photochemical smog. The expanded use of fossil fuels also built up the concentration of carbon dioxide in the atmosphere, contributing to global warming.

Persisting anomalies of climate, such as drought, and the impacts of extreme weather events have long been the cause of many deaths and the destruction of property. Droughts in India, China and Africa during the latter part of the 19th century contributed to famine and disease, and millions of people died as a consequence. They were also a powerful stimulus for meteorologists to examine the links in the climate throughout the tropics. Better understanding of the climate system, the nature of climate variability and its extremes, and early warning of dangerous weather and climate events now provide tools for improved planning and response.

Benefits from research

In the early part of the 20th century major advances were taking place in understanding the large-scale motions of the atmosphere. Thanks principally to Vilhelm Bjerknes and his colleagues in the 'Bergen School' in Norway, models of air masses and the structure and behaviour of cyclones and anticyclones were developed. The synoptic climatology on which the models were built became a basis for weather forecasting in middle and high latitudes that persisted for nearly three-quarters of a century, until the development of computers and numerical weather prediction.

At the same time the first inklings of larger-scale patterns in the weather systems of the tropics were beginning to emerge from studies into causes of the fluctuating strength of the Indian monsoon from year to year. It would not be until later in the century that the true importance of these insights would be realized. The so-called El Niño events of 1972–73 and 1982–83 were to confirm fluctuations in the ocean circulation, especially those of the tropics, as being a major source of global climate variability on seasonal to interannual timescales, and to stimulate new directions of climate research.

With the development of air transport, knowledge about the behaviour of the upper atmosphere became an essential requirement for safe and efficient operations. Scientists, such as Carl-Gustaf Rossby, were drawing on data provided by regular balloon soundings of the atmosphere to provide a physical explanation for the large-scale horizontal wave patterns of airflow in the upper atmosphere and their importance for the development and decay of surface weather systems.

Technological advances

The development of computers has transformed our ability to handle massive amounts of climate information that are now regularly generated and held in archives. Computers carry out vast numbers of calculations at high speed for research and applications. The computer models of the atmosphere, first developed for weather forecasting, have been further developed and coupled with models of the oceans, the terrestrial biosphere and the polar ice to generate future climate scenarios.

In parallel, new technologies have emerged to view the climate system and our environment. The advent of the Space Age, with orbiting and geostationary satellites, has provided climatologists with eyes in the sky. These continually view the Earth's weather and monitor changing vegetation patterns, anomalies of the oceans' circulation and a range of other physical and chemical measures linked to climate variability.

The clearing of the Earth's forested areas to make way for agriculture and settlements over time has affected the hydrological cycle and the Earth's radiation balance.

Sustainable development

Our developing appreciation during the 20th century of the fragility of the environment in which we live, including the sensitivity of the climate system to human impact, has thrown the political spotlight on a range of climate-related issues. The greenhouse effect and global warming, the destruction of the ozone layer, vulnerability to natural hazards, protection of land and water resources, and renewable energy are a few of the areas where climate will be increasingly important in the 21st century.

The themes throughout this book highlight the science, technology and applications of climate that are important for sustainable use of the Earth's resources and the protection of the climate system.

Antarctica – charting the coldest climates

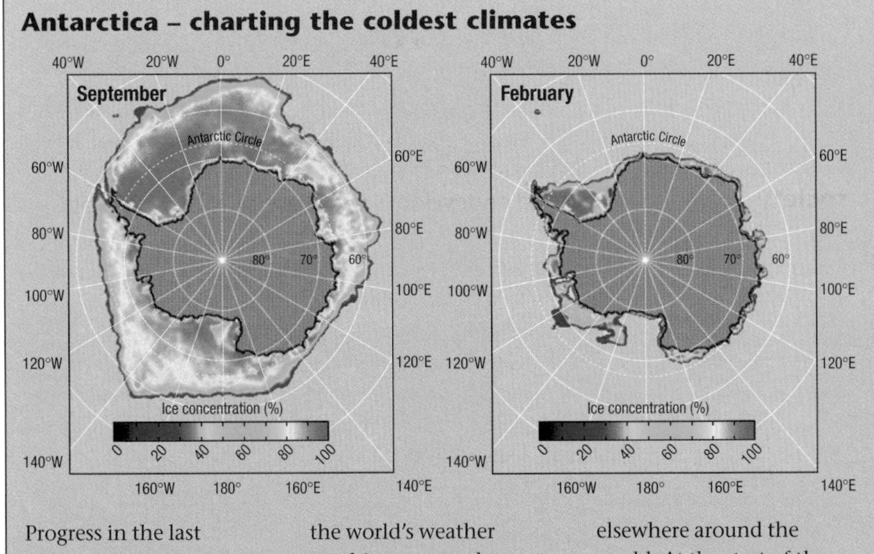

Progress in the last 100 years in understanding the role played by Antarctica in the world's weather machine encapsulates many of the advances in meteorology seen elsewhere around the world. At the start of the 20th century, intrepid Antarctic explorers were discovering the harshest climates on the planet. By the end of the century the Antarctic ozone hole loomed as the clearest example of the impact of human activities on the global climate.

Early studies discovered that Antarctica experienced the lowest temperatures on the surface of the Earth (the all-time low observed at Vostok, a Russian research station, in 1983 was measured at –89.24°C; at the limits of instrumental capability). Around the edge of the continent bitterly cold winds (known as katabatic winds) drained off the ice caps, at speeds over 150 km/h. Since this early work scientists have discovered many of the reasons for these extreme conditions. They have come to understand why the ice pack reaches its maximum expanse in September and minimum in February, and how measurements of the glaciers and ice domes can tell us something about the history of the Earth's climate.

1.2 What is climate?

Climate is so central to every aspect of our lives that we give little thought to what precisely it is. To appreciate fully all the reasons why the climate affects so many features of our existence, we need to define what we mean by climate.

The word climate came from the Greek word *klima*, which means inclination and refers to the angle that the Sun's rays are inclined above the horizon. The ancient Greeks knew that the climate was cooler if the inclination of the Sun was low. Thus climate is hotter close to the equator and colder at higher latitudes, but we now know that what determines the climate of a location is far more complicated. Factors such as the distance from the sea, altitude and overall general circulation of the atmosphere are also important in determining local climates.

Weather or climate?

Mark Twain observed that "Climate lasts all the time weather only for a few days", while Robert Heinlein put it as "Climate is what you expect, weather is what you get". Both encapsulate the essence of the difference between weather and climate. At the simplest level the weather is what is happening to the atmosphere at any given time. Climate is a measure of what to expect in any month, season or year and is arrived at using statistics built up from observations over many years.

The statistics often concentrate on averages, but it is the extreme events, including lengthy spells of abnormal weather (e.g. droughts, floods, heatwaves and severe winters), that generate the most interest. While dangerous, and often deadly, these rare extremes are still part of normal climate for many parts of the world. We will therefore make a particular effort in this book to explain how our perceptions of the climate are dominated by experiencing occurrences of the extremes.

There is another unclear boundary to explore: the difference between climate variability and climate change. It follows from the definitions of weather and climate that the concept of climate change is about shifts in meteorological conditions lasting many years. These changes may involve a single measure or parameter, such as temperature or rainfall, but usually accompany more general alterations in weather patterns to, for example, warmer, drier and sunnier conditions over sustained periods that result from compensating shifts in weather patterns around the world. Generally, changes are linked to an overall warming or cooling of the globe.

The distinction between climate variability and climate change inevitably has a degree of arbitrariness and depends on the time frame considered. As is shown in the box opposite, it is possible to spell out these differences. Detecting climate fluctuations of the many forms described poses major challenges, especially when considering the relatively small changes that have occurred in recent history. Nevertheless, the important point is that when we talk about climate variability and climate change we must be clear that we are dealing with two different concepts. This will become particularly obvious in grappling with the question of whether the incidence of

The climate system is the sum total of the processes and interactions of the Earth's atmosphere, oceans, land surfaces, ice sheets, and its flora and fauna that are driven by the incoming radiation from the Sun, which in turn is balanced by heat radiated back to space.

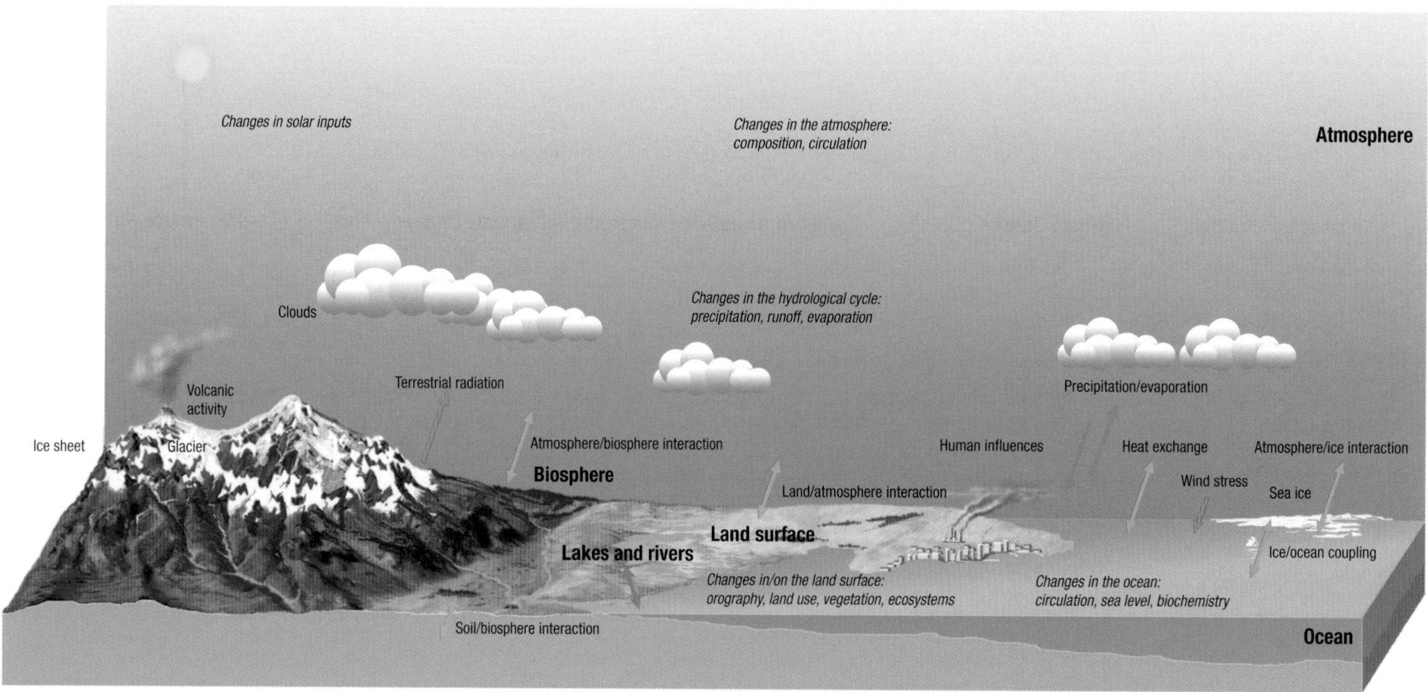

Weather is manifest in many ways, few more dramatic than the lightning discharges associated with violent convection, as shown in this 1902 photo.

extremes is changing and just what this might mean for the future.

Another challenge in the consideration of climate change and variability is the matching of different records. Data from ice cores, pollen trapped in deep layers of peat, tree-ring widths and tiny creatures in the ooze at the bottom of the oceans can tell us a lot about climatic conditions long ago. Interpreting this information is rarely easy, though. While no one can fail to see the broad features of the last Ice Age in the records, the effects of sunspots, for example, lurk deep in the statistics, and are surrounded by controversy. It is this mixture of the well established and the unknown that makes the subject so hard to pin down, and the reason it has continued to fascinate climatologists over the years.

The annual cycle of weather determines the local climate and natural ecosystems. Some climates limit the potential for productive land management.

Climate variability and climate change

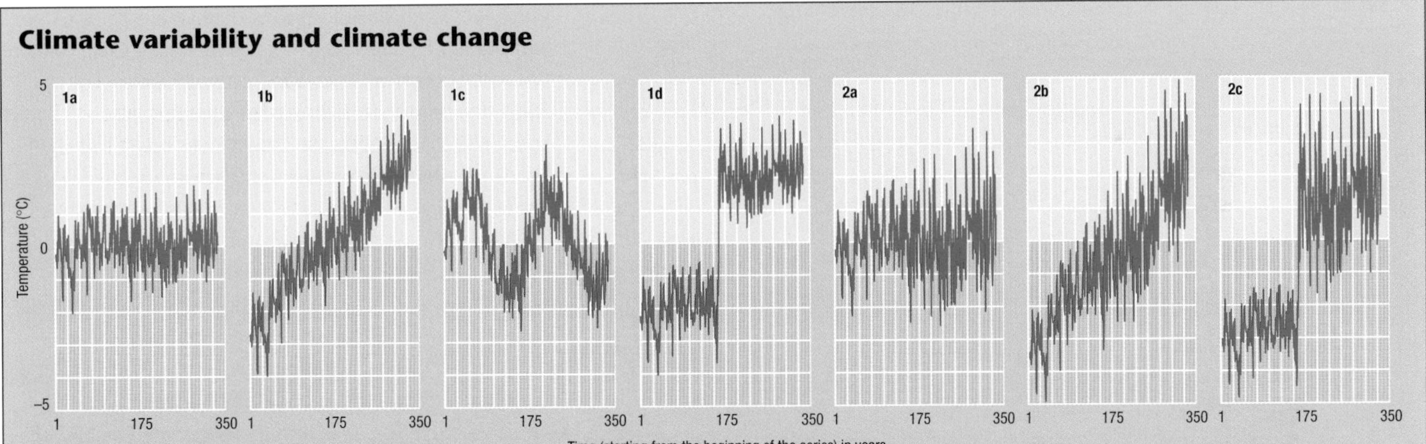

Any set of meteorological observations can be represented as a time series (see 1a). If, over the full period of the measurements the average values for shorter periods are effectively the same, the series is said to be stationary. The parameter in question may, however, fluctuate considerably from observation to observation: every winter in Argentina is not exactly the same; the onset date of the monsoon in India is different each year. This fluctuation about the mean is a measure of climate variability. In time series 1b, 1c and 1d the same example of climatic variability is combined with examples of climatic change. The combination of variability and a uniformly increasing trend is shown in 1b. In curve 1c the variability is combined with a periodic change in the underlying climate, and in curve 1d the variability is combined with a sudden rise, which represents, during the period of observation, a fundamental change to the climate system.

The implication of the forms of change shown in the above set of time series is that the level of variability remains constant while the climate changes. This need not be the case, and time series 2a–c present the implications of variability changing as well. Series 2a presents the combination of the amplitude of variability doubling over the period of observation, while the average climate remains constant. Although this is not a likely scenario, the possibility of the variability increasing as the climate warms (as in series 2b) might be more likely. Similarly, the marked increase in variability following a sudden rise in temperature (series 2c) could be a consequence of climate change. These concepts of climate variability and climate change will be explored in this book.

1.3 The challenge of measuring global climate

Climate data and information are increasingly used for making decisions that benefit us. Improved methodologies for agriculture, managing water resources, eradication of pests and diseases and building safer communities are some of the applications. Evidence of global warming has extended our interest in climate. In the past, climate data were distinctly atmospheric, but today data from the entire earth system, including the land, air, sea and ice, are needed.

Greater than 50% coverage for temperature observations, 1891 through 1910

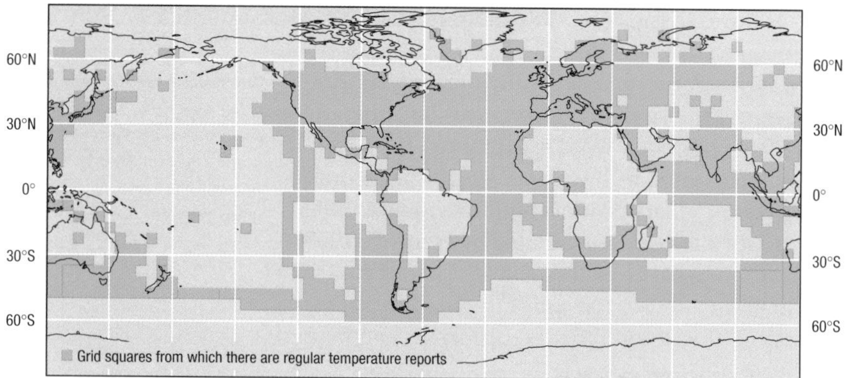

Greater than 50% coverage for temperature observations,1970 through 1999

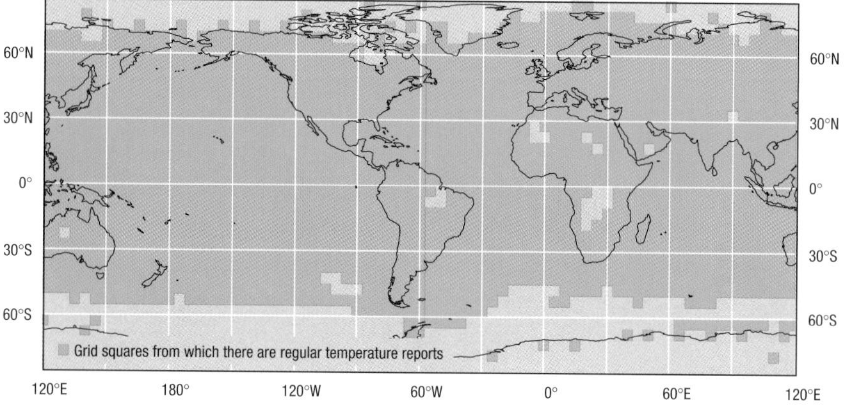

At the beginning of the 20th century there were large gaps in the global network for regularly observing and reporting surface temperature, but by the end of the century these were mostly filled, except for the inhospitable polar regions.

How human activities are dominated by climate has driven our search for better knowledge of how it functions. It started with how daily weather events affected agriculture and has extended now to how wider features of the global climate can influence commerce and trade around the world. For much of recorded history this learning was encapsulated in folklore and indigenous knowledge systems. Sayings of the "red-sky-at-night ..." variety and observed links between animal behaviour and seasonal weather all helped our forebears to handle the uncertainty of weather.

A century ago, directors of observatories and meteorological offices wrote letters to one another that were transported by train or ship, reporting on droughts, floods and emerging famines. They had no way of detecting early signs of extreme weather patterns in the global climate, because they had no way of rapidly building up a picture of what was happening. That could only be done retrospectively. A century on, global patterns, including phenomena like El Niño and La Niña that last seasons and more, are mapped day by day, as they emerge. The 20th century was the time when global climate observing, using an array of tools including land networks, ships, aircraft and satellites, came of age.

Early weather and climate networks

The basis for modern weather and climate networks was laid during the later part of the 19th century as governments realized that mapping weather systems could provide valuable public information, especially for rural decisions and safety of shipping. Government officials and volunteers in many countries around the world took on duties as weather observers. In many cases the weather observing was closely linked to the duties of the local telegraphist and coded observations were quickly sent to national headquarters and exchanged internationally. The observations, systematically carried out at fixed times of the day, were recorded in field books that were regularly forwarded to headquarters for the compilation of climate statistics. At the beginning of the 20th century there were quite extensive meteorological observing networks covering the lands of both hemispheres, although with significant gaps.

At sea the challenge of building up a complete climatic picture was more demanding. Nevertheless, from the middle of the 19th century in many parts of the world, arrangements were made to obtain systematic records from ships. The collection of these data and their meticulous analysis, which was much more complicated than

Measurement of basic meteorological elements, such as temperature and rainfall, has varied little over the century. Volunteer observers have made an important contribution to many meteorological services.

for the fixed land stations, provided guidance to mariners on a range of conditions at sea. Information gathered included air and sea surface temperature, air pressure, wind speed and direction, wave height, the extent of sea ice and visibility. All of these were used initially to assist navigation and improve safety, but have since proven invaluable for research into the climate system.

New technologies have made more observations possible and so the nature of the measurement systems has evolved over time. The organization of national weather services and the use of telegraphy enabled more data to be used quickly to build up a better picture of current weather patterns and recent climate. From the beginning of the 20th century wireless telegraphy was being exploited to provide real-time information of weather at sea. Analysis of the data from land and sea enabled the identification of storms and the provision of warnings of dangers to come.

It was not until the 1950s and 1960s, however, that some national weather services started to record these data in computer-compatible forms. Now, with the looming threat of global warming, the world wants to know more about how the climate varies and what exactly is changing. Early records are vital in helping to answer such questions. The huge amounts of data recorded earlier by hand therefore had to be digitized before they could be analysed by computers. This task is still continuing. A recent example of this work of data recovery is the digitization of millions of Japanese observations of sea surface temperatures collected by the Kobe Observatory in Japan.

The demands of aviation

The advent of aviation transformed the demand for measurements. During the 1920s aircraft were used to measure conditions in the lower atmosphere.

Safety for the developing aviation industry was a major impetus to increase meteorological observations above the ground. This bi-plane (1934) was equipped to measure weather elements.

Research ships such as the Canadian Coast Guard ship *Quadra* have been important for exploring the currents and structure of the oceans.

Then, during World War II, as aircraft were flying higher, pilots discovered exceptionally strong winds at 6 to 7 km altitudes. Sometimes, strong tailwinds blew them far from their targets without their being aware, or on other occasions, flights would be into the teeth of headwinds up to 300 km/h, perhaps ending tragically. These first encounters with the jet stream and the realization that it was linked with large-scale atmospheric circulation patterns controlling many features of the weather, was an eye-opener to meteorologists. Its implications for safe air travel created an urgent demand for routine observations of conditions aloft. This accelerated the development of new technologies to measure the upper atmosphere, and regular balloon-borne measurements (radiosondes) became a part of meteorology. Although initially collected to support aviation safety, these observations also became a valuable resource for studying the climate.

Weather satellites

The next great step in observing the world's weather came with the launch of the first weather satellite in April 1960. Within a few years orbiting satellites were providing daily coverage of global weather patterns, and geostationary satellites, rotating synchronously with the Earth over the equator, were providing full-disk pictures every 30 minutes. Now satellites measure a whole host of different features of the climate system, including the temperature throughout the depth of the atmosphere and of the surface of the land and the oceans. Everything, from the extent of polar snow and ice, through the wind and wave conditions at sea, to the vegetation state from equator to poles, is routinely observed. Satellites even measure aspects of global pollution and variations in the ozone layer. It has, however, taken some effort to derive information from the satellite data that can be used to measure climate change, because of drifts in the performance of the on-board instruments and the slightly changing characteristics of successive instruments. Nevertheless, the advent of weather satellites has without doubt transformed our perspective of the climate of our planet.

1.4 Climate patterns

Climate is not just a matter of statistics; it affects many aspects of our lives depending where on the globe we live. It is therefore equally important to define what aspects of the climate matter to us, both in human terms and to all other forms of life.

Understanding how we adapt to the climate in any part of the world requires identification of what aspects of the climate matter. Equally, we need to appreciate how flora and fauna around the world reflect the climatic conditions where they can thrive in their many different forms. The way species adapt to their specific types of climate can be used to classify similar climate patterns. This basic information is vital to anyone involved in making decisions about the best way to come to terms with the climate in any part of the world. Furthermore, the fundamental knowledge of how species respond under different climatic conditions is also vital when we come to consider the implications of any future changes in the climate.

How the human race has learned to cope with almost the complete range of the Earth's climate is a measure of our adaptability. We have ventured into the hottest, most arid deserts, the most inhospitable, icy regions of the poles and atop the highest mountains. In different ways, many other forms of life have also evolved to be able to survive in the most extreme climatic environments. The extraordinary range of adaptations made by flora, fauna and humans, and the limits are central to our understanding of the control that climate can exert over life on this planet.

Even on a small scale there can be a wide range of climate conditions. On the Island of Réunion in the southern Indian Ocean the orography is particularly important because of its impact on temperature and rainfall distribution. Annual mean temperature (upper) tends to vary with elevation, while rainfall (lower) ranges from more than 6 m per year in the windward southeast, to less than 1 m per year to the lee of the interior mountains. These factors have been combined to define Réunion's basic climate zones.

Temperature
°C
21–29°
19–26°
16–24°
14–22°
12–20°
10–18°
8–15°
0 10 km

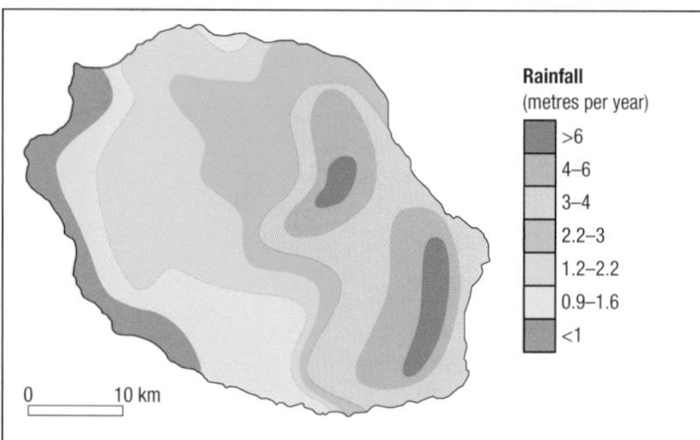

Humans have ventured into all the Earth's climates and have learned to survive independently virtually everywhere, such as in the cold of Siberia (top) and in monsoon-prone Viet Nam.

Climate and natural selection

Travels of exploration into the New World by European scientists in the early 19th century led to the recognition that different plants and animals are restricted to fairly narrow climate regimes. The German naturalist Alexander von Humboldt and his French botanist assistant Aimé Bonpland travelled extensively through South and Central America from 1799 to 1804 collecting samples and making notes of what they found. Von Humboldt then spent more than 50 years researching and writing extensively about his findings. A major conclusion, among many, was the link between climate and vegetation species.

The British naturalist, Charles Darwin, spent the years from 1831 to 1836 travelling on the *HMS Beagle* that was carrying out a scientific expedition around the world, but principally around the coast of South America and the Galapagos Islands of the eastern equatorial Pacific Ocean. Darwin also researched and published his findings, including his theory of natural selection, which holds that species adapt to their environment through the reproductive success of individuals with the most appropriate range of randomly variable characteristics. Adaptability to climate variability and change has therefore been fundamentally important for the sustainable development of all living species.

Climate classification

Later in the 19th century, attempts were made to classify global climates. The most often used classification scheme is that of Vladimir Köppen, first presented in the early 1900s, and revised frequently since. Köppen concluded that the five main kinds of plant formations must represent five principal climate classes. All the lesser formations such as the bushlands of the maquis and the chaparral represented subdivisions of one of the main climatic types. In *Why big fierce animals are rare?*, Paul Colinvaux wrote: "Any contemporary atlas will contain, next to the map of the world's vegetation, a map of the world's climate ... The two will match because they are the same map."

There are many ways of classifying the climate. Most atlases have maps of temperature and precipitation around the world, and some may contain maps of atmospheric pressure, prevailing winds, ocean currents and extent of sea ice throughout the year. Many countries need more detailed classifications for various reasons. For example, the average dates of the first and last killing frost are of value to farmers and growers, as is the average length of the frost-free growing season. In colder places, the number of days below freezing affects building design. The number of degree-days below or above some reference value (e.g. 18°C for heating and 22°C for cooling) provides a measure of the energy demand for heating, air conditioning or refrigeration in homes and offices.

Vladimir Köppen (1846–1940)

As director of the meteorological section at the Deutsche Seewarte (naval observatory) at Hamburg, Köppen had a crucial influence on developments in weather observing and forecasting in Germany and the rest of Europe. Meticulous archiving of daily weather reports from 1875 provided the foundations for climate studies that were to lead to the publishing of the Köppen Climate Classification. The five-part classification defined the boundaries between vegetation types primarily on temperature and precipitation into:

A: Tropical rain forest
B: Hot desert flora
C: Temperate deciduous forest
D: Boreal forest
E: Tundra.

Subdivisions of the primary classes are based firstly on precipitation and secondly on temperature variations.

Köppen's original map has been significantly modified and refined with the advantage of additional data over time. Between 1937 and 1980 Glenn Trewartha (1896–1984), Emeritus Professor of Geography at the University of Wisconsin, USA, published variations of the map below in five editions of his textbook *An Introduction to Weather and Climate.*

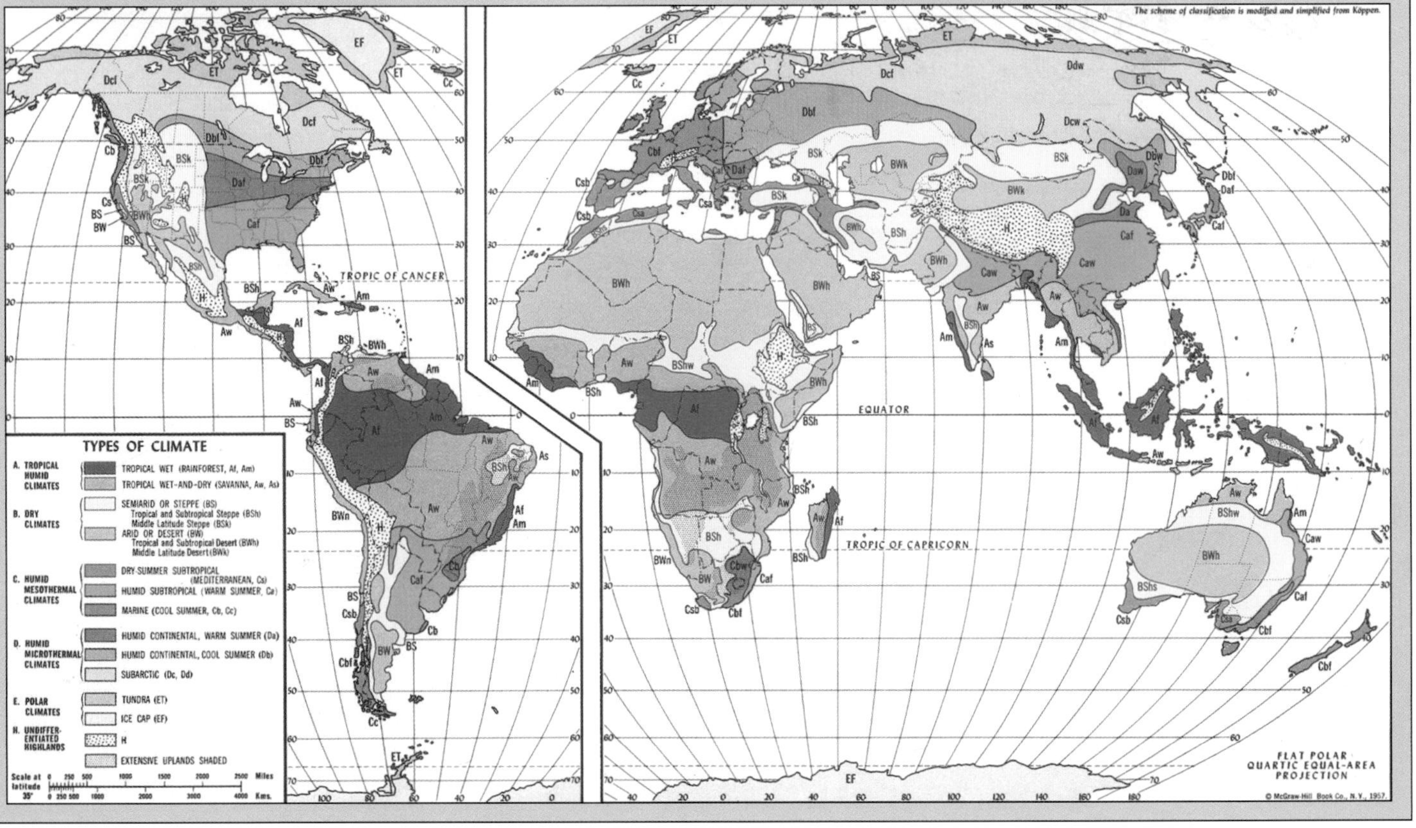

The scheme of classification is modified and simplified from Köppen.

TYPES OF CLIMATE

A. TROPICAL HUMID CLIMATES: TROPICAL WET (RAINFOREST, Af, Am); TROPICAL WET-AND-DRY (SAVANNA, Aw, As)
B. DRY CLIMATES: SEMIARID OR STEPPE (BS) – Tropical and Subtropical Steppe (BSh), Middle Latitude Steppe (BSk); ARID OR DESERT (BW) – Tropical and Subtropical Desert (BWh), Middle Latitude Desert (BWk)
C. HUMID MESOTHERMAL CLIMATES: DRY-SUMMER SUBTROPICAL (MEDITERRANEAN, Cs); HUMID SUBTROPICAL (WARM SUMMER, Ca); MARINE (COOL SUMMER, Cb, Cc)
D. HUMID MICROTHERMAL CLIMATES: HUMID CONTINENTAL, WARM SUMMER (Da); HUMID CONTINENTAL, COOL SUMMER (Db); SUBARCTIC (Dc, Dd)
E. POLAR CLIMATES: TUNDRA (ET); ICE CAP (EF)
H. UNDIFFERENTIATED HIGHLANDS: H
EXTENSIVE UPLANDS SHADED

Scale at latitude 35°: 0 250 500 1000 1500 2000 2500 Miles; 0 250 500 1000 2000 3000 4000 Kms.

TROPIC OF CANCER
EQUATOR
TROPIC OF CAPRICORN

FLAT POLAR QUARTIC EQUAL-AREA PROJECTION

© McGraw-Hill Book Co., N.Y., 1957

1.5 Variability and the character of extremes

Statistical analyses of events, be they sporting achievements, stock market prices or climate series, generally concentrate on the most outstanding aspects. Attaching real meaning to these exceptional events requires a disciplined approach to the statistics.

It is in our nature to want to see every extreme weather event as record-breaking. As Piglet observed in A.A. Milne's *Winnie-the-Pooh,* "It rained and it rained and it rained. Piglet told himself that never in all his life, and *he* was goodness knows *how* old – three, was it, or four? – never had he seen so much rain. Days and days and days". Measuring how extreme events impact upon both human activity and the world around us is a complex process. Our adaptability only goes so far. At some stage, conditions deviate too far from the normal. Then there is a dramatic breakdown and the fabric of human societies is destroyed: droughts kill crops and livestock; snowstorms paralyse transportation systems and damage energy supply systems; and floods wash away homes, kill people and drown agricultural land. Longer-term changes can alter the entire balance of ecological systems. To measure these impacts, we must address a few statistical questions from the outset, in order to take a consistent approach in this book.

Climate and statistics

With adequate measurements of the climatic conditions covering many years, it is possible to define what is considered normal and what is an extreme event for any part of the world. Data gathered over the 30-year period from 1961 to 1990 define the latest Normals used for climate reference. At any given time of the year, an extreme high temperature might be defined as one that occurs only once in every 30 years. A cold winter or hot summer can be specified in a similar way, or in terms of the number of days below or above defined exceptional values. This means that when there is a succession of extremes, or more extreme events over a period

How climate change affects us

For a temperature record that is stationary, the statistical parameters can be presented in what is often referred to as a 'bell curve'; the peak of the curve provides the mean and the width of the curve is a measure of the variance. If the climate undergoes a warming without any change in the variance then the whole bell curve moves sideways (left). The consequence of a shift to a higher mean is that there are fewer cold days and more hot days, and a higher probability that previous record high temperatures will be exceeded. If, however, there is an increase in variance but no change in the mean the bell curve becomes fatter and lower (middle). The consequence is that there are more cold and more hot days and a high probability that previous records for both the coldest and hottest days will be broken. If both the mean and the variance increase then the bell curve both shifts sideways and becomes lower and fatter (right). The effect is for relatively little change in the frequency of cold weather or the occurrence of extreme low temperatures, but a big increase in hot weather and previous record high temperatures being exceeded far more often.

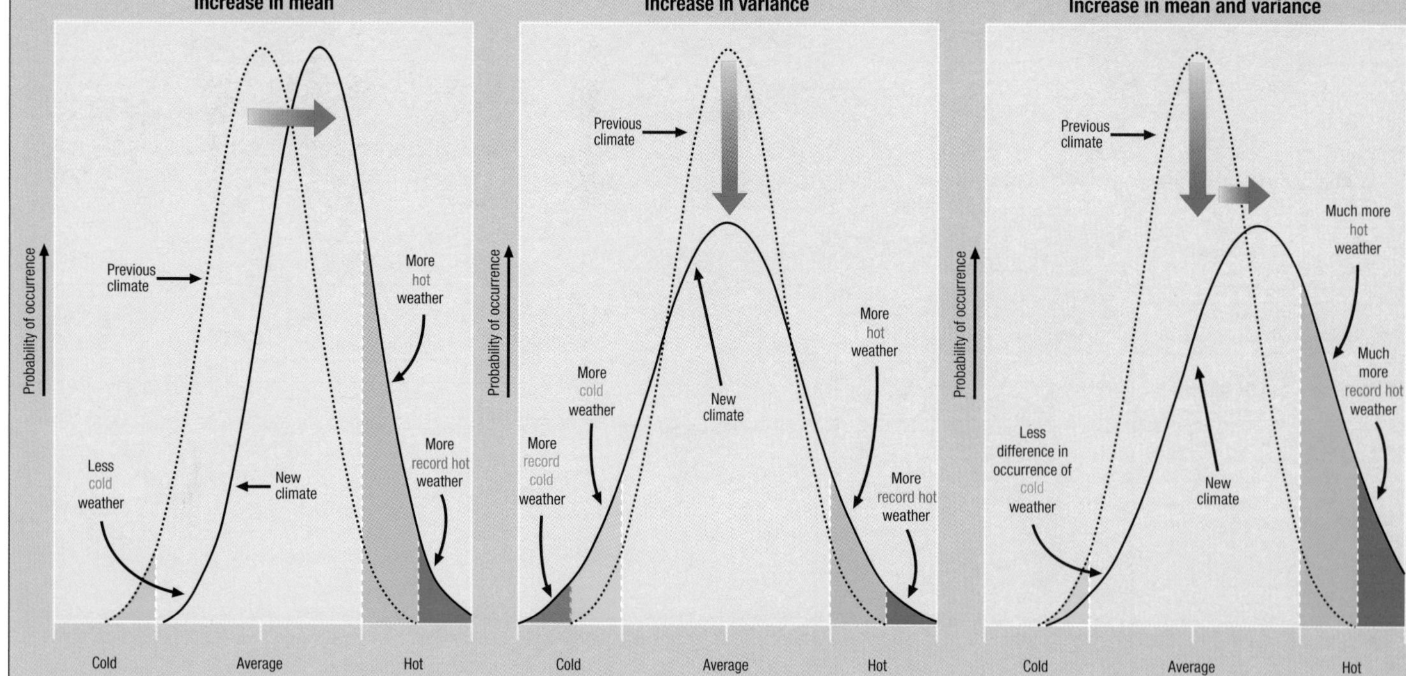

(e.g. season), it is possible to estimate whether they seem to be part of the normal expectation for the locality, or are so unlikely that they can only be explained in terms of some more radical shift in the climate. The basic properties of any data series, for example temperature, can be defined in terms of the mean over time and the amount of variance about the mean.

Other meteorological variables exhibit more complicated statistical properties. For instance, rainfall is episodic. In many parts of the world, much of the annual rainfall falls in a short rainy season. In addition, most of that rain may be concentrated in a few heavy falls, and small shifts in the large-scale weather patterns from year to year may significantly alter the amount and distribution of seasonal rainfall. More complex techniques usually are needed to interpret variations in rainfall.

Many historic observatories previously in garden settings, such as in Montreal, Canada, have become completely surrounded by buildings. To compensate for these changes more recent data have to be adjusted by comparing them with nearby rural data.

Can we trust the statistics?

Handling the statistics of rare events is of particular importance because we are so vulnerable to extreme weather. There is, for example, a special need to have reliable measurements over as long a time as possible, in order to build up a reliable set of statistics for what are inherently rare events. One complicating factor can arise from changes at the site where the measurements were made and in the instruments used. One well-known site problem results from the expansion of urban areas around an observatory. Urbanization changes many characteristics of the local climate, notably the replacement of cooling trees with concrete and asphalt that heat up during the day but cool only slowly at night. Buildings and structures also change the ground-level wind flow and create eddies. Urbanization can also lead to rapid rainfall runoff and increases in flash flooding.

In rural areas the smaller shifts that may occur are more difficult to detect. For instance, the growth of trees around a farmstead that maintains a weather station alters the local wind flow and temperature patterns, and so reduces extreme wind speeds and the incidence of frosts (where they occur). The trend in the observations reflects the change of the microclimate of the farmstead while the general climate may not have changed. Greater challenges arise when it comes to interpreting widely differing measurements of destructive storms. For instance, it was only with the advent of weather satellites that a reasonably complete and consistent record of tropical storms could be maintained. Earlier records depended on shore-based and shipping observations, plus an increasing number of aircraft measurements from the 1940s, all of which relied on different types of equipment and provided only partial mapping of storms around the world.

All of this means that any long climatic series must be subjected to close scrutiny to ensure that what appear to be significant changes are a real part of the larger climate and not due to changes in the equipment, observing practices or the site itself. As we will see throughout this book, the history of the study of climate change has been principally a matter of painstaking and scholarly detective work to establish the reliability and applicability of the data.

1.6 Implications of variability

Climate fluctuations have an impact on human existence in many ways. The capacity of societies to adapt to extremes is finely tuned to the local climate and how much it normally fluctuates.

It is easy to lose sight of just how much our lifestyle is attuned to the local climate. Winter snowfall and intense cold are part of normal life in Siberia or Canada, but much lesser events bring London, England or Washington, DC to a standstill. Despite the fact that many societies have developed systems to handle their normal climate extremes, droughts, floods, heatwaves and severe winters still cause enormous economic damage and human suffering in many parts of the world. Just how damaging these extremes are depends on many features of the societies affected.

Communities that are frequently hit by disruptive weather events have generally developed procedures to recover quickly, but even in communities with highly developed infrastructures, in unusual events, such as this snowstorm in Geneva, Switzerland, citizens have to take special action.

The worst matters much more than the bad

The reason it is so hard to predict how much impact an extreme event might have is because most extreme events have a disproportionately greater impact than the more frequent lesser events of the same type. Much of the fabric of any society is designed to handle a degree of day-to-day variability, so up to a certain threshold the consequences of certain types of adverse weather are generally small. Once the conditions exceed this critical level, however, the damage to life, limb and property escalates rapidly.

Flash floods become much more dangerous when rainfall accumulation exceeds the design flow of the drainage system. If 50 mm in a couple of hours is likely to cause serious damage in a city, 100 mm in the same time will do many times more damage. In many middle to high latitude cities snow and ice will disrupt traffic and public transportation systems. Although there may be few disruptions in mild winters, in colder winters they may be severe and frequent, even though the difference in average temperature may be only a few degrees. Similarly, the wind and storm surge damage caused by tropical storms will rise rapidly as the wind exceeds a threshold value of first significant damage. What is more, in any part of the world the threshold for damage may differ appreciably due to different standards for construction.

When rainfall or winds exceed the design criteria of the local infrastructure, the ensuing destruction can escalate rapidly. Several days of heavy rains in Brazil, for example, resulted in severe flooding of the Piracicaba River at São Paulo on 2 February 1983.

Economic analysis

The variability of climate affects so many aspects of our lives that it has many economic consequences. Changes in temperature or rainfall can lead to changes in demand for weather-sensitive goods and services or alter the supply of other goods and services. The changes in demand are manifested in the increased sales of anything from ice cream in hot weather to electricity, natural gas and oil during cold winters. On the supply side, both droughts and excessively cold wet summers can reduce agricultural production, or a late killing frost can destroy entire crops (e.g. coffee, grapes or soft fruit) in a locality. These changes in demand or supply can be linked to shifts in prices with obvious consequences for producers and consumers.

Historical impact

The impact of climatic change on human history became the subject of fierce debate during the latter part of the 20th century. Some historians argue that fluctuations in the climate were wholly inconsequential in the course of history. Conversely, a few have gone so far as to suggest that climate change is an unrecognized mainspring that has controlled the outcome of many events. The essence of the debate is establishing precisely where on the spectrum of impact various events lie. These issues will be explored in this book, which will present examples of where extreme weather or more sustained shifts of the climate have had measurable consequences, and how these in turn have combined with other factors to influence social, political and economic development.

The historical impact can be found at many levels, from the devastation of local communities, through disasters that can alter the economic and social progress of nations, to subtle but nonetheless significant changes in global developments. Flash floods can sweep away villages, which have existed for centuries, and that are never rebuilt in the same vulnerable spot. Hurricane *Mitch* devastated the economy of Honduras in October 1998, killing over 10 000 people in that country and neighbouring Nicaragua. It will take many years and massive international aid to get the region fully back on its feet again. Bangladesh's society is regularly buffeted by floods that so often swamp much of the country, and in the past by the awful loss of life that occurred when storm surges driven by tropical cyclones swept in from the Bay of Bengal.

More subtle effects include how the course of World War II was influenced by the exceptionally bitter winter of 1941–42 that stalled the German advance on Moscow. Then, in the aftermath of the War, the paralysis of much of Europe by the savage cold of February 1947 had a profound effect on both the USA's approach to foreign policy relating to the defence of the West, and to European thinking about economic union. More recently, concern about global warming has been a major factor in many aspects of international environmental policy.

Does climate still matter?

The waves of innovation that swept across the developed world during the 20th century have had a strong influence on our level of vulnerability to climate. These range from basic domestic technologies, such as refrigeration and air conditioning, through a whole range of transportation developments from automobiles to jet aircraft, to various forms of electronic equipment which enable us to manage the complex infrastructures of modern society. These technologies have enabled industrialized nations to construct massive networks to supply food, water, natural gas and electricity; climatic information has been necessary to plan these infrastructures and to ensure that effective services are maintained even during exceptional events. Although these systems are occasionally disrupted by bouts of extreme weather, they have driven economic growth, improved public health, and given us greater mobility. In many respects it can be claimed that much of the developed world is less vulnerable to climate than it was in 1900 but only because of the climate knowledge applied and early warning systems in continuous operation to underpin technology and allow adaptation to mitigate the impacts of extreme events.

If this view leans towards technological optimism, it is as well to remember the environmental problems facing major cities in the late 19th century. During one eight-day hot spell in New York in August 1896, not only did human mortality rise sharply with the records showing 'heat stroke' as the cause of death for 617 people, but thousands of domestic animals also died. Over 1000 horses died on the streets, and the facilities for removing their bodies were so inadequate that many remained for days where they fell. The scientific journal *Nature* recorded that some of the busy streets "were strewn with dead horses like a field of battle".

Electrical and electronic tools and appliances have enabled communities to adapt to climatic variability more effectively, but they are particularly vulnerable to damage from extreme weather events and their loss adds significantly to the overall cost, such as when the Red River flooded Grand Forks, Minnesota, USA, in 1997.

Very rare severe weather events can destroy everything in their path. The tropical cyclone that struck Orissa, India, in 1999 left 10 000 dead, and millions homeless and without food.

1.7 Where people live

We are all familiar with our local climate, so sometimes find it hard to envisage living elsewhere, or to conceive of the consequences of our climate changing appreciably.

As a general rule, most people come to enjoy the climate where they live. As the American historian Carl Becker wrote in 1910, "We who live in Kansas know well that its climate is superior to any other in the world, and that it enables one, more readily than any other, to dispense with the use of ale". Nonetheless, however much we have come to terms with our own climate, we must always face the challenge of climatic variability, and possibly will have to confront more radical climate change.

The argument that the developed world is becoming increasingly independent of climatic variability has to be countered by one interesting observation. Despite a marked decline in mortality and social disruption associated with extremes, the monetary consequences of climate variability are increasing. One reason for this is because, with rising disposable income, many people are investing in desirable properties in more vulnerable locations, such as close to the seashore, in the flood plains of rivers, or high in mountains. Combined with having more possessions, the insured losses incurred as a result of extreme weather events in these parts of the world are rising steeply.

In many parts of the developing world, the challenges are different. Crowded cities with inadequate services are becoming increasingly susceptible to weather disasters. In particular, building in flood-prone areas, including shanty towns without adequate early warning services and infrastructure for evacuation, increases the vulnerability of people, especially to flash floods and mudslides. Here, it is the human suffering that matters most when adverse weather strikes. Areas that are most likely to experience catastrophic inundation should be uppermost in our minds, as disaster scenes from China, Madagascar, Mozambique and Venezuela late in the 20th century have shown so clearly.

Out of sight but not out of mind

Concentrating on where people live does not mean that we should think only in terms of our backyard. The message from the 20th century is that we cannot ignore the global nature of climate. With a global perspective, we see why certain types of anomalous events occur and how they are often balanced by opposite anomalies in other parts of the world. Discoveries, including that changes in the temperature of the sea surface in the equatorial Pacific (the El Niño) are of vital importance to people in the tropics, and the fact that changes in the extent of polar pack ice are part of wider changes, matters to us all. The study of the real impact of the climate on people goes hand in hand with assessing how the often distant links between various components of the climate (referred to as teleconnections) act to produce the fascinating variety of climate fluctuations that are part of our daily lives.

Broad patterns

What are the broad features of climate that are important to people and their activities? The annual temperature cycle dominates the lives of those living outside the tropics and defines the seasons throughout the year. The seasonal variation of precipitation is equally important. While there is rainfall throughout the year in parts of the tropics and many temperate regions, in other areas there are marked rainy and dry seasons that control many aspects of life irrespective of the annual temperature cycle. Failure of summer rains in India or the Sahel can be a calamity, while a relatively dry summer in Scotland or southern Chile is no hardship, and may even be seen as a benefit. Late spring

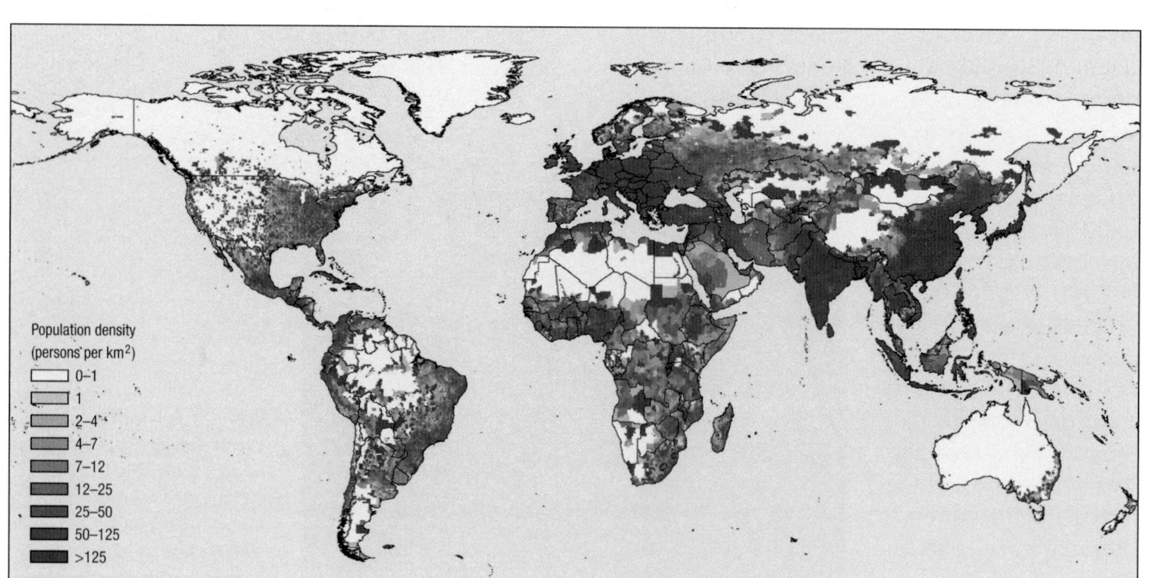

People have tended to settle where overall the climate and lands provide adequate food, freshwater and tolerable living.

frosts are normal in Minnesota, but can devastate the coffee crop in Brazil. In temperate zones a complete range of variations can disrupt peoples' lives, ranging from cold winters, snowstorms, and spring frosts to summer heatwaves, droughts and severe thunderstorms, as well as damaging floods at almost any time of the year.

In the populous parts of the tropics, where there is a marked cycle in annual rainfall, two factors matter most. The first is the progress of the main rainy season, which for much of the region is associated with the summer monsoons across Asia, Africa, Australia and the Americas. The second is a significant risk in many places of destructive tropical storms (variously known as tropical cyclones, hurricanes or typhoons in different places). In these regions it is the variation in the amount of seasonal rainfall, often linked to the number and intensity of tropical storms each year, that is important. However, to people living in any region, the changing seasons may seem a mystery. As Jawaharlal Nehru wrote in 1939 of the coming of the monsoon to Bombay, "A few showers came. Oh, that was nothing, I was told; the monsoon has yet to come. Heavier rains followed, but I ignored them and waited for some extraordinary happening. While I waited I learnt from various people that the monsoon had definitely come and established itself. Where was the pomp and circumstance and the glory of the attack, and the combat between cloud and land, and the surging and lashing sea? Like a thief in the night the monsoon had come to Bombay, as well it might have done in Allahabad or elsewhere. Another illusion gone".

New frontiers

The assumption that many aspects of agriculture are well tuned to the global climate does not apply when people have opened up new lands. In places like Argentina, Australia, the Great Plains of North America and the steppes of Central Asia, the extension of grain production into semi-arid lands where there were no historical rainfall records sometimes led to major difficulties. Success in a series of good years could be rapidly followed by disaster in dry years. What is more, at times there was even an antagonistic reaction to well-informed criticism of misguided land use. Politicians attacked the renowned Australian geographer, Griffith Taylor, in the 1920s for pointing out that much of the country was arid or semi-arid, and that contemporary land-settlement policies would be ill fated. His books were, for a time, even banned in Western Australian schools.

The failure to take adequate account of local climate has bedevilled projects well into the 20th century. The problems have been greatest where people sought to exploit a region where they knew little or nothing about the local conditions, as in the case of the failure of the 'Groundnut Scheme' in the late 1940s in what was then Tanganyika Territory in East Africa. This experience has been repeated elsewhere with varying outcomes. The expansion of grain production by the former USSR into the 'virgin lands' of central Asia during the 1950s was somewhat more successful, but erratic rainfall led to frequent poor harvests in the 1960s and 1970s, demonstrating the importance of climatic information. Only by having adequate data on how the climate can fluctuate from year to year is it possible to make a realistic assessment of the prospects for introducing any new form of agriculture.

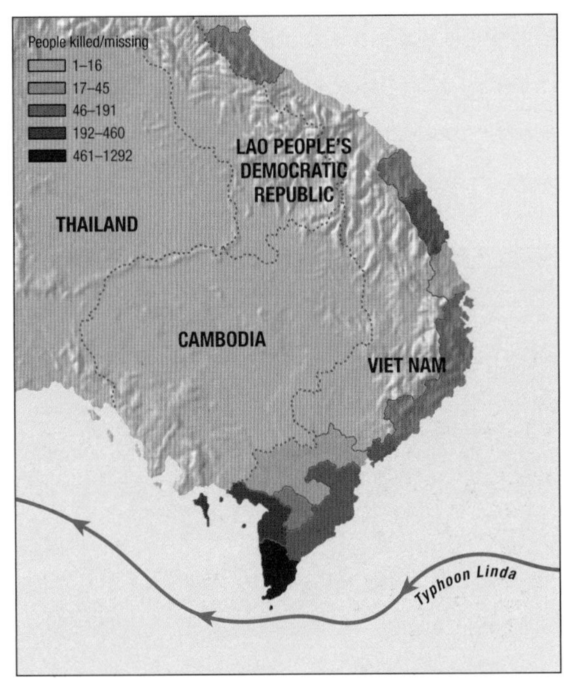

Over many parts of the world, such as Viet Nam, the climate and soils of the low-lying coastal margins are very productive and encourage settlement. However, there are ever-present hazards such as destructive typhoons. Typhoon *Linda* in 1997 was one of the worst weather disasters in that area in over 100 years, affecting hundreds of thousands of people.

Where much of the convenient land is low-lying, even major cities are severely threatened by floods.

1.8 What people think about their climate

Climate is not just a matter of definitions and statistics. How we come to terms with the everyday and the exceptional events where we live is an essential part of dealing with climate, and the better we understand our climate the better we cope.

The concept of a 'White Christmas' stems from winter scenes such as this in Randa, Switzerland.

These people appear to be thoroughly enjoying their day on the frozen Zuiderzee in 1929. The long-term records, however, show that this was one of the coldest winters in the 20th century in the Netherlands.

The French author Marcel Proust observed "I never allow myself to be influenced in the smallest degree either by atmospheric disturbances or by the arbitrary divisions of what is known as time", but then he spent most of his life in bed! For the majority of us, our own climate is a very real part of our lives, and is the subject of a peculiarly ambivalent attitude. Rather like a pet dog, we are likely both to defend its unwelcome attributes stoutly, while complaining from time to time that it fails to meet our expectations.

This personalized outlook on climate tends also to view any unpleasant event as being way outside past experience. Fanned in part by media hype, every big storm, flood, heatwave or snowstorm is seen as having exceptional characteristics. Somehow, it seems to give us reassurance to be told that a particular event was the most extreme in the last five years. In many instances, however, unpleasant weather is nothing more than part of the normal fluctuations that make up climate. We continually need to put current events into a longer-term perspective, to remind ourselves of what has occurred in the past and what we are likely to have to face in the future.

Are our memories reliable?

If you ask elderly people about the climate in their childhood they will often have clear memories of certain features being more pronounced. As the 15th century French poet François Villon said, "Mais où sont les neiges d'antan?" (Where are the snows of yesteryear?). Depending on our location, our memories are often of snowy winters, balmier springs, long hot summers or sunlit autumns. Unfortunately, these recollections may have more to do with how our memories embellish features of long ago and little to do with real climate change.

Actions to address real climate change must also address popular perceptions. For instance, the image of a 'White Christmas' has spread from the snowier parts of Europe and eastern North America to be a common cultural theme in many places, including parts of the Southern Hemisphere. Thus the lack of such snowy festive seasons is sometimes falsely portrayed as evidence of recent warming. Winter temperatures have indeed risen in middle latitudes of the Northern Hemisphere, but not as much as our memories might suggest. In the Swiss Alps there is little evidence that the amount of snow on the ground on Christmas morning has changed appreciably since the early part of the 20th century. At an altitude of 500 m it is a rarity, but at 1500 m it is a near certainty.

Another factor is that nearly half the world's population now lives in urban areas. The most obvious effect of urbanization is that the temperatures in large cities are getting higher, so it is not surprising that people who live in them perceive that snowfall is less common than it used to be or that the heatwaves seem to be more intense.

The role of the media

The media's handling of weather events influences our perceptions appreciably. The reason coverage has risen sharply in recent decades has much to do with the heightened interest of viewers, listeners and readers; perhaps the reverse is also true. While the evidence of global warming during the 20th century is clear, the case for a comparable rise in extreme weather events is not so well established, in spite of media reports to the contrary.

The popular assumption about the weather becoming more extreme is not surprising. Advances in recording technology mean that visually exciting images of destructive events around the world can be widely reproduced and transmitted quickly and efficiently. We can thus see weather disasters unfolding on our television screens, presented with everything from false-colour satellite images of hurricanes to video film of tornadoes or flash floods in full flight. These pictures are not only gripping but also convey accurately the scale of destruction and human misery associated with extreme events. Not surprisingly, awareness of our own vulnerability to such events is heightened.

Another factor is more directly related to increased public awareness of the scientific debate about global warming. As this issue has assumed greater importance since the 1970s, it has become a thread connecting the many severe weather events. Although many people may not understand how so many different forms of weather disasters could all be due to the same cause, links are often drawn between global warming and the occurrence of heatwaves, droughts, floods and hurricanes.

The media bring us graphic images and details of current extreme weather events from around the world but are also the source of information about the weather conditions expected in the local area. Media training for meteorological personnel helps to ensure that the right information is made available in ways that meet the needs of the local community.

Indigenous knowledge

In many cultures there is abundant folklore about the behaviour of the local climate and even about specific weather phenomena. Some of these are exceedingly widespread and have been known for centuries. The appearance of a halo round either the Moon or the Sun is rightly seen as a sign of approaching rain. The Zuñi people of Southwest USA observe that: "When the Sun is in his house, it will rain soon." Longer-term predictions are also based on experience. For instance, deep snow cover is seen as a boon for over-wintering grain in both Sweden ("Much snow, much hay.") and the Russian Federation ("Wheat is as comfortable under the snow as an old man is under his fur cloak.").

Farmers in the Andes of Peru and Bolivia make planting decisions based on observation of the properties of high clouds. For centuries they have observed the apparent brightness of the stars in the constellation of Pleiades around the southern winter solstice. If the visibility is poor because of thin high clouds they delay the planting of their potato crop because they expect drought conditions which reduce yields. Recent studies have shown that thin high clouds and reduced visibility occur in El Niño years that are associated with low rainfall in this part of the Andes.

Similarly, the behaviour of animals and plants is widely seen as an indicator of forthcoming weather. Although many of these observations relate to how flora and fauna are reacting to past and current weather, and there is little evidence that apart from seasonal behaviour these past patterns are a good guide to future conditions, folklore is a deeply held part of many peoples' lives. In the USA, geese flying south is clearly a sign of approaching winter, while geese flying north portends spring. In contrast, claims that the early arrival in Britain of waxwings is a sign of an impending cold winter is more likely to be a reflection of the weather or food supplies in Scandinavia, where they spend their summers.

Indigenous knowledge systems are widely applied in other parts of the world in the planning of agricultural activities. These include the behaviour of types of trees and birds. However, the success and failure rates are not always well documented since they have frequently been applied by communities that disseminate their information orally rather than by written methods.

1.9 The possible impact of human activities

The whole issue of the possible impact of human activities on the climate during the 20th century has undergone a complete turnaround in scientific attitude; from being little more than a curiosity on a local scale before the 1950s to become a dominant focus of climate studies on a global scale.

A ship's 'cemetery' now stands where decades ago there was a harbour, as the Aral Sea has now receded tens of kilometres away.

To the end of the 19th century there was only a limited perception of the possibility of climate change and there was little concept that human activities could have a measurable impact on the global climate. This view was to be shaken by emerging evidence of major glacial periods in the past million years and calculations by the Swedish chemist Svante Arrhenius in 1896 that changing the atmospheric concentration of carbon dioxide could substantially affect the Earth's temperature. He drew on a proposal made by the French mathematician Joseph Fourier in 1827 that the Earth is kept warm because the atmosphere traps heat, and observations by the British physicist John Tyndall in 1859 that gases such as water vapour, carbon dioxide and methane absorb infrared radiation.

Grand schemes

In Australia there have been various schemes to divert rivers into the heart of the country, and even flood with sea water the extensive area that is below sea level, to ameliorate the climate. One proposal from the end of the 1930s, Dr J.J.C. Bradfield's Inland Irrigation Scheme, named the Bradfield Scheme after its notable proponent, claimed that diversion of the rivers would lead to "evaporation – 100 inches (2500 mm) per year – from such a water surface ... would cause a fall of rain of 4 inches (100 mm) over 500 000 square miles (1.3 million km^2) of the dry inland. That rain, after refreshing the vegetation would evaporate and fall again as rain". This rosy view, effectively of meteorological perpetual motion in a favoured locality, was not uncommon in the early decades of the century.

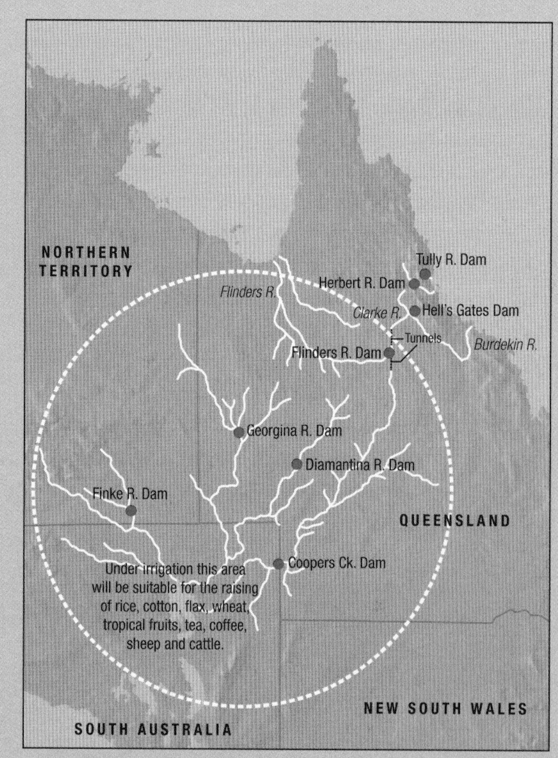

Climatic interventions

The public attitude towards human activities affecting the climate was peculiarly ambivalent for much of the 20th century. On the one hand, the possibility of inadvertently altering the global climate was largely ignored. On the other hand, there was a strong assumption that with the right effort humans could conduct large-scale experiments to alter regional climates to improve environmental conditions. For instance, in Australia there have been various proposals to bring water into the arid heart of the country, such as the Bradfield Scheme described (left).

As late as the 1960s, some scientists were pressing for schemes to divert the rivers in the Russian Federation southwards to create huge areas of additional fertile land on the steppes, or even going as far as wanting to melt the ice in the arctic by covering it with soot in summer to absorb sunlight, and thereby bring a benign climate to polar regions. Environmental disasters like the decline of the Aral Sea and more general concerns about global warming have completely altered our thinking about such forms of climatic intervention. These days few countries would contemplate such schemes.

Weather modification

Another scientific development, which exerted considerable influence on meteorological thinking about the altering of climate, was the rise of weather modification activities. The first scientific efforts to modify clouds occurred soon after World War II. By then an accepted theory of precipitation had been developed, which proposed that the essential first stage was the formation of ice crystals on dust and other particles in the atmosphere (freezing nuclei). It was argued, therefore, that introducing more nuclei into clouds would improve the efficiency of precipitation. The apparent success of early experiments using dry ice or silver iodide crystals led to a rush to judgement.

Within a few years operational programmes, principally to suppress hail, were being conducted in France, Italy, Kenya, the Russian Federation, Switzerland and the USA. Many meteorologists were deeply suspicious of the claimed successes and it slowly emerged, with sophisticated statistical analysis and carefully designed experiments, that the results did not live up to early expectations.

The fundamental limitation with the initial work was that nobody could be certain what might have happened without seeding. Only 'double-blind' statistical trials could provide the answer. These required that neither the people seeding the clouds nor those making the measurements of precipitation knew whether or not the flares used to seed the clouds actually contained seeding material. This was essential to ensure no bias in selecting clouds or interpreting their precipitation patterns. Only when all the measurements had been made was the identity of the seeding and non-seeding runs revealed. That experience serves as a useful reminder of both the complexities of the climate system and the need for scientific rigour in drawing the conclusions that have attributed global warming to human activities.

The threat of global warming

In spite of the marked warming of the global climate observed between 1920 and 1944 the standard view at the time on climate change was neatly summarized by T.A. Blair in his authoritative textbook *Climatology, General and Regional* published in 1942. He declared, "We can say with confidence that the climate is not influenced by man except locally and transiently". This view had its roots in 19th century European observations that showed lower temperatures in the middle of the century but a recovery in the latter half. The prevailing view that the Earth's climate is stable was sustained, in part, by an interruption in the rise in global temperatures during the 1950s and 1960s, especially in the Northern Hemisphere, which led to the assumption that the earlier warming had been a temporary phenomenon.

Nevertheless, there was a growing group of scientists concerned with the potential for human-induced global warming. An initiative of the International Geophysical Year for scientific research in 1957–58 was to establish careful measurements of carbon dioxide on the Mauna Loa mountain observatory in Hawaii, far from the industrial centres of the world. These, and other later observations around the world, confirmed the build-up of carbon dioxide in the atmosphere and formed the basis for concluding that human activities were altering the atmosphere. Studies have now provided estimates of how much of the carbon dioxide produced by human activity has been absorbed by the oceans and biosphere, and how much has remained in the atmosphere. By 1985 the evidence was sufficiently compelling that an international meeting of scientists in Villach, Austria declared that human activities were leading to global warming and if not controlled would lead to dangerous climate change.

Aircraft fitted with hygroscopic flares spew chemically laden smoke into areas of developing clouds in an attempt to increase rainfall in otherwise semi-arid areas.

Better science, more powerful computers and more realistic numerical models in the 1970s and 1980s allowed atmospheric scientists to start tackling the problem of global warming in earnest. The growing confidence in the performance of computer models for predicting the weather a few days ahead reassured scientists that climate models were a sound tool. The fact that this research coincided with a second surge in global temperatures also lent additional urgency to the work. There is now a broad consensus that the use of computer models to develop future scenarios of climate change is an essential part of formulating international policy on controlling the future emissions of greenhouse gases.

Svante Arrhenius (1859–1927)

A Nobel prize winner, the Swede Arrhenius is probably best known for his contributions to physical chemistry. Generations of students have grappled with the Arrhenius equation that describes the temperature dependence of the rate constants of chemical reactions. He also made major contributions to electrolytic dissociation and immunochemistry. In climatology his fame rests on an inspired analysis of how carbon dioxide can contribute to warming the Earth. In 1896 he published a paper, which estimated that a doubling of the concentration of carbon dioxide in the atmosphere would lead to a 5°C warming of the Earth's surface, a figure he subsequently reduced to 4°C. Without the aid of modern supercomputers, or a detailed knowledge of the infrared absorption characteristics of carbon dioxide he managed to produce a figure that turns out to be well within the range covered by the latest estimates.

1.10 Technological advances

Technology has been a driver for developing our understanding and capabilities. Climatologists have been at the forefront in making use of computers, remote sensing including radar, satellites, communications systems and the Internet. The applications of technology for research and provision of better services have accelerated since about 1950.

Instruments have been developed with sufficient robustness, while maintaining sensitivity, to make measurements in the most inhospitable environments, such as here, at the summit of Nevado Sajama, at 6542 m the highest peak in Bolivia.

Our growing knowledge of the climate has been built on the development of new technologies. This has led to more accurate and varied measurements of a wide range of parameters on an increasingly comprehensive scale. These measurements can now be communicated around the world more rapidly and made available to more people quickly so that they can be put to good use.

One hundred years ago the communication of some meteorological observations relied on semaphore signals and messages picked up in baskets by passing trains. In the 1970s, it took nearly three months to compile the data and analyse the seasonal climate patterns across the globe. Now supercomputers analyse and model the global climate and personal computers in many private homes enable us to see the products of these massive computations swiftly via the Internet. Telephone warning systems (pagers) alert people to imminent severe weather. Better still, these developments are likely to continue apace for many years to come.

Hardware

Advances in measurement technologies have occurred as a continuously evolving process. Now automatic weather stations (AWS) can measure and report every few minutes from remote localities. Electronic sensors can be put on balloons, buoys, planes and ships, and on a wide range of satellite systems enabling a near-global and almost continuous weather and climate watch. The operational programmes of the World Meteorological Organization and the specialized observations for the internationally coordinated World Climate Research Programme are all built on these systems.

A good example of the internationally coordinated and cooperative efforts to advance climate science is the array of equipment set up to observe El Niño/Southern Oscillation (ENSO) events, which has been operating in the tropical Pacific since 1994. This has transformed our ability to measure and forecast these events. This observing array provides data on surface and sub-surface temperatures, wind speed and direction, sea level and current velocity. During 1997–98 the array produced a much more timely and accurate picture of the major ENSO warming event and was an important factor in early warning as the event developed.

Modern radar systems track hurricanes and typhoons as they approach land and locate their rotating destructive eyes with an accuracy of one kilometre or so. They can measure both rainfall and wind speed and direction in severe thunderstorms. More sensitive radars are being developed that are likely to increase our knowledge of the properties of these storms. This offers the prospect of being able to provide more detailed and timely warnings of damaging conditions. An immediate benefit of radar, particularly in the USA, is to provide better warnings of tornadoes. The success of these warnings and the public willingness to respond and

The analysis of radar signals provides detailed information about the structure and motions of severe storms and is vital for providing early warning.

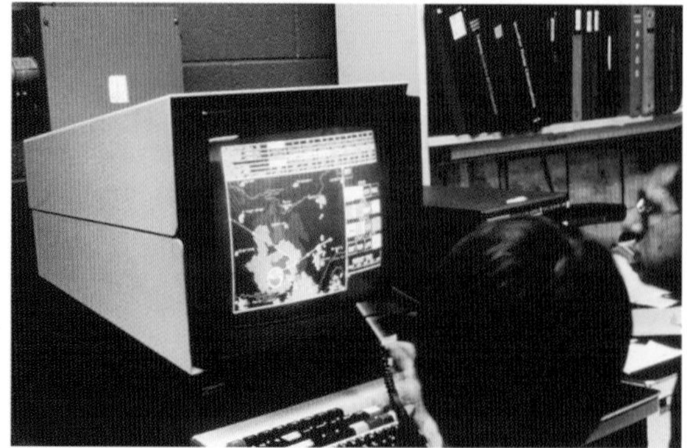

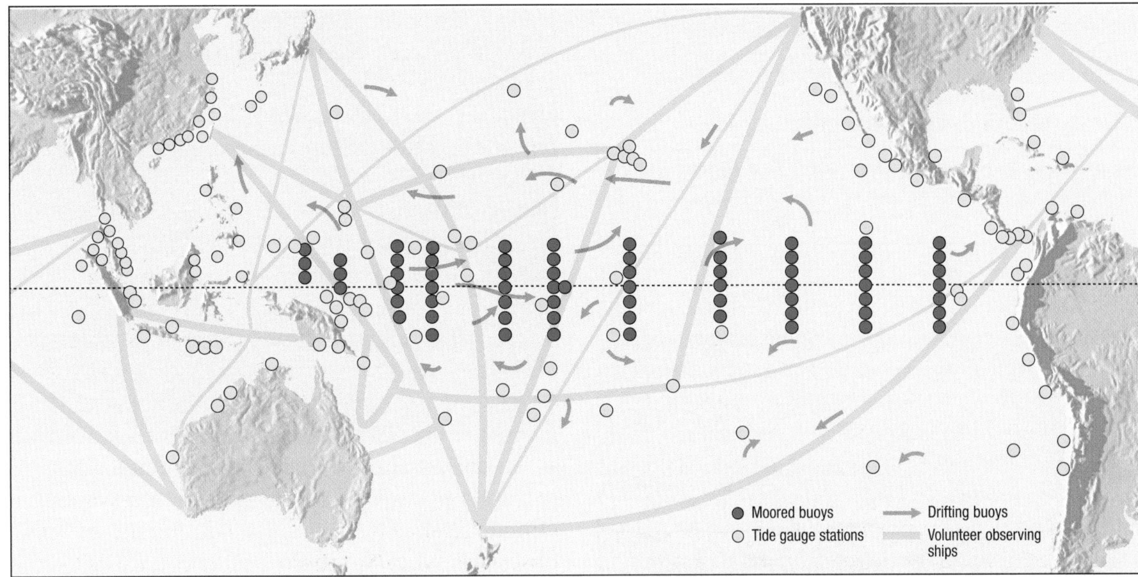

The Tropical Ocean Global Atmosphere (TOGA) project developed a comprehensive array of moored buoys, drifting buoys, tide gauges and shipboard measurements for the Pacific Ocean. Collectively these systems measure the temperature and movement of the ocean to a depth of 500 m or more, provide standard weather data, and measure waves and changing sea-surface topography. The data, relayed by various satellite systems, support further research into the El Niño phenomenon and provide early warning of developing events.

take shelter has greatly reduced the loss of life in these devastating storms. Associated work in anticipating flash flooding from heavy rainfall is another major potential exploitation of radar and linked AWS technology.

These systems offer the potential of making improved observations of the climate and using them both for better forecasting work and improving our understanding of the physical processes involved. It is not simply a matter of new technology. Better measurement techniques have to be combined with the hard work and dedication of keeping observing systems going day in and day out. This work involves both high standards of maintenance of the equipment and proper observing methods. This scrupulous effort is what enables us to build up a more reliable picture of the many physical processes (e.g. ocean–atmosphere interactions and land–atmosphere energy exchange) that control our climate and its fluctuations. Continuing progress will also depend on sustaining the right amount of effort to maintain existing systems, which provide a vital part of our growing knowledge base.

Exploitation of data

A challenge that arises from the accumulation of more and better observations is to make effective use of them. From both scientific and practical viewpoints, the data are used many times over, including to:

- monitor current weather patterns and their rate of change;
- start numerical weather prediction and climate models;
- monitor the variability of climate, especially on monthly, seasonal and annual timescales;
- build local climate statistics;
- verify past forecasts;
- establish regional and global climate archives for research and applications; and
- diagnose physical processes.

The vast increase in the flow of data and data management have become central issues in making better weather and climate forecasts. Also, as we gather increasingly more measurements, the collection, quality control and assimilation processes are becoming important components of the whole computerized modelling and prediction exercise. A greater amount of effort is being devoted to making the most effective use of the available sources of data and to developing new sources to fill prevailing gaps.

Advances in seasonal forecasts in the tropics and subtropics have been built mainly around knowledge of the El Niño/Southern Oscillation. These predictions have enabled countries to improve the planning of their agricultural policies and practices on the basis of whether the coming growing season is expected to be wetter or drier than normal. To be fully effective, however, this information must be transmitted to users, for example farmers who make the planting decisions, in time for them to take action. If the benefits of the new forecasts are to be fully realized, this poses major communication challenges, particularly in less developed countries.

Ongoing maintenance and calibration are essential functions in the management of observing systems.

1.11 Our increasing vulnerability

As the global population increases and settlements expand, many communities are becoming more vulnerable to weather extremes and fluctuations in the climate. How we minimize this increasing vulnerability will depend on making more effective use of weather and climate services.

Where populations have increased there has always been a tendency to exploit marginal land while growing conditions have been good. In subsequent bad years this can lead to dreadful retribution. In Europe, during the 12th and 13th centuries, favourable climate conditions led to both population growth and widespread moves into marginal land. A famine during 1315 and 1316 killed huge numbers of people and led to many marginal settlements being abandoned. Over the centuries, this process has been repeated in many places and continues to this day in more vulnerable parts of the world.

Coastal plains

A recent example of increasing vulnerability can be seen amongst the growing number of communities located close to the sea. Already, almost half the world's population lives within 100 km of the coastline. In the developed world, people have moved to locations close to the sea because, for a variety of social reasons, they are regarded as desirable – perhaps by those seeking a milder climate or to enjoy a beach-side lifestyle. This movement in many countries has contributed to a steep rise in the economic costs of damage from both winter storms and summer tropical storms. In the developing world, the fertile soils of coastal plains and near-shore fisheries are vital as food sources for many communities. However, the risks of storm surges and inundation are an ever-present danger in the exploitation of these valuable resources. These problems will become even greater in the future if sea level rises further, as predicted.

The world's poor in crowded urban areas in floodplains and low-lying coastal lands are particularly vulnerable to weather extremes such as flash floods.

Flood plains

In some instances the increased vulnerability to extreme events is a direct product of action that has been taken by societies to cope with more normal weather fluctuations. A good example of this is flood control. In many countries massive civil engineering works have channelled major rivers and controlled runoff water from urban areas. At the same time the flood plains have been developed for commerce, housing and industry and high-technology agriculture. Rainfall and runoff are managed efficiently for most weather and/or seasonal occurrences, but beyond threshold rates of rainfall accumulation these management systems fail and the parts designed to be protected are flooded. Instead of mitigating impacts, the failure of the system exacerbates the problem and the damage and disruption of

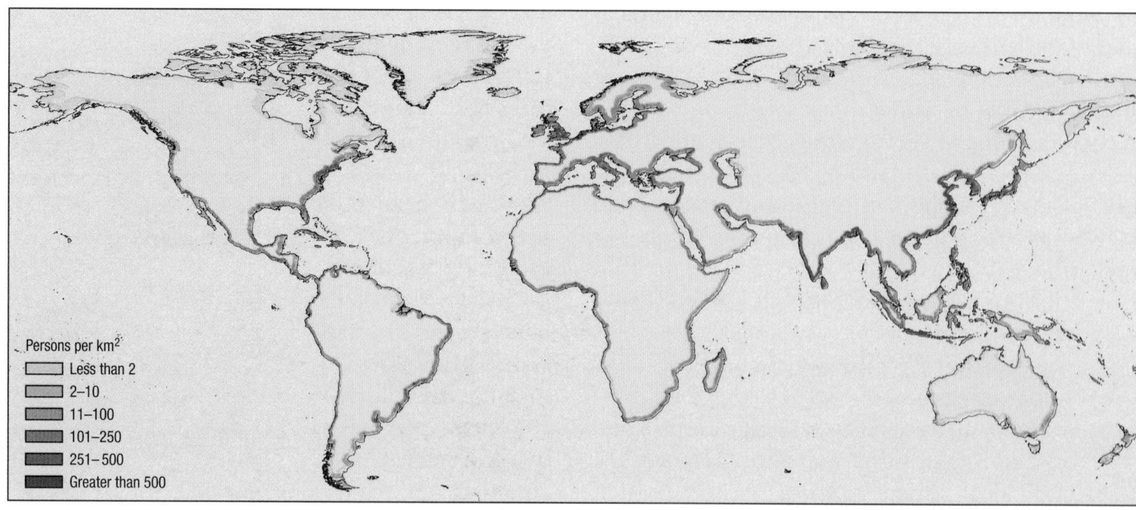

In 1995, roughly 2.2 billion people (then approximately 40 per cent of the world's population) lived within 100 km of the coastline.

flooding increase. For example, it is estimated that the time taken for water to flow down the Rhine River has halved from 10 to five days during the 20th century and this was an important contributing factor to the record-breaking floods in December 1993 and January 1995.

In less developed parts of the world the threat of flooding is part of normal life. In Bangladesh much of the country is inundated most years during the summer monsoon rains, but the consequences can still be dangerous for many communities. Elsewhere, although less frequent, large-scale flooding is a serious threat as the floods in Mozambique demonstrated in March 2000.

Arid regions

The most lengthy and damaging drought episode of the 20th century afflicted much of the region along the southern fringes of the Sahara during much of the last three decades of the century. During the same period the number of people in this vulnerable region rose rapidly. In Niger the number of people in drought-prone areas rose from 250 000 in 1960 to 3 million in the 1990s. Similarly, in Ethiopia the number of people classed as food-insecure, estimated at 2 million in 1980, had risen fourfold by the mid-1990s.

Deforestation

In many mountainous parts of the world, the consequences of deforestation have been considerable. They have greatly increased the vulnerability of local communities to heavy rainfall, which leads to flash flooding and landslides, with substantial impacts on downstream communities.

Cities

The number of people living in urban areas has risen by more than a factor of 12 since the beginning of the 20th century and now stands at nearly 3 billion. By 2030 this figure is expected to increase by another 2 billion to represent some 60 per cent of the world's population. Almost all of this increase will occur in the less developed world. Both past changes in the climate of cities and changes that will take place in the next few decades are therefore of vital importance to over half the Earth's population. Urbanization can alter the local climate in a wide variety of ways. Some changes are beneficial in parts of the world (e.g. warmer winter nights, less snowfall and the abatement of strong winds) while some are not (e.g. higher daytime maxima in summer heatwaves and the trapping of noxious pollutants during these episodes).

Many of the largest cities are in warmer parts of the world, consequently heatwaves and rising air pollution exacerbated by global warming and urban heating have serious health implications. In addition, the drift of rural poor to urban areas, many of whom dwell in shanty towns in locations that are especially vulnerable to flash flooding, has made matters worse. The loss of life as a result of the torrential rains from Hurricane *Mitch*, which hit Honduras and Nicaragua in October 1998, and the floods and landslides of December 1999 in Venezuela, provide tragic examples of this distressing trend.

During the 1998 El Niño episode, Thailand suffered one of its worst droughts in 40 years. At the time the country also was in economic crisis. To supplement their diet, these Thai farmers are looking for fish in what is left of a dried-up lake.

Making better use of services

The common message that emerges from these examples of increasing exposure is that disasters are largely a product of vulnerability. In principle, we should be able to mitigate the impact of a natural hazard by better planning. This does, however, require major investment in systems that minimize the need to exploit vulnerable areas and maximize the protection we install. These actions are essential to avert damage when extreme weather strikes and will involve not only fully exploiting existing knowledge, but also developing improved forecasting services to enable more people to receive early warnings and to take precautionary action.

Urban growth, if not adequately addressed, will expose ever-larger numbers of people to health risks, such as photochemical smog.

1.12 Changing sensitivity to climate

Whether communities are becoming more sensitive to climatic variability depends on a range of changing socio-economic factors, as well as how exposed they are to certain types of extreme events.

The balance between the advantages gained from advances in technology and the increasing vulnerability of various aspects of modern life to certain weather extremes varies greatly from place to place. Broadly speaking, in the 20th century the economic cost of climatic variability rose while the loss of life dropped. These shifts reflect better early warning of dangerous events and improved infrastructure response strategies to mitigate many of the food, water and health impacts in the aftermath of disasters. However, the death toll from coastal storm inundation, flash flooding and landslides still remains distressingly high, especially in the developing world.

Which climate disasters cause greatest hardship?

Overall, the climate extremes that matter most are droughts and floods. Although there are many weather extremes that cause great hardship, mortality and damage to property (e.g. hurricanes, storm surges, tornadoes and freezing weather), the extent and sustained impacts of too much or too little rain have the most profound consequences. The very nature of drought is difficult to handle, as it is a slow insidious process. Often disaster seems to creep up before anyone realizes its onset. To minimize this impact requires early warning in order that defensive action can be taken (e.g. increased planting of drought-resistant crops, reducing stock numbers or holding back reserves of water).

Better long-range forecasts, particularly those on monthly to seasonal scales, are enabling some people to respond to the threat of drought, and are clearly demonstrating their value. Sadly, in some parts of the world, both the limited capacities to exploit these advances and the growing numbers of people dependent on regular harvests are inhibiting full realization of the potential benefits. In terms of the global impact of droughts, as reflected in both shortages and prices, grain commodity markets appear to be less sensitive to drought than they were in the 1970s. This reflects a complex response of all aspects of the food-supply chain to increasing knowledge of what creates drought. Better monitoring of weather developments and the progress of crops in the fields, and improved global transport for food grains, are enabling markets to adjust to the global variability of climate and respond more systematically to the effects of regional droughts. Also, the management and distribution of international aid have improved significantly so that while famine and death tolls associated with drought still occur all too often, there has been a marked drop in the overall loss of life due to drought.

International aid, assisted by assessments of the extent, severity and likely duration of a drought, is providing relief and mitigating the worst impacts. In the past, in many parts of the world, drought was accompanied by much greater famine, disease and loss of life.

Millions of people in China have perished in flooding catastrophes over time, and countless numbers of homes and businesses have been destroyed. In the 1990s alone, floods repeatedly struck central, eastern and southern regions of the country, causing the loss of thousands of lives, US$ billions in losses and severe hardship.

Our changing sensitivity to floods is also complicated. Although many developed countries have installed increasingly elaborate defences, the scale of damage remains awfully high. This is not just a matter of accelerated runoff. Pressure to exploit the flood plains means that many expensive new installations are sited right by rivers. More important is the fact that flooding of homes, offices and factories causes much greater economic loss than in the past. The proliferation of electrical and electronic appliances results in far more costly damage during flooding, as does the damage of other valuable household possessions.

In the developing world, the situation is starker. Flood plains offer much of the most valuable agricultural land and exploitation is often vital in providing adequate food. In countries like China and Bangladesh, for example, this poses the daunting challenge of devising flood control strategies to control adequately the mighty rivers that flow through their agricultural heartlands.

Public health

The impact of the weather on our health has changed appreciably during the last 100 years or so. At the beginning of the 20th century, in many parts of the world, mortality rates were highest in summer. That was because so many young children died during hot weather from gastrointestinal infections. Even in generally cool, cloudy England some 2500 children died during the heatwave of August 1911. Infant diarrhoea associated with unclean water supplies remains a major killer, and reducing this cause of death is a major public health challenge. In spite of progress, and the widespread introduction of air conditioning in many warmer parts of the world, heatwaves remain major killers.

Although the health consequences of heatwaves are a widely publicized aspect of extreme weather, winter chill is also a big killer. This is because the effects of both respiratory and cardiovascular diseases are exacerbated by cold weather. At the simplest level, when we get cold our blood thickens and this means that people who already have heart disease or partially blocked arteries are more likely to suffer complications. Consequently, many countries in the middle latitudes of the Northern Hemisphere show an increase in mortality during the winter half of the year. As the climate warms, fewer people should die of extreme cold because of more frequent, milder winters, but this may be offset by a rise in summertime mortality.

In higher latitudes an extended period of extreme cold can lead to rationing of food and heating reserves, and can contribute to ill health and a rise in the death rate.

Transportation

The increasingly global nature of transportation means that the costs of moving people and goods around the world are becoming more sensitive to disruptive weather events. Whether local commuter services into our major cities or intercontinental air services, the increases in traffic place ever-greater emphasis on the need for systems to run on schedule. As planners seek to add increasingly more journeys into overcrowded networks, it is inevitable that fog, heavy rain or snow and ice will continue to disrupt traffic on those networks from time to time.

Is there an index of climate change?

Changes in our sensitivity to various features of the climate are a complex mix of technological and social developments. This complexity makes it difficult for people to develop an informed view about claims that things are either going from bad to worse, or getting better. Climatologists must therefore strive to produce reliable information on all the changing sensitivities. Explaining what current changes in the climate mean to local communities is not easy. The fact that the global average temperature has risen by about 0.6°C over the last 100 years may now be indisputable, but what has it meant to our lives? In short, very little in the developed world, because the rate of change has been relatively slow and our advancing technologies have allowed us to adapt for the most part to any local changes – so far. In the developing world the picture is more complicated because population and other pressures have often played a greater role in increasing vulnerability to climate hazards than any increased risk that might be attributable to climate change.

What people really want to know is whether they are experiencing more heatwaves, droughts, floods, cold spells, etc., and this also is not an easy question to answer. First we must define what really matters to people, as this is crucial in ensuring that they will support the politically difficult decisions that may be needed to avert the worst effects of climate change caused by human activities.

An objective assessment of how the climate is changing must define the outcomes in terms of people's local experience. Not only must such an assessment measure whether climate is changing but also how it relates to the predictions for global warming. Attempts to develop an index of climate change have used temperature and precipitation statistics, because they are the changes that people notice most and they have been recorded for longer than other measures (e.g. winds, sunshine and cloud cover, or humidity).

The Intergovernmental Panel on Climate Change has addressed the problem by compiling a series of climate change indices. No one index will be completely adequate but a range of indices will allow comparison of sensitivity factors between different locations. Such indices cover factors like accumulated degree-day departures from the seasonal mean temperature (for each season), accumulated heating and cooling degree-days (departures below and above reference temperatures requiring artificial heating or cooling respectively) and frequency of temperatures above and below critical values (numbers of hot or cold days respectively). Similarly, indices for moisture relate to total seasonal precipitation, annual water deficiency (taking account of local evapotranspiration) and frequencies of heavy precipitation and sequences of dry days.

One of the biggest problems in applying these indices in many countries is the lack of historical daily data that are available. Some countries urgently need assistance in managing their observing networks and in managing historical data that may not yet be in a form suitable for computer processing.

1.13 Climate since the last interglacial

Much of our knowledge about how the climate can change has been built on the study of the huge changes that occurred during the last ice ages. Although those may seem remote events, they are important messages for the present.

Air bubbles trapped in the layers of snow laid down each year over the glacial domes of Greenland and Antarctica provide a history of the chemical composition of the atmosphere over the past several hundred thousand years.

At the beginning of the 20th century, scientists deduced from the available geological evidence in Europe and North America that there had been four major glaciations in the last million years. It was, however, assumed that the current interglacial would last a long time and climate change was of little or no concern. In the 1950s, thinking started to change. Cesare Emiliani, at the University of Chicago, published a set of papers on the properties of fossil shells of the tiny creatures found in the sediments of the tropical Atlantic Ocean and Caribbean Sea. Using the reversal of the Earth's magnetic field 700 000 years ago as a marker, he was able to show that there had been seven glacial periods since, occurring every 100 000 years or so.

That new picture of the most recent ice ages was not immediately accepted. Since the 1960s, the worldwide programme of collecting cores of the solidified ooze at the bottoms of the deepest oceans has yielded huge amounts of information about past climates. Other sources of information, including Antarctic and Greenland ice cores and pollen records from the parts of Europe and North America that had not been covered by ice, have confirmed Emiliani's conclusions. It is now agreed that glacial periods occurred more frequently during the Pleistocene than early theories suggested. These cold periods have been interspersed by shorter, warm interglacials. Within each glacial period there were substantial fluctuations in the climate with temperatures ranging from extremely cold to nearly as warm as during the interglacial periods.

More detailed analyses, both of ocean sediment and ice core data for the period covering the last major glacial period, show the striking suddenness of some of the changes that have occurred. The dramatic warmings that are revealed by the Greenland ice core data appear to coincide with characteristic layers in the ocean sediment data, known as Heinrich Layers. The sediment layers are thought to have been produced by debris carried out into the North Atlantic by a surge of icebergs resulting from the sudden collapse of part of the ice sheet covering North America. The last glacial was not a period of unremitting cold, but exhibited marked fluctuations on timescales from years to millennia.

The value of new observations

The availability of ocean sediment data shows how new evidence can transform our view of the past. A

The ice core recovered at Vostok, East Antarctica, extends to a depth beyond 3300 m and reveals how carbon dioxide, methane and air temperature changed over a period greater than 400 000 years that included four major glacial periods.

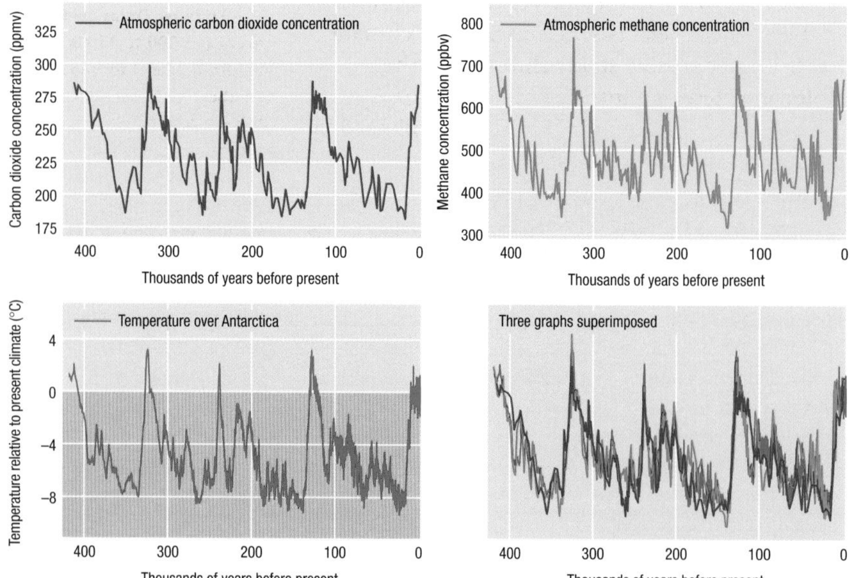

Milutin Milankovitch (1879–1958)

A Serbian astronomer at the University of Belgrade, Milankovitch dedicated his career to developing a mathematical theory of climate. The Milankovitch Theory states that as the Earth travels around the Sun, cyclic variations in three elements of the Earth-Sun geometry (orbital eccentricity, changes in obliquity and precession) combine to produce seasonal and latitudinal variations in the amount of solar energy that reaches the Earth. The variation in solar energy control led the onset and decline of the major glacial epochs.

century of work on glacial deposits on land had been a frustrating experience. The changing rates of deposition of debris and disturbance of soil layers by subsequent events muddled things considerably. In contrast, the marine record is often continuous and more easily dateable. The development of vessels that could retrieve cores of sediment from the bottom of the deepest oceans was a significant advance. It was thus possible to measure the conditions in the oceans over millions of years. The detailed measurement of various properties of the cores made it possible to build up an accurate chronology of past Ice Ages, and put the continental evidence into context.

The product of all this work is an increasingly ordered picture of the waxing and waning of the ice sheets during the last million years or so, and lends far greater insights into the workings of our climate. In particular, the expansion of the ice sheets every 100 000 years or so provided clear support for the theory that the Ice Ages were driven by variations in the Earth's orbital parameters. This theory, which was enumerated in great detail by the Serbian astronomer Milutin Milankovitch, explains a great deal about the factors controlling our climate, and how relatively small variations in certain physical parameters can produce major shifts.

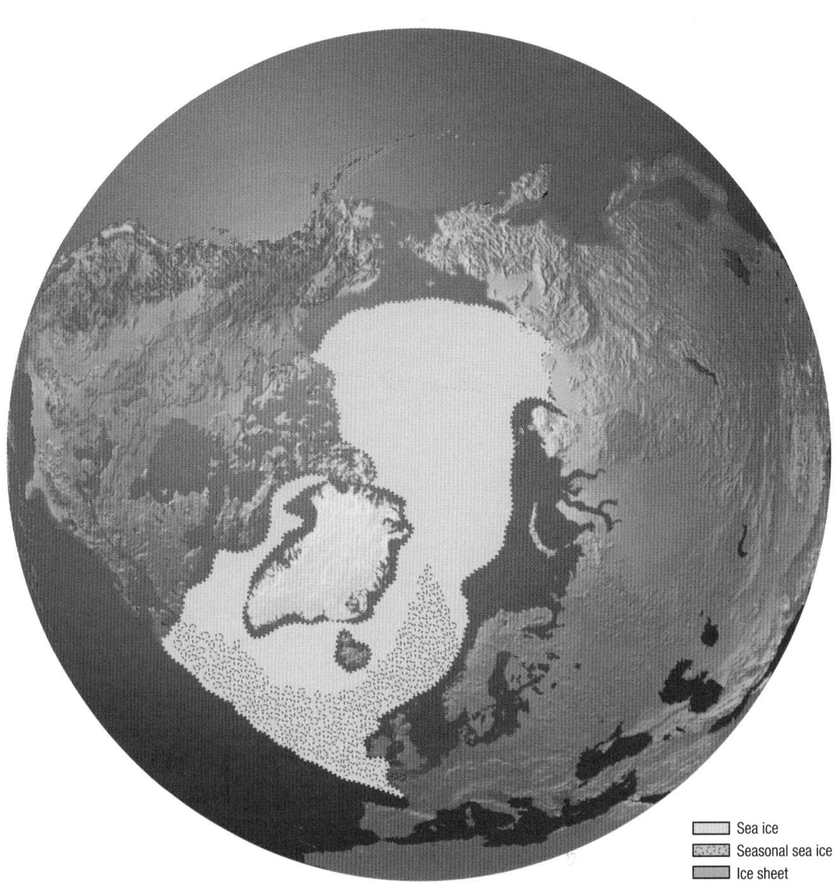

Twenty thousand years ago, large expanses of North America and Northern Europe were covered by huge ice sheets and the sea level was significantly lower. All that now remain of these ice sheets are the Greenland ice dome and smaller areas of permanent mountain ice.

The depths of the Ice Age

Just how much the global climate can change can be gauged from the depths of the ice sheets up to 3 km thick, that were laid down during the last glacial period, as recently as 21 000 years ago. These ice sheets covered most of North America as far south as the Great Lakes, all of Scandinavia and extended from the northern half of the British Isles to the Urals of the Russian Federation. In the Southern Hemisphere much of Argentina, Chile and New Zealand were under ice, as were the Snowy Mountains of Australia and the Drakensbergs in South Africa. The total volume of ice locked up in these ice sheets has been estimated to be about 90 million km^3. As a result, the average global sea level then was between 90 and 120 m lower than at present. Today, the volume is about 30 million km^3.

The global average near-surface air temperature during the last Ice Age was 4°C to 5°C lower than at present. Over the ice sheets of the Northern Hemisphere, temperatures were around 12°C to 14°C cooler. Temperatures in the tropics at the time are less well understood. Recent studies suggest that the tropics might have been about 3°C cooler.

Then, about 15 000 years ago, a dramatic warming started. Although the emergence of the Earth's climate from the last Ice Age over the period of 15 000 to 10 000 years ago may seem to have little immediate relevance to current concerns about climate change, this is not the case. Increasing evidence in recent decades shows that this period was marked by a variety of sudden changes which raise important questions about the stability of the global climate. The eventful transition between glacial and post-glacial climates ushered in a more settled period covering the last 10 000 years (the Holocene). Apart from a sudden short sharp cooling about 8200 years ago, which may have been linked to the final collapse of the ice sheet over North America, the climate has been remarkably stable throughout the Holocene. Research into the causes of this combination of dramatic fluctuations followed by relative calm is shedding new light on our understanding of the causes of current climate change and is helping us to develop credible scenarios of future climate developments.

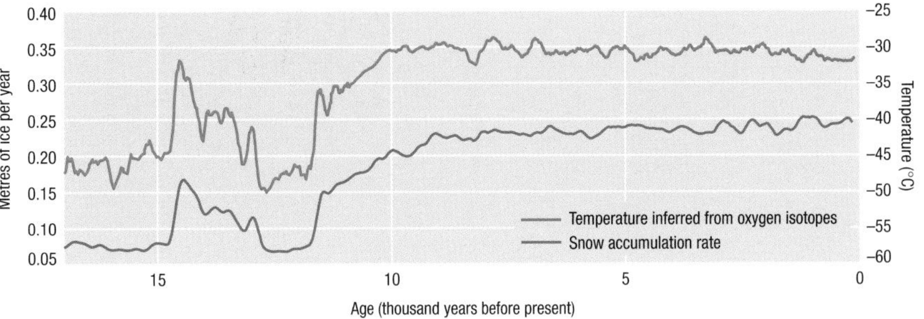

The Younger Dryas cold event (13 000 to 11 700 years before present) terminated in an abrupt climate shift. In this ice core record from central Greenland, a local rise of about 10°C and a doubling in annual precipitation volume occurred in the span of a decade.

1.14 Climate from historical accounts

The references to prevailing climate contained in historical writings have been difficult to interpret. How a more accurate picture of changes that have occurred has been established from these sources is an important part of how our overall understanding of the global climate system has advanced.

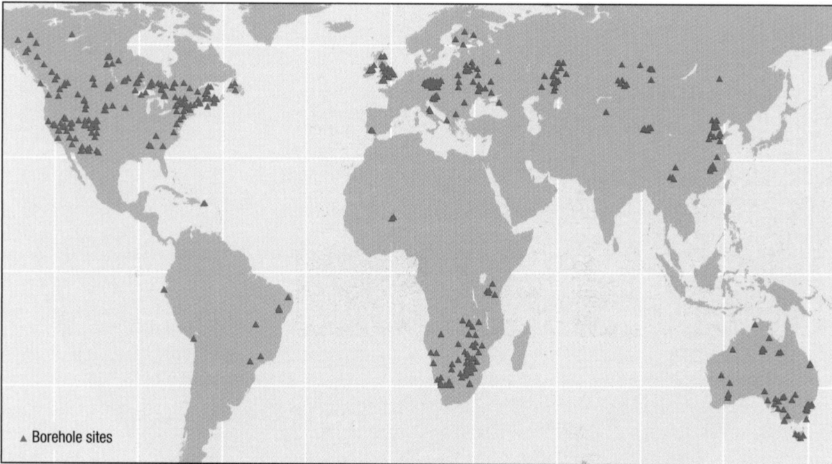

More than 600 boreholes, drilled into the Earth's surface, have provided data used to assess the Earth's temperature variations in past times.

The Central England Temperature (CET) time series (annual values expressed as anomalies from the 1961–90 period average) are overlain by borehole temperatures for the period 1600–2000. Shown here is the average of data from 26 boreholes. The borehole data reflect well the observed near-surface air temperature variations in the region.

Our knowledge of changes in the climate over the last few thousand years was transformed during the 20th century. A lack of appreciation that changing climate was a possibility can be attributed to, in part, an interpretation of classical literature, which appeared to describe the climate in similar terms to current experience. Studies of written works from Classical Greece through to the early years of the 20th century had concluded that there had been no change in the climate since the fifth century BC. These analyses were based on descriptions of the fertility of the country, the nature of streams and rivers, and the dates of sowing and harvests. This view changed gradually during subsequent decades, thanks principally to the painstaking work of several scholars who produced a variety of evidence to show that the climate had indeed changed on almost every timescale and in every part of the world.

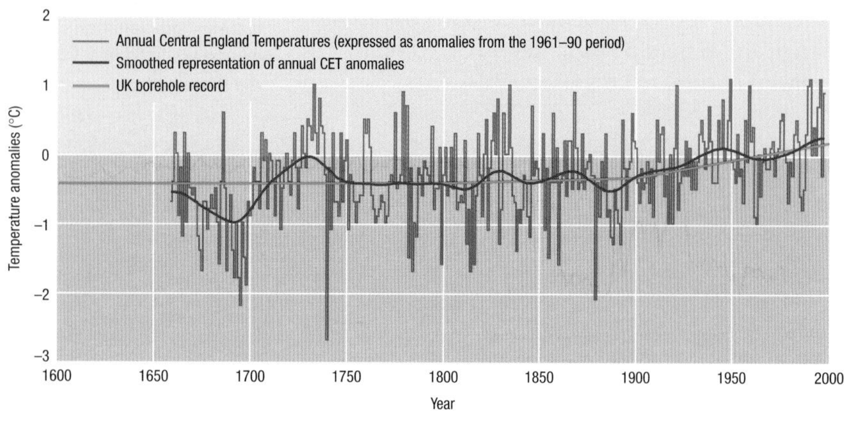

The broad picture is that by around 6000 years ago the post-glacial warming trend reached a peak. This was mainly a Northern Hemisphere summer phenomenon. On the basis of evidence of pollen from trees, the average summer temperature in middle latitudes of the Northern Hemisphere was 2°C to 3°C warmer than at present, largely because of the increased levels of summer sunshine that peaked, as a result of the Milankovitch effect, about 9000 years ago.

Around 5500 years ago, in the early stages of the development of the ancient civilizations of Egypt and the Middle East, the climate began to cool gradually and become drier. These changes were small compared with the sudden shifts at the end of, and during, the last Ice Age. Nevertheless, some of the changes were profound. The Sahara dried up. Desert formed where giraffes, elephants and antelope had roamed. More generally, there is evidence of a decline in rainfall in the Middle East and North Africa setting in around 4000 years ago. At the same time the poleward extent of the treeline across the Canadian Arctic and Siberia started to shift southward. This trend towards cooler, drier conditions continued until near the end of the first millennium AD but was punctuated by warmer periods. There is also considerable evidence of mountain glaciers around the world expanding around 2500 BC, but then receding to high elevations around 2000 BC. Further glacial expansions occurred around 1400 BC to 1200 BC, and between 200 BC and around AD 500–800. In between these colder episodes the climate was warmer with glaciers receding to higher elevations around the world.

The medieval warm period

Although there is no sudden change in the amount of historical evidence on the climate at around AD 1000, much points to the fact that many conditions in northern Europe and around the North Atlantic became warmer during the 9th and 10th centuries. In part, this is inferred from the expansion of economic and agricultural activity throughout that region. Grain was grown farther north in Norway than is now possible. Similarly, crops were grown at levels in upland Britain that have proven uneconomic in recent centuries. The Norse colonization of Iceland in the 9th century and of Greenland at the end of the 10th century is also seen as evidence of a period of more benign climate in this region.

There is little doubt the warmer conditions over much of northern Europe extended into the 11th and 12th centuries, but the geographical evidence is complicated. Tree ring data from northern Europe show warm periods between

The Grindlewald Glacier in Switzerland is an example of the worldwide retreat of mountain glaciers since the Little Ice Age was at its maximum in the middle 17th century. The painting (far left) shows the glacier around 1820, and the photograph on the left was taken in 1974.

AD 870 and AD 1100, and again from AD 1360 to AD 1570. Similar measurements made on trees growing in the northern Urals of the Russian Federation, however, show no evidence of the earlier warm period, and indicate that the warmest times there were the 13th and 14th centuries and into the late 15th century.

The Little Ice Age

In comparison to the uncertainties of earlier centuries, the evidence of a cooler period which started in the 14th century and became more marked between the mid-16th and mid-19th centuries, is widely assumed to be generally more representative of conditions. The 17th century was the coldest in Europe and China, whereas the 19th century was the coldest in North America. Conclusions about other regions can only be drawn when sufficient new evidence emerges. Throughout Europe, the popular image of the late 17th century is of bitter winters with frozen waterways, together with periods when cold wet summers destroyed harvests. Frost fairs on the Thames in London were frequent, but changes to London Bridge (including rebuilding between 1825 and 1835) affected the tidal limit. This meant that the river then only froze upstream of the new tidal limit in the 1962–63 winter, the coldest since 1740. Changes in a local regime due to human activity are just one of the many factors that must be taken into account, therefore, when considering the significance of historically recorded extremes of climate. Widely known as the Little Ice Age, the 17th century period has been closely studied by climatologists for many years. This growing body of work is showing that, as with all aspects of climatic change, the real situation is frequently more complicated than the early studies had suggested.

A large volume of 'proxy' data has been amassed in recent years, allowing much greater understanding of the climates of this period. Using a combination of tree rings, ice cores, coral growth, historical documents and a few long instrumental records, it has been possible to build up a reasonably comprehensive picture of the climate variations in the Northern Hemisphere since around AD 1000. The conclusions include that:

- the five centuries before were colder than the 20th century, and the coldest decade globally (1601–10) was, averaged over the decade, about 1.0°C below the 1961–90 average; and
- the 20th century was the warmest in the last millennium, with the 1990s the warmest decade by far, and 1998 the warmest year.

Northern Hemisphere (NH) near-surface air temperatures were reconstructed for the past 1000 years using palaeoclimatic records from tree rings, corals, ice cores, lake sediments, etc., along with historical and long instrumental records. The 20th century is revealed as unusually warm in the millennium period.

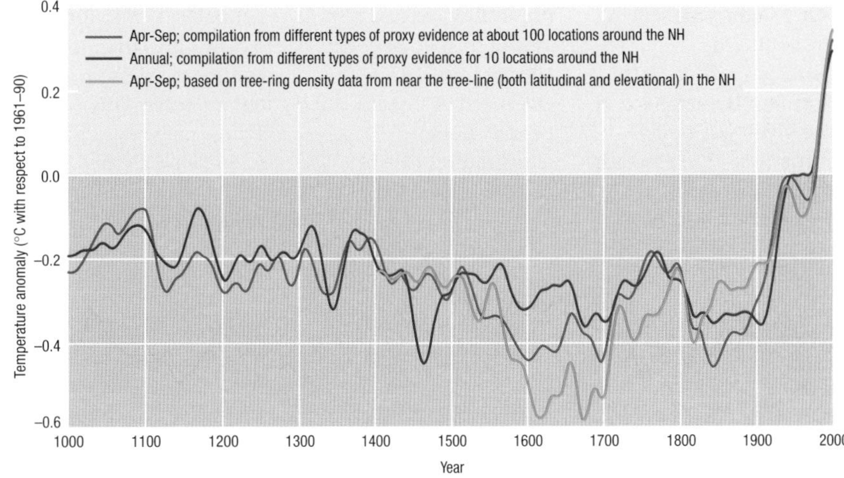

1.15 20th century climate change

The global climate warmed appreciably in the 20th century. A detailed examination of the changes shows significant differences in how fast temperatures have risen during the century in different regions and where the most marked rises have occurred. These variations provide important insights into the possible causes of global warming.

Before we discuss how much the Earth has warmed, we need to decide what we mean by global temperature. This is not as easy as it might seem, because as we will see at various places throughout this book, maintaining adequate measurements of the surface temperature around the world requires an instrument network that covers all the many climate regimes of the world. Even when the land surface observing networks were at their peak during the latter part of the 20th century, global coverage was inadequate; in addition there remain significant observation gaps over the oceans. Nevertheless, the best analysis of observations made on land and at sea indicates that the annual average near-surface air temperature of the world, calculated over the 1961–90 reference period, is approximately 14°C (14.6°C in the Northern Hemisphere and 13.4°C in the Southern Hemisphere).

There is an annual cycle in the Earth's near-surface air temperature because of the thermal inertia of the oceans and the fact that there is a higher proportion of land in the Northern Hemisphere than in the Southern Hemisphere. The relatively larger land mass in the Northern Hemisphere means that the hemisphere warms more in summer than the Southern Hemisphere does in its summer and cools down more in winter. As a consequence, the global mean near-surface air temperature rises to 15.9°C in July and falls to 12.2°C in January.

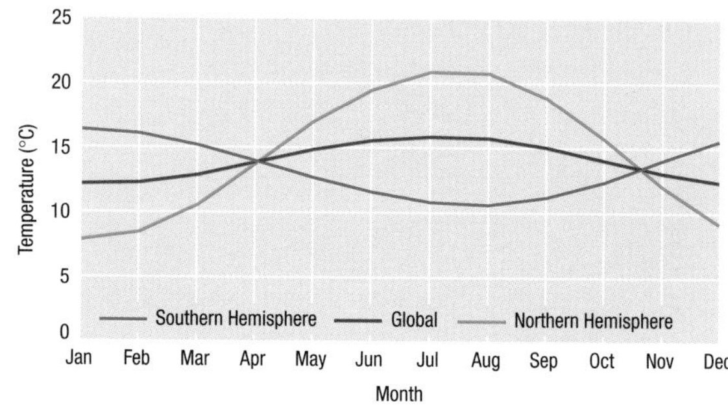

Hemispheric and global near-surface mean air temperatures in degrees Celsius, based on the period of the 1961–90 reference Normals, show a seasonal cycle.

Where has it warmed?

The distribution of near-surface warming has not been uniform across the globe. The most pronounced warming in recent decades has been over much of the northern continents and has been restricted largely to winter and spring. A few areas show cooling over the northwest Atlantic and middle latitudes of the North Pacific Oceans. The effect of the overall warming on extreme temperatures has shown up most noticeably in the Northern Hemisphere winter. For instance, in China the number of cold outbreaks and cold days

How much has it warmed?

The globally averaged annual mean temperature at the end of the 20th century was about 0.6°C above that recorded at the end of the 19th century, the rise being slightly greater in the Southern Hemisphere than in the Northern Hemisphere. The seven warmest years in the 1856–2000 records were recorded in the last decade with 1998 the warmest year, probably not only of the 20th century but also of the whole millennium.

The greatest warming of the 20th century was during the two periods 1901–44 and 1977–99. Over each of these two periods global temperatures rose by about 0.4°C. In between there was a slight cooling, which was more marked in the Northern Hemisphere. In recent decades there have been greater increases in night-time minimum temperatures than in daytime maximum temperatures. This change may be associated with an increase in cloudiness. Surface observations suggest that around the world cloud cover increased between 3 and 9 per cent during the 20th century and, although there is some uncertainty as to the accuracy of these measurements, the increases have been greater over the continents than over the oceans.

The time series show anomalies of annual near-surface air temperatures over land from 1856 to 2000, relative to the 1961–90 reference Normals, for the Northern Hemisphere, the Southern Hemisphere and the globe.

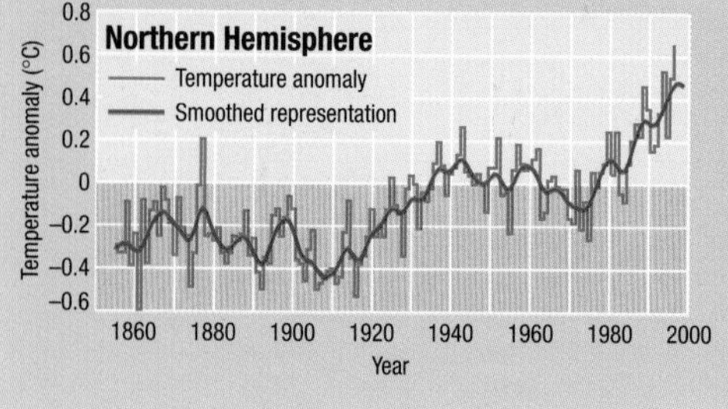

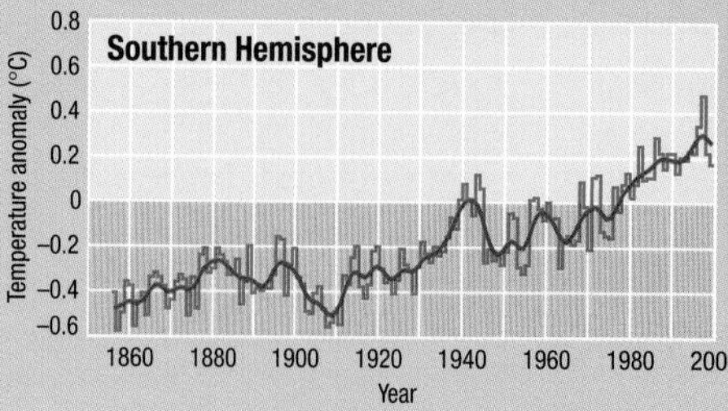

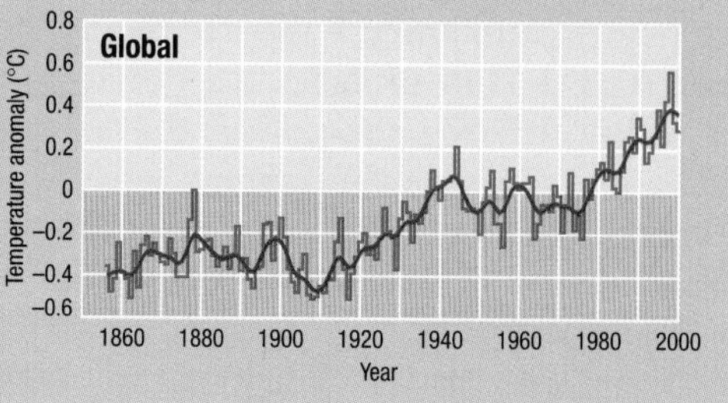

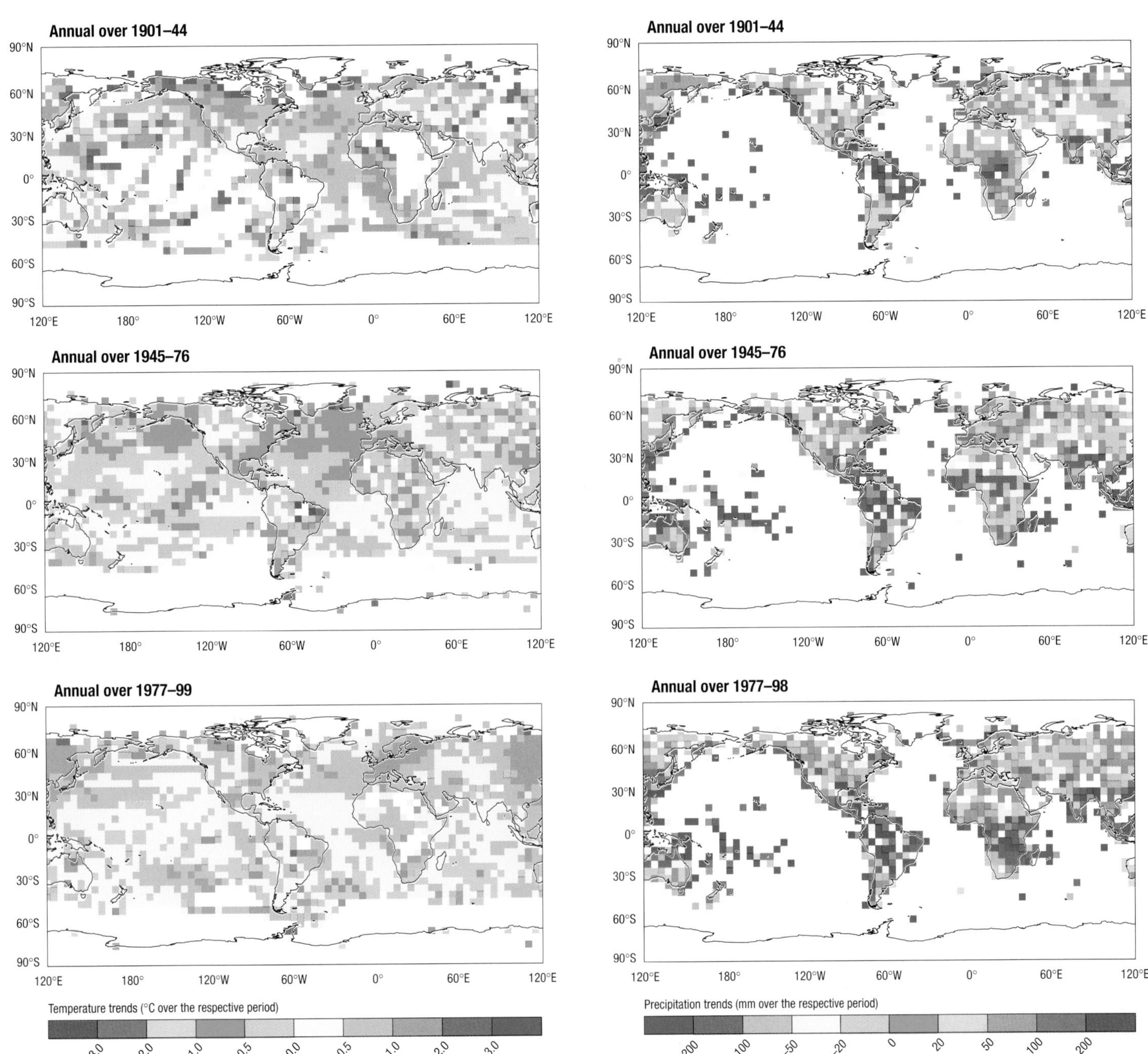

has declined sharply since 1950 and a significant rise in minimum temperatures has occurred. In contrast, there is no significant trend in mean maximum temperatures for the country as a whole, and in eastern China there has been a summer cooling.

Precipitation changes

Precipitation is much more variable in both space and time than temperature. Moreover, reliable long-term records exist only over land, and even here the coverage is far from complete. Estimates of annual total precipitation over the land areas of the globe (except Antarctica) show a slight increase during the first half of the century, followed by a smaller decline since then. An important factor in the interpretation of the data is the marked year-to-year variability, especially in the early part of the record, which raises questions about the accuracy of these observations. Precipitation has increased over land in high latitudes of the Northern Hemisphere during much of the 20th century, especially during winter. A step-like decrease of precipitation occurred after the 1960s over the subtropics and tropics from Africa to Indonesia. Both these changes in precipitation occurred as temperatures in these respective regions were increasing. There are, however, some specific regions that appear to be getting wetter (e.g. central and eastern USA, Argentina and the Amazon Basin).

The maps of near-surface air temperature trend (left) show consistent regional patterns over many parts of the Earth during the two periods of global warming (1901–44 and 1977–99) and the period of slight global cooling (1945–76). Regional precipitation trends over land areas (right) for almost the same periods do not show the same coherence.

2.1 The Earth's energy budget

All aspects of the Earth's climate – the wind, rain, clouds and temperature – are the result of energy transfers and transformations within the atmosphere, at the Earth's surface and in the oceans.

To understand how the Earth's climate system functions, we need to know about how the Sun's energy drives the whole system. At the most fundamental level these processes are all about how much solar energy is absorbed by the Earth and how this energy is re-radiated at longer wavelengths to space in the form of infrared terrestrial radiation. Over time the Earth's climate system remains largely stable because the energy received is equal to that lost. Much of our knowledge of the climate system is built on discoveries made at the end of the 19th century about the basic physical laws of emission and absorption of electromagnetic radiation and how these related to incoming solar radiation and outgoing terrestrial radiation. The other part of our understanding of the radiation 'budget' comes from the discovery of how atmospheric gases and other components of the climate system (e.g. the Earth's surface, clouds, aerosols and particulates) absorb and re-radiate radiation. At the surface green vegetation absorbs more of the Sun's

The Earth's energy budget

The Earth's response to sunlight depends on the properties of this solar radiation. It is at its most intense near the middle of the visible portion of the electromagnetic spectrum. Almost 99 per cent of the Sun's radiation is contained in the wavelength range 150 to 4000 nm. Some 9 per cent falls in the ultraviolet, 45 per cent in the visible and the remainder at longer, including infrared, wavelengths. When there are clear skies most of the solar radiation reaches the Earth's surface, although the shorter ultraviolet radiation is absorbed by oxygen and ozone in the stratosphere and some of the longer wavelengths are absorbed by water vapour and carbon dioxide at lower levels.

The Earth's surface and atmosphere radiate energy primarily in the mid-infrared range (4000 to 50 000 nm). The amount of this long-wave terrestrial radiation emitted increases with temperature. The principal atmospheric gases (nitrogen and oxygen) do not absorb appreciable amounts of infrared radiation, therefore the radiative properties of the atmosphere are dominated by certain trace gases that do, notably water vapour, carbon dioxide and ozone.

The effect of the radiatively active trace gases on terrestrial radiation is complex. Each species has a unique set of absorption and emission properties known as its molecular spectrum. This means that the atmosphere absorbs terrestrial radiation strongly at certain wavelengths but is transparent at others. The wavelengths where the atmosphere is transparent act as a 'window' because terrestrial radiation emitted by the surface and clouds at these wavelengths is lost to space without attenuation. We need to know precisely in which wavelengths each gas absorbs and emits infrared radiation to understand how the climate is altered by variations of the concentration of these gases.

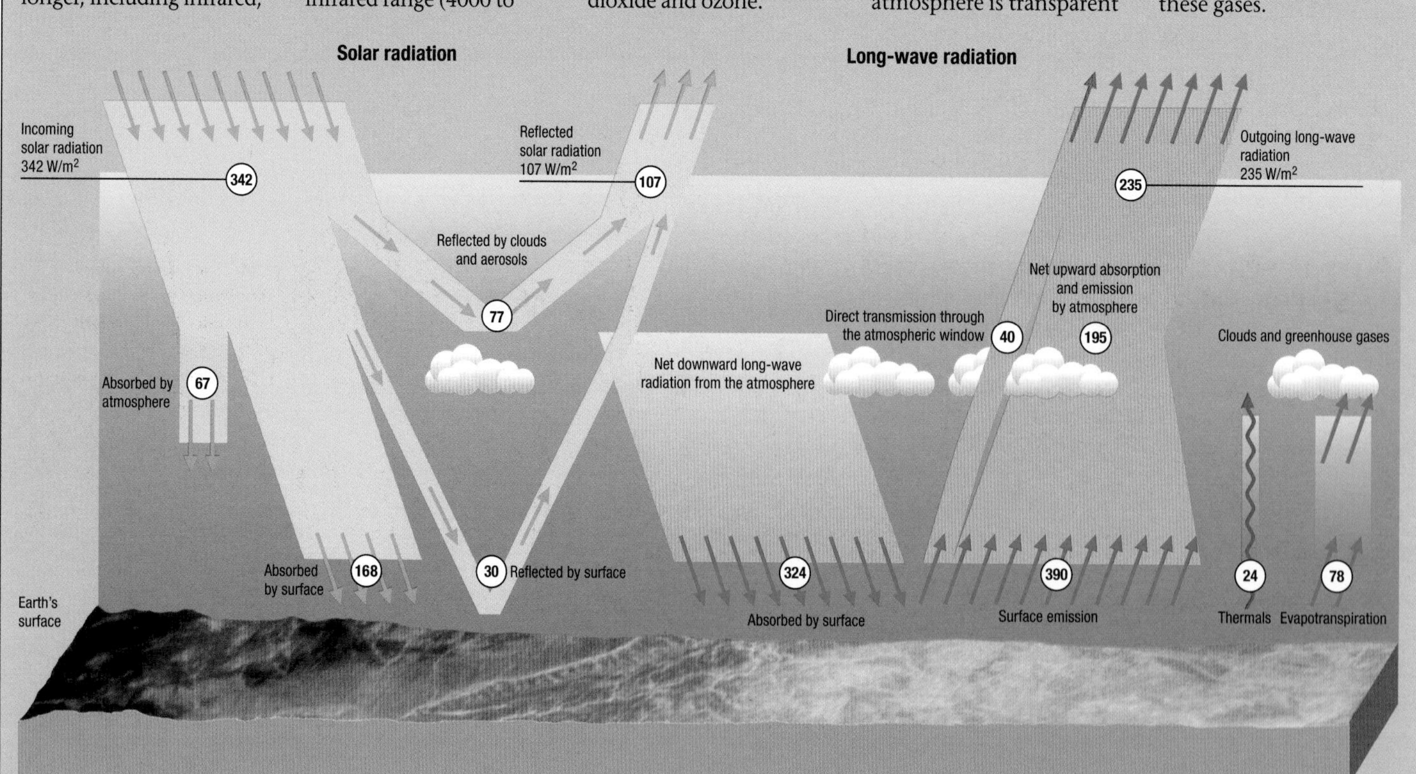

radiation than does bright snow, for example, and determining the detail for the many and various surfaces of the Earth is a huge task. The complex structure of clouds makes scientific study of their interplay with incoming and outgoing radiation a real challenge as well.

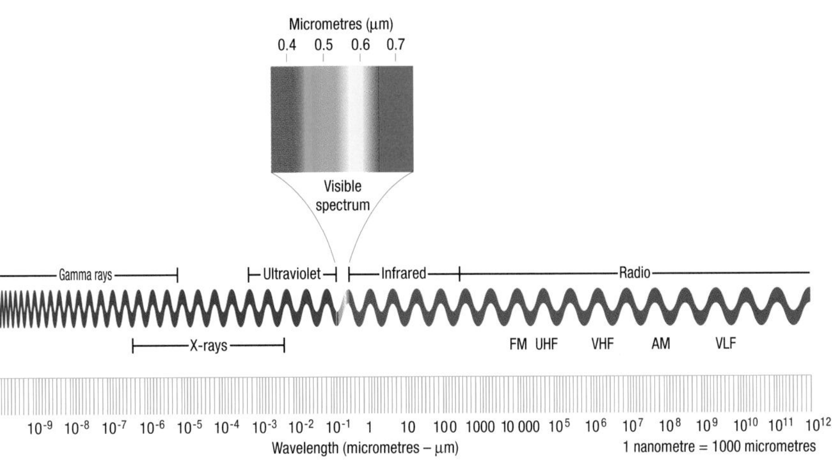

Radiation is the transfer of energy by electromagnetic waves. This radiation ranges from very long radio waves to ultra-short gamma rays. The Sun's radiation is concentrated in the ultraviolet, visible and near-infrared wavelengths while outgoing terrestrial radiation is concentrated at longer mid-infrared wavelengths.

Solar and terrestrial radiation

The Sun emits radiation as if its surface temperature were about 5700°C. However, the intensity of its radiant heat decreases with distance and so the Earth is bathed in an average solar influx of 1370 watts per square metre (W/m^2). As the Earth is spherical, each square metre receives on average one fourth of this, or about 342 W/m^2. We can calculate how much energy the Earth will radiate to space for any given temperature to balance this incoming energy. In so doing, we have to allow for the fact that the Earth only absorbs about 70 per cent of the radiation from the Sun (the rest is reflected back into space). In theory, the Earth's average surface temperature might be expected to be around –19°C, much lower than the observed value of about 14°C. The reason for this difference is the Earth's atmosphere, which itself absorbs and re-emits energy. This naturally occurring process is commonly known as the Greenhouse Effect, which plays a significant role in setting the Earth's surface and atmospheric temperatures.

More in at the equator, more out at the poles

There is one other essential feature of the Earth's radiation budget. The majority of incoming solar radiation is absorbed at low latitudes, whereas even in mid-winter, the polar regions emit considerable quantities of infrared radiation. This means that, averaged over the year, there is a net inflow of energy in the tropics and a net outflow of energy in high latitudes. To balance these differences, the atmosphere and the oceans must transport energy towards the poles. This equator-to-pole or meridional radiation imbalance is the fundamental driving force of the climate system.

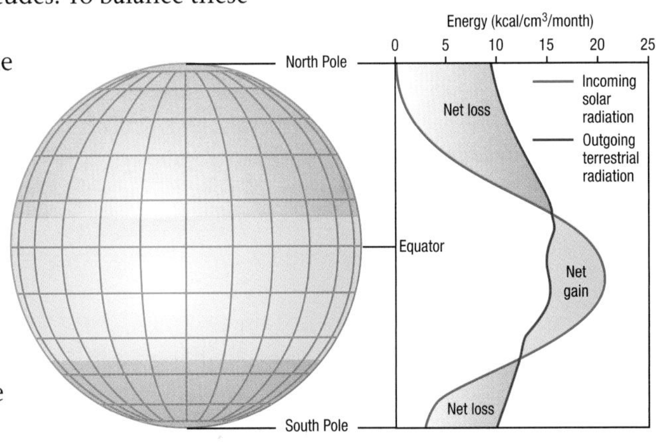

Most incoming solar radiation is received at low latitudes. At high latitudes the amount of outgoing terrestrial radiation exceeds the amount of incoming solar radiation, particularly in winter. To balance these differences, energy is transported polewards by the atmosphere and the oceans. This diagram shows the pole to pole radiation imbalance averaged around each latitude for the Northern Hemisphere winter half of the year (October to March).

The greenhouse effect

The impact of the naturally occurring, radiatively active gases can be calculated in terms of the contribution they each make to the warming of the Earth. This calculation shows that water vapour contributes 21°C, carbon dioxide (CO_2) 7°C and ozone (O_3) 2°C. These figures show that water vapour is the most important greenhouse gas. However, because of the abundance of water, there is virtually no direct human influence on the amount in the atmosphere. In contrast, the ability of human activities to alter the concentrations of CO_2 and O_3 in the atmosphere highlights the importance of addressing the specific activities in question, especially the burning of fossil fuels and the release of chlorofluorocarbons (CFCs). In addition, other products stemming from human activities (e.g. methane, oxides of nitrogen and sulphur dioxide) also modify the radiative properties of the atmosphere.

As the concentrations of radiatively active gases in the atmosphere increase, the amount of solar energy absorbed at the surface remains almost unchanged, but the lower levels of the atmosphere absorb more outgoing terrestrial radiation. This warms the lower atmosphere a little, which then emits a little more infrared radiation, both upwards and downwards. As incoming and outgoing radiation are balanced, the surface and the lower atmosphere warm, and the upper atmosphere cools. In consequence, the surface warms up and the atmosphere, as a whole, effectively radiates from a higher and cooler level.

2.2 Circulation patterns and global climate

Understanding the global climate depends on knowing how the transport of energy drives the motions of the Earth's atmosphere and oceans. Building up knowledge about atmospheric circulation, ocean currents and how water vapour moves through the climate system is central to this understanding.

East winds
West winds
Trade winds
Meridional circulation
East winds
Trade winds
West winds
East winds

Near the surface the circulation of the troposphere is marked by semi-permanent regions of low pressure where the air is usually rising and regions of high pressure where it is descending.

Painstaking research has revealed many of the essential features of atmospheric and ocean circulation patterns, such as the jet streams, Hadley cells, the Gulf Stream and other major ocean currents. There is still much to be learned about the dynamics of the oceans and ocean–atmosphere interactions (see 2.5).

The circulation of the atmosphere is a process in three dimensions. In the equatorial regions the planet is girdled by an extensive belt of intense convective activity and rising air, known as the intertropical convergence zone (ITCZ), that generally shifts with the seasonal latitude of maximum solar heating. During the northern summer the ITCZ develops into the more sustained expansion of the monsoon over much of southern Asia. Also, activity around the ITCZ can erupt, now and then, in the form of tropical cyclones. Similar processes operate in the Southern Hemisphere during the southern summer, but the definition of the ITCZ is less pronounced and there the number of tropical cyclones is only about half the northern figure.

To the north and south of this tropical hive of activity, the continental deserts mark regions of high atmospheric pressure, subsiding air and low rainfall. Further poleward still, the middle latitude depressions swirl endlessly around the globe, often steered by concentrated cores of strong westerly winds aloft known as jet streams. These 'rivers' of air are usually found between altitudes of 9 and 12 km. Wind speeds are at a maximum during winter and often average near 180 km/h, although peak speeds can exceed twice this value. Jet streams can be very turbulent and hazardous for aircraft, with speeds changing rapidly over small height intervals ('wind shear').

In the Southern Hemisphere, this procession continues unabated throughout the year with only a modest shift to higher latitudes in the austral summer. In the Northern Hemisphere middle latitudes there is a crescendo of activity in the winter half of the year, with a marked lull during the summer. In polar regions the annual melting of snow and ice in the summer contributes to increased cloud cover. In both hemispheres, the variation of the polar pack ice is substantial. As well, during the winter, much of the northern continents are blanketed by snow, the extent of which varies markedly from year to year.

The vertical motion in the atmosphere associated with these climatic patterns is central to understanding the processes at work. Latent heat energy is released into the atmosphere as rising air cools and moisture condenses to form clouds and rainfall. In the tropics, particularly in the ITCZ, the air primarily rises buoyantly within short-lived convective clouds (called 'hot towers') and rainfall is often very intense. When the rising air reaches an altitude of around 12 to 15 km and virtually all the moisture has been extracted it spreads out. Descending air on each side of the ITCZ creates zones of dry, hot air that maintain the deserts in the subtropical regions of the world. At the surface the Trade winds flow back towards the ITCZ. First interpreted by George Hadley in 1735, this basic circulation pattern now bears his name (the Hadley cell).

As they swirl around in middle to high latitudes, depressions cause huge quantities of air to rise over wide

Satellite observations of the elevation of the world's oceans can be analysed to provide a measure of surface currents. Colour is used to show ocean topography and arrows show the speed and direction of ocean currents. This map shows how currents move clockwise around higher regions in the ocean in the Northern Hemisphere.

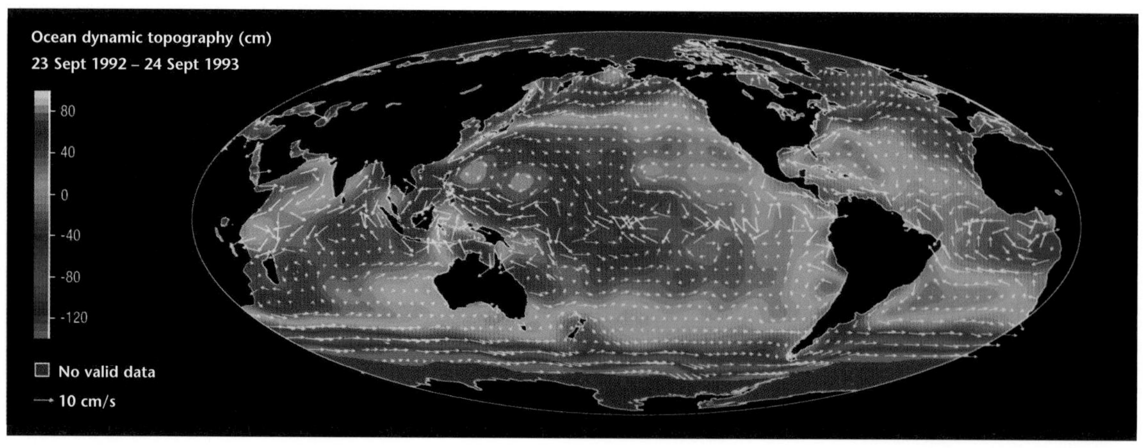

areas to form extensive layer clouds. Rainfall is generally spread more uniformly and is less intense than in the tropics, but can accumulate locally to produce high totals during storms. The air over polar regions also descends and creates the cold deserts of the Arctic and especially the Antarctic.

Gulf Stream

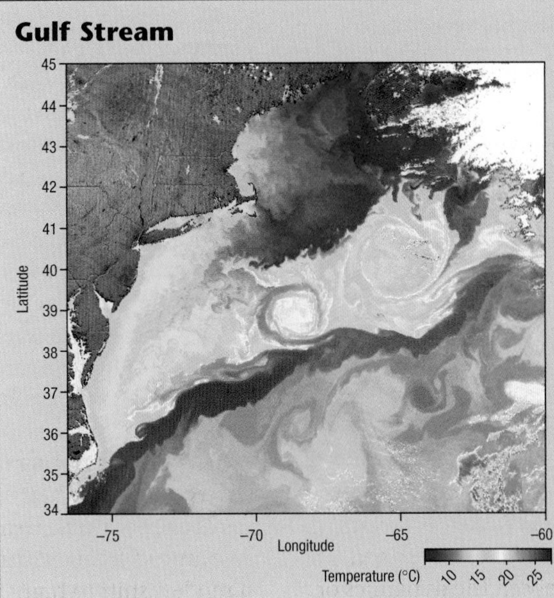

The Gulf Stream is the western boundary current in the North Atlantic. It sweeps warm waters out of the Gulf of Mexico and up along the southeast coast of North America. It then flows eastward and partly northward towards Europe, where it arrives still warm as the North Atlantic Drift. This flow of warm water maintains surprisingly high winter temperatures in northern Europe. At the same time, this flow is offset by deep cold water flowing from Greenland down to Newfoundland. Satellite images of sea surface temperatures provide clear pictures of the structure of the Gulf Stream, with a sharp boundary between the warm waters flowing from the tropics and the icy waters from the Arctic. The flow also contains swirls and eddies of warm and cold water that can be tracked for months.

Ocean circulation

Ocean circulations transport roughly the same amount of energy towards the poles as does the atmosphere. The basic form in both hemispheres is a basin-wide gyre with wind-driven westward flow in low latitudes close to the equator and poleward-directed currents along the western margins. Beyond about 35°N and 35°S the major currents sweep eastward carrying warm water to higher latitudes. This pattern is seen most clearly in the North Atlantic and North Pacific in the form of the Gulf Stream/North Atlantic Current and the Kuroshio/North Pacific Current. To balance the poleward flow there are returning currents of cold water moving toward the equator on the eastern sides of the ocean basins. In the Southern Hemisphere, because of the virtual absence of land between 35°S and 60°S, the ocean gyres are linked with a strong circumpolar current around Antarctica. There are also regions of significant vertical motion associated with these global ocean circulations.

Hydrological cycle

The continual recycling of water between the oceans, land surface, underground aquifers, rivers and the atmosphere (the hydrological cycle) is an essential part of the climate system. Ice requires much energy to melt (latent heat of fusion) and water needs even more energy to evaporate (latent heat of vaporization), so the cycling of water through the atmosphere by evaporation and its subsequent precipitation is a significant mechanism through which energy is transported throughout the climate system.

How much and where water is absorbed into the atmosphere, the nature and form of the clouds it creates, and how quickly it is precipitated out again, are all processes of the climate system. Over the oceans, the temperature of the surface is generally the dominant factor in evaporation, with wind speed playing a secondary role. However, the fierce winds associated with tropical cyclones and other storms markedly accelerate evaporation from the ocean and feed more energy into the weather systems. Over land the presence of living matter complicates the issue considerably. The amount of water vapour passing into the atmosphere over land is a combination of evaporation of available moisture directly from the soil and transpiration from plants, defined as evapotranspiration. The amount of water vapour in the atmosphere affects its radiative properties and is a critical factor affecting the climate system.

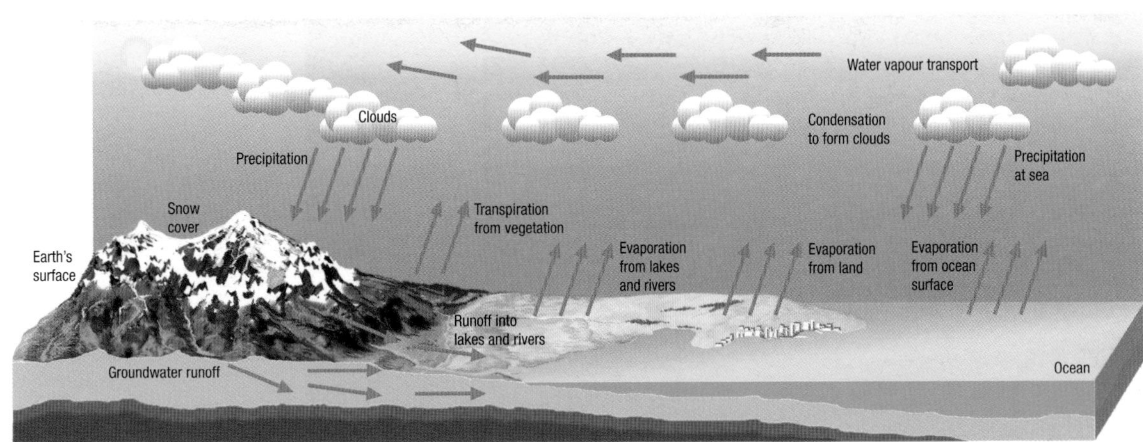

Water vapour is continually evaporating from the oceans, freshwater sources and soils and through transpiration from plants, forming clouds and eventually precipitating as rain or snow. This precipitation often returns quickly to rivers and oceans as surface runoff, but can remain for some time as soil moisture in groundwater, or as ice or snow.

2.3 The framework to measure the climate

To measure all the aspects of the Earth's radiation budget and the processes of the climate system requires a comprehensive measurement system. Many components of this system were initiated during the 19th century and further developed during the 20th century, but there are still large gaps to fill in order to ensure an adequate measurement framework.

Scientific curiosity initially drove development of meteorological instruments and ongoing measurement of various climate elements. By the early part of the 19th century there were many observatories making and recording meteorological observations. Instruments of quite different construction and measurement scales were used in various parts of the world, producing data and assessments of the local climatology that were often not directly comparable. However, success in constructing rudimentary weather maps from these observations and the advent of telegraph led, in the middle 19th century, to an expansion of meteorological networks for weather forecasting and a move to standardize observing instruments and methods. The weather networks that evolved have had an emphasis on rapid collection of essential elements for forecasting, particularly temperature, pressure, humidity, wind speed and direction, and precipitation. Selected stations were also equipped with an expanded set of instruments to measure such elements as solar radiation, evaporation, sunshine hours, soil temperatures and daily minimum grass temperature (for frost occurrence).

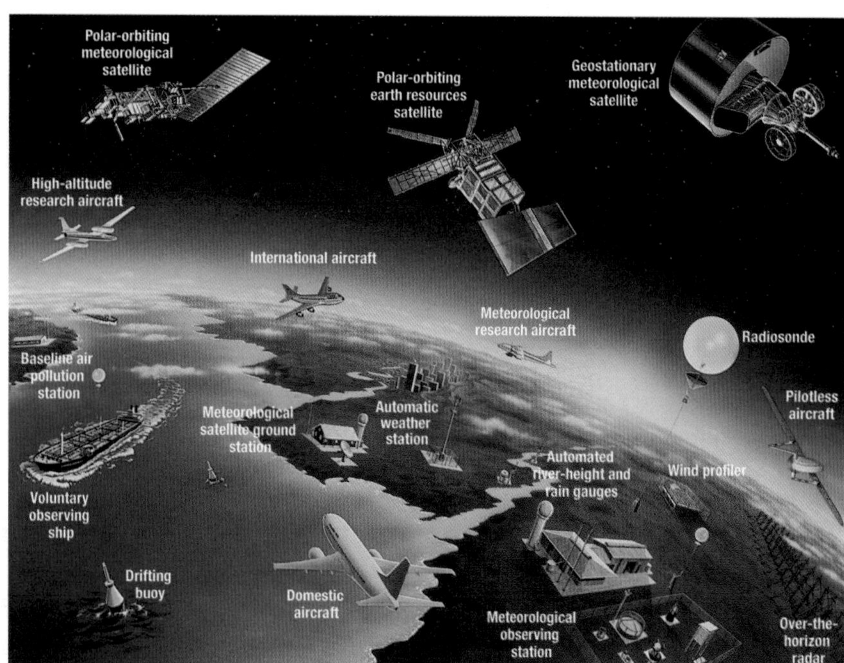

Monitoring the Earth's weather and building up the best possible picture of the climate involves a wide range of different observation systems, on land, at sea, in the air and from space.

Precipitation is the most variable of climate elements and its adequate analysis requires a much denser station network than other basic elements. The growth of networks and the dedication of volunteer observers is truly remarkable. Farmers, in particular, recognized the value of good precipitation records and many are avid volunteers. It is not unusual to find farming families who have made daily observations, from generation to generation, for more than 100 years.

One of the great strengths of basic climate records is that, for many elements, the instruments and methods of observations changed little for most of the 20th century. Only in the last decade did automated instruments begin to be deployed to retrieve measurements from remote regions, to modernize infrastructure, or to offset the increasing cost of staff to read instruments routinely day and night, seven days a week.

National climate archives

The volume of records held in the archives of many National Meteorological Services is truly awe-inspiring. The data were traditionally recorded in field books, each covering a month of observations, and many stations have data covering more than a century, which adds up to a huge amount of paper. Analysis of data in such manuscripts is a slow and tedious task. It became common at the end of the month to extract a few relevant statistics for each station. Average and extreme temperatures, monthly precipitation totals, strongest wind gusts, and number of days with frost and thunder were typical statistics computed and often published. In the 1930s, with the guidance of the International Meteorological Organization (IMO), countries began exchanging monthly climate summaries. Also for the first time, they published 30-year climate 'Normals' (statistics covering the standard period 1901–30) that formed the basis for comparison of regional climates and the identification of climate anomalies.

National Meteorological Services have been at the forefront of analysing data by machines. This began with the use of punched card systems in the 1950s and progressed through storage on magnetic tapes to directly accessible computer databases. Compared to the tedious search through paper records it is now a trivial task to interrogate the computer database to check for extremes, identify trends and compute complex climate statistics.

Constructing the global picture

A century scale data set was released in 1997 and consists of monthly surface observations from some 7000 land stations around the world. The Global Historical Climatology Network, Version 2, is one of the most complete of the digitized climate data sets available. Collecting and checking all these data was a laborious process, even with the assistance of computers, and required a great deal of detective work. For example, in order to get a long time series for a location it was often necessary to combine data from several nearby stations collected over different, but preferably overlapping, times. Without adjustment to compensate for different elevations (for example) at these sites, such merging can lead to biases and pose challenges for interpreting longer-term trends. Special statistical techniques were developed to form homogeneous series from merged data sets. The work also had to take account of differing national and regional procedures for generating apparently the same statistic. For example, in some countries the monthly average temperature is computed from daily maximum and minimum temperatures while in others it is computed from hourly or three-hourly temperatures. Each method produces a slightly different result.

Even where comprehensive records were kept of how and where the measurements were made (called 'metadata'), there remained demanding tasks in generating homogeneous time series for identifying real long-term trends. Changes to the instrument specifications and the introduction of automated systems can create their own characteristics. For example, sensitive thermistors used in automated systems are more responsive than traditional mercury-in-glass thermometers . The heating effects of urbanization have been another major challenge, and the movement of weather stations from downtown sites to airports in the nearby countryside has further complicated the analysis effort. Even far from the cities one has to be on the lookout for subtle inhomogeneities that have crept into the data because of site shifts over relatively short distances and the growth of surrounding vegetation. Effective climate research relies on meticulous attention to and correction for all non-climatic, especially human-induced, effects embedded in the archived data.

The Global Climate Observing System (GCOS)

Just how much work is involved in current international efforts to improve measurements of the climate is exemplified by the Global Climate Observing System. Originating from the 1990 Second World Climate Conference, and sponsored by the World Meteorological Organization, the Inter-governmental Oceanographic Commission of UNESCO, the International Council for Science and the United Nations Environment Programme, its priority is to ensure that data needs are met for seasonal-to-interannual climate prediction and for the detection and attribution of long-term climate trends. GCOS is being built on the existing systems for observing the atmosphere and the Global Ocean Observing System (GOOS) sponsored by IOC, ICSU, WMO and UNEP.

GCOS aims to fill important geographical gaps and expand the coverage to fully encompass the oceans and land surfaces. Space-based and *in situ* networks measure sea surface temperature, winds, waves, salinity, sea level, sea ice properties, surface and subsurface currents and other properties. For example the Argo Programme, a component of the Global Ocean Data Assimilation Experiment (GODAE), is a GOOS pilot project to introduce a global array of some 3000 ocean buoys to undertake large-scale sampling of the temperature and salinity of the ocean from the surface to a depth of 2000 m.

Observing stations must be maintained in all parts of the world to ensure an adequate coverage of surface conditions. These range from the balmy warmth of Samoa in the central Pacific to the Arctic chill on Bear Island.

As the oceans cover more than two-thirds of the Earth's surface, research vessels, like Japan's *Keifu Maru,* play an essential part in gathering data about the climate at sea.

2.4 Our accumulated knowledge of climate

Weather and climate knowledge is used in a wide variety of everyday services. From daily forecasts and reports, to statistical advice on one-in-fifty-year extremes, which need to be considered when building bridges or dams, there is abundant evidence of how much we know about the weather and our climate and how they are changing.

When people pick up their daily newspapers and look at the weather maps showing what the atmospheric conditions are expected to be that day, it is easy to overlook how much effort has gone into producing this standard service. First, there are the huge numbers of measurements that have been made around the world. These are the raw material for generating forecasts. Second, the powerful computer models that produce the forecasts do a mind-boggling number of calculations, and provide forecast products that can be used with confidence almost anywhere in the world. Third, all these data can be processed further and interpreted to develop information that is useful to the community. These activities rely on knowledge of both the physics of weather and climate and on the accumulated experience of the meteorological profession. Only by being able to appreciate the capabilities of the complex models and how their output relates to the detailed behaviour of the weather and climate can the forecasts be made more useful to the general public.

Some people might argue that the example of weather forecasts in a climate context is a double-edged sword. The fact that forecasts are inevitably sometimes wrong is clear evidence that even at short timescales we do not yet know all the answers. However, this criticism loses sight of the real progress that has been made in forecasting on longer timescales, especially for some critical areas of the globe. This progress will be further explored in this section.

Historical records that provide graphic images of past weather disasters contain useful information about climate change. This woodcut shows a flood that struck the Bristol Channel in western Britain in early 1607.

The Archival Climate History Survey (ARCHISS) project

Vital data and information, including documentary evidence, that pre-date official meteorological services can be found in such diverse locations as museums, colonial and company records, and even the Vatican archives. The ARCHISS project aims to locate and retrieve such information, and strives to make it available in digital form. This project, jointly sponsored by the World Meteorological Organization, UNESCO, the International Council for Science, and the International Council of Archives, has completed a number of successful data retrievals. In 1996–97, for example, a significant amount of serial instrumental data from the period 1753 to 1894 for about 30 stations was retrieved and digitized from the archives of Mexico and Cuba.

Our accumulated knowledge of climate

Much of what we know about climate and how it is varying is based on bitter experience. The patient collection of data, day in and day out, forms the basis for many insights on how the climate works as well as how it can change. When it comes to exploring our vulnerability to the varying climate, however, it is the extreme events that provide the most important messages. Extreme events will form a significant focus of the rest of this section and much of the next one, because they have a disproportionate impact on the lives we lead, and so represent a vital aspect of climate variability and change.

Historical documentary sources

For times before instrumental measurements of the climate were made, some of the gaps can be filled in with documentary records of events that relate to the weather. These include direct references to the weather, agricultural records, wine harvest dates and phenological records. The latter consist of compilations of annual records such as leaf opening, flowering, fruiting and leaf-fall together with climatic observations. Phenological calendars were used long ago in both China and the Roman Empire. A lot of work was done on this subject up to the mid-20th century, especially in Europe, but fell out of fashion. Recently there has been a revival of interest in the subject in both Europe and North America as part of efforts to identify the effects of global warming on flora and fauna.

Proxy data

Where we do not have instrumental observations or documentary records, much of our information about the climate has been obtained from what are known as proxy data. These are obtained by analysing a wide variety of materials whose properties are influenced by the surrounding climate. Tree rings, ice cores, ocean sediments, coral growth rings and pollens are the best known examples. Proxy data rarely ever provide a direct measure of a single meteorological parameter. For instance, the width of tree rings is a function of temperature and rainfall over the growing season, and also of ground water levels reflecting rainfall in earlier seasons. Only where the trees are growing near their climatic limit can most of the growth be attributed to a single parameter (e.g. summer temperature). For other records (e.g. analysis of the pollen content in lake sediments, or the creatures deposited in ocean sediments), drawing climatic conclusions depends on knowing the sensitivity of the plants or creatures to the climate and how their distribution might be a measure of the climate at the time.

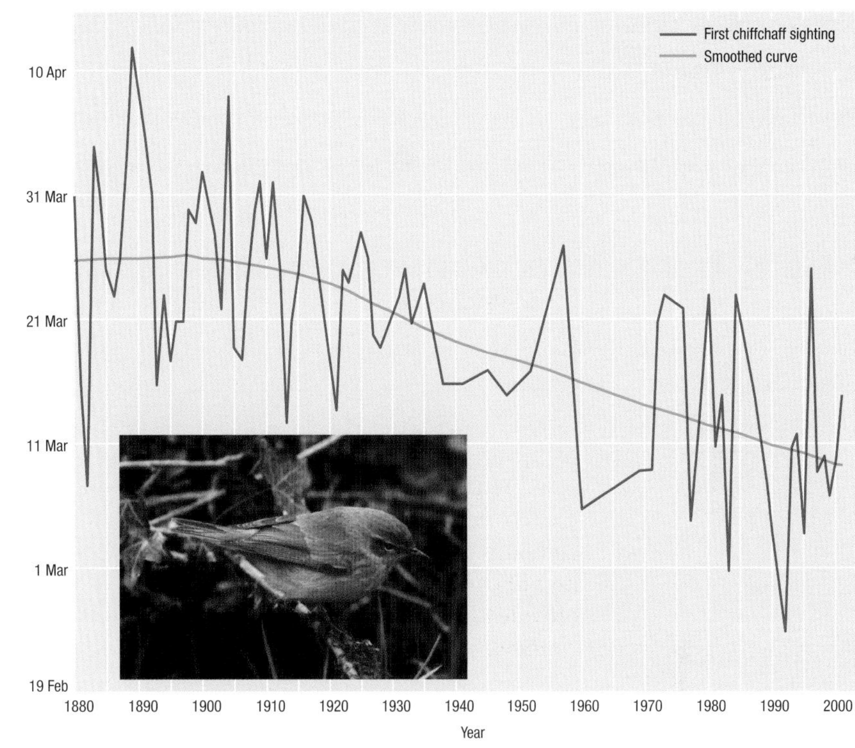

The timing of the return of migratory birds to their breeding grounds depends on the temperature. In Hertfordshire, southern England, chiffchaffs were returning some two weeks earlier each spring at the end of the 20th century than they were at the beginning of the century.

Obtaining proxy records

The search for proxy data involves working in many different parts of the world, whether drilling on coral reefs (top left) or studying the layers laid down in the glaciers of Peru (top right). The samples then have to be carefully handled, as in the case of ice cores (bottom left) and prepared in a suitable form for measurement as in the case of tree rings (bottom right). These processes involve building up a huge database of each form of proxy record to establish reliable statistical measures of the causes of the observed changes and the part played by climate in these changes.

2.5 On the frontiers of climate research

There are many features of the climate that remain a closed book to us. Filling these gaps is not simply a matter of making more standard observations. We have also to come to grips with those areas that have been largely out of reach. Nowhere is this more the case than in our gaps in understanding of how the atmosphere and the ocean interact and in what drives the motions of the deep ocean.

The atmosphere and the oceans are intimately coupled on every physical scale and all timescales, and it is not simply a question of one driving the other. As our knowledge of the behaviour of the climate has improved the importance of atmosphere–ocean interactions has become increasingly more evident. Underlying variations in sea surface temperatures, in particular those lasting from months to decades, are a major factor in driving climate variability. Over the tropical Pacific,

The Great Ocean Conveyor

It was only in the 1980s that Wallace Broecker of the Lamont–Doherty Earth Observatory of Columbia University, USA first pointed out how changes in the circulation of the world's oceans could explain some of the great mysteries of past climate change. Most important of these postulations were sudden shifts in the behaviour of what he called the 'Great Ocean Conveyor'. The broad features of this circulation of the world's oceans Broecker likened to a great conveyor that transports warm, low-salinity water from the tropical Pacific and Indian Oceans round South Africa to the North Atlantic near Iceland. During the northward transport evaporative heat loss leaves surface waters cooler, saltier and more dense. At temperatures near freezing point the very dense waters sink to the ocean bottom and flow back towards Antarctica, and eventually to the Indian and Pacific Oceans. This global circulation is particularly dependent on the salinity and temperature of water in the North Atlantic and could explain sudden changes in the climate at the end of the last Ice Age. Many of the details remain to be described, but what is so important about the observations is that they raise the spectre that large changes in the thermohaline circulation might be triggered by global warming.

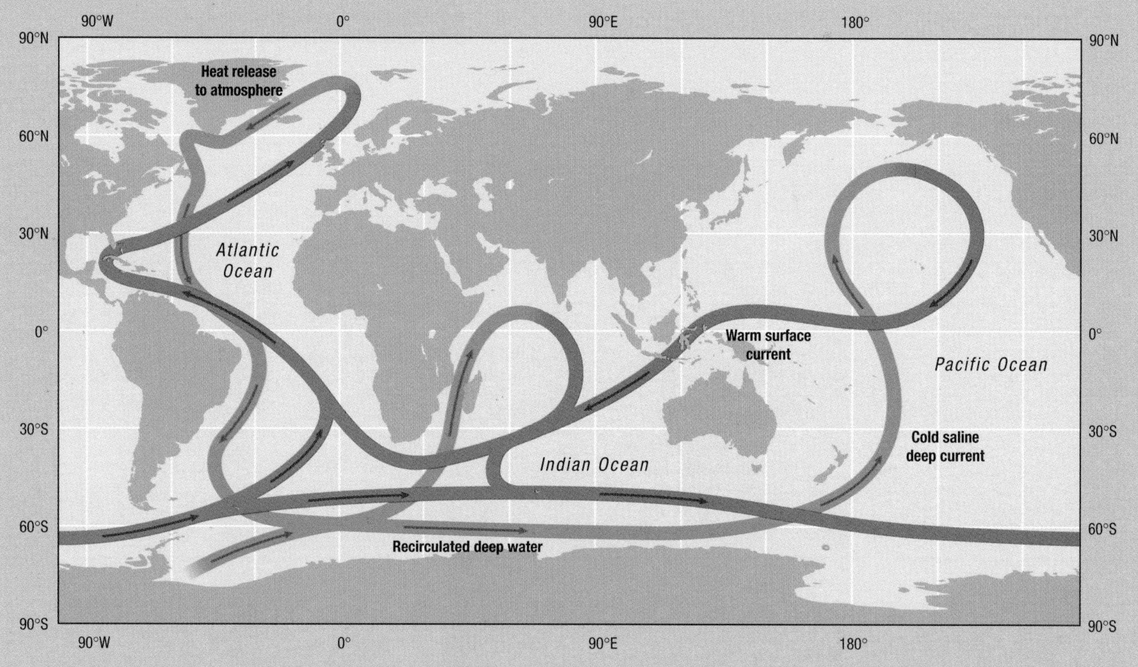

Understanding feedback processes

A major challenge in understanding climate change is to quantify various feedback processes. These arise because, when one climatic variable changes, it alters another in a way that influences the initial variable that triggered the change. If this circular response leads to a reinforcement of the impact of the original stimulus, then the whole system may move dramatically in a given direction. This runaway response is known as positive feedback. In the case of the climate, an example of this behaviour might be the effect of a warming leading to a reduction in snow cover in winter. This, in turn, could lead to more sunlight being absorbed at the surface and yet more warming, and so on.

The reverse situation occurs when the response tends to damp down the impact of the initial stimulus to produce a steady state. This is known as negative feedback. An example of this type of climatic response is where a warming leads to more water vapour in the atmosphere, which produces more low clouds. These reflect more sunlight into space, thereby reducing the amount of heating of the surface and so tending to cancel out the initial warming. Both of these forms of behaviour permeate the processes of the climate system and represent a major part of what we do not know about climate.

the events known as El Niño and La Niña are the offspring of ocean–atmosphere coupling. The challenges scientists are now confronting include whether these fluctuations can be predicted months and even years in advance, and whether the behaviours observed in the Pacific may have counterparts in the other major ocean basins.

Thermohaline circulation

The major process driving the formation of the deep ocean currents is known as the thermohaline circulation. Warm water is usually less dense than cold water and the saltier the water, the more dense it becomes. The density of seawater increases with declining temperature right down to its freezing point at around –2°C. Further, the salinity of the surface layers of a body of water increases through evaporation, but it is reduced by the addition of freshwater from rainfall or runoff from rivers and melting of the ice sheets.

Over much of the oceans a layer of less dense water caps the colder denser water beneath. There is a sharp transition, known as the thermocline, between the surface water and the colder water below. If surface water becomes denser than the deeper layers through evaporation and cooling, it can sink to great depths. However, this occurs in large quantities to great depths in only a few places, notably in the northern North Atlantic and around Antarctica. More detailed aspects of the circulation and how it changes with time are only just beginning to emerge from present research, notably through the World Ocean Circulation Experiment, which ran from 1990 to 1997.

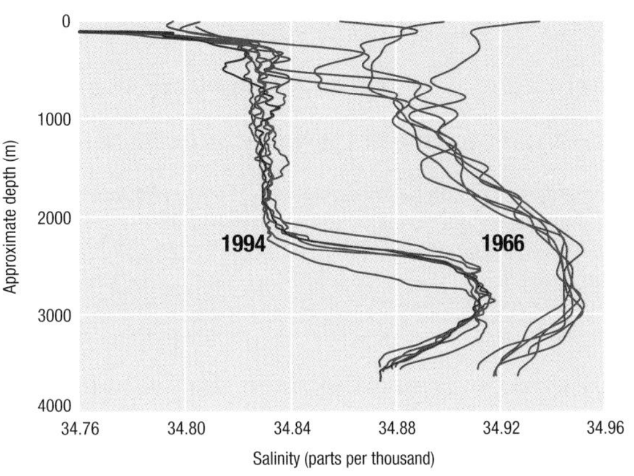

Vertical salinity profiles taken in the central Labrador Sea show that the salinity at all depths down to 3500 m decreased between 1966 and 1994, probably because a large volume of freshwater was exported out of the Arctic through Fram Straight and into the Labrador Sea in the late 1960s.

Henry Stommel (1920–92)

The American Henry Melson Stommel, senior scientist at Woods Hole Oceanographic Institute, is widely recognized as the world's most original and important physical oceanographer. He loved observational work at sea, helped design new instruments, taught at Yale, Harvard and at the Massachusetts Institute of Technology, wrote prolifically, and inspired others. Professor Stommel's research interests included the Gulf Stream, the Kuroshiro and Antarctic Circumpolar Currents, eddies, the thermocline, the thermal structure of the oceans, deep water convection, interactions between the oceans and the atmosphere, and climate. His fundamental contributions are largely responsible for the modern concepts of ocean circulation.

Is the climate chaotic?

A chaotic system is one whose behaviour is so highly sensitive to its current state that it is impossible to define that state with sufficient accuracy in order to make a precise prediction of a future state. Even quite simple systems can exhibit chaos under some conditions. The atmosphere is a turbulent fluid and the chaotic behaviour of weather systems means that forecasts lose much of their skill beyond a week to 10 days.

Although the atmosphere can be chaotic on a day-to-day basis, the same need not apply to longer-term averaged conditions. We know that the climate in any particular part of the world at any given time of the year stays within relatively narrow limits: the temperature hardly ever rises above –20°C at the South Pole, or falls below 20°C in Singapore. The oceans strongly influence the behaviour of the atmosphere, and knowledge of slowly varying characteristics is being used to make useful predictions of seasonal weather in some regions. In addition, the much longer variations associated with the ice ages can be largely explained in terms of changes in the Earth's orbital parameters. These results imply that some features of the climate have a level of predictability well beyond the limits that apply to weather forecasting.

There are, however, many examples of the climate behaving in an apparently unpredictable manner. Between 12 900 and 11 600 years ago, for example, when temperatures in the North Atlantic region slumped back to depths comparable to the last Ice Age, and then showed a much shorter sudden decline of about 3°C some 8200 years ago. Both these events have been linked with features of the collapse of the huge Northern Hemisphere ice sheets occurring at the time. This suggests that while there are occasions when the climate behaves in a chaotic way, for the most part the regularity of seasons suggests it is not strongly chaotic.

2.6 Which events matter?

The threats to human security from weather and climate variability come in many forms. Some, such as tornadoes, are swift and deadly while others, such as drought, are slow and insidious but in the long run can be far more damaging. The development of meteorology has been driven to a large extent by a desire to learn from the past and to minimize future human and economic disasters.

In the simplest terms, the events that matter most are those that do the greatest harm to people, their possessions, their livelihoods and the environment around them. It is important, though, to look beyond the basic and often grim mortality and economic statistics, and to examine not only the wider implications of events but also what society as a whole can do to mitigate the impact of future extremes. A combination of forecasts, sensible planning and wise precautions can prevent us suffering heavily from disasters in the future, and this is the essential benefit of an improved understanding of our climate. In terms of assessing which extreme events matter most, we will first consider those events that have driven us to find out more about the climate. These events were not necessarily the greatest disasters, many of which are covered elsewhere in this book, but those which exercised a critical influence on our thinking about how to prevent us suffering the same fate again and again.

For those in peril on the sea

The climatic events that have the greatest impact on the largest number of people are droughts and floods. As much of the early work on meteorology was conducted in the industrialized countries of Europe and North America, it was inevitable that extreme events in these parts of the world would have a significant influence on the direction and urgency that climate studies subsequently would take.

Disasters at sea were a major reason for early decisions to set up national weather forecasting services. A devastating storm hit the Anglo-French fleet in November 1854, in the Crimea. It destroyed 13 transport ships laden with food and clothing for the troops and forage for the horses. This catastrophe caused more hardship, disease and loss of life during the subsequent winter than did military action, and triggered initiatives to set up both the British and French weather services. Similarly, the storms that hit the Great Lakes of North America in 1868 and 1869, sinking or damaging over 3000 vessels and killing 500 people, led to the establishment of the US Weather Bureau and the Canadian Weather Service. In a different way, the sinking of the *Titanic* in 1912 also had a profound impact on the provision of meteorological services.

The realization that we were almost completely in the dark when considering many types of extreme events was unfortunately compounded by a tendency to turn a blind eye to the potential threat until a disaster of sufficient magnitude struck. In many cultures the disastrous impacts of severe events were termed 'Acts of God', implying that nothing could be done about extreme events. The enquiry following destruction of the Tay Bridge in Scotland on 28 December 1879 by gale-force winds causing the loss of an entire train and all on board, identified that natural disasters could be mitigated and led to much tighter rules on the standards of construction of bridges. Sometimes, though, it took longer for the message to sink in. For instance, storm surges have been a scourge of the communities around the North Sea for centuries. London suffered serious flooding with loss of life due to storm surges in 1881 and 1928. However, it was not until after a catastrophic flooding in

The destruction of the Tay Bridge across the Firth of Forth in Scotland by a mighty storm in December 1879 provided a graphic illustration of the need for engineering design to take adequate account of the characteristics of extreme weather events.

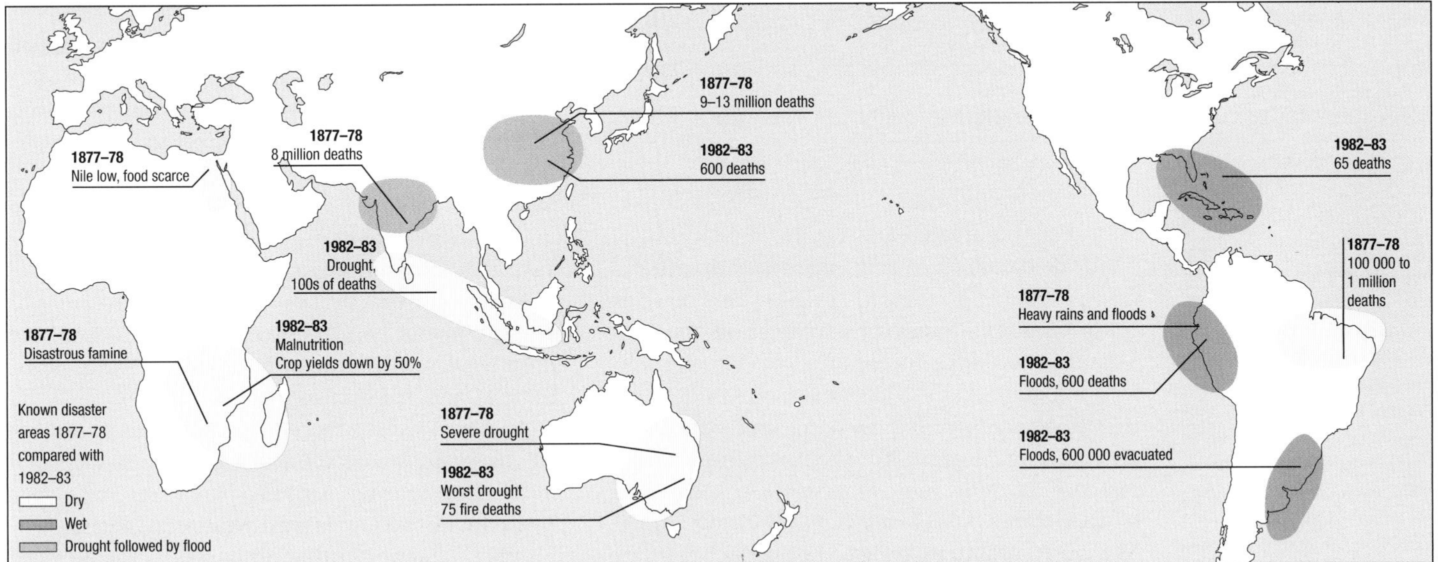

The consequences of the El Niño event of 1877–78 included famines and disasters around the world. A similar pattern of impacts occurred during the event of 1982–83, just over a century later, but the mortality was far lower. Improved national infrastructure, health and services accounted for most of the vast difference in mortality.

1953 that major work was set in hand, both to understand the nature of storm surges and to protect the city from further inundation.

Warfare also played a part in the development of studies of intense storms, notably in understanding the behaviour of wind and waves. Unexpectedly heavy swell during military landings (Operation Torch) on the Atlantic coast of North Africa in November 1942 proved a near disaster. These dangerous conditions were generated by depressions tracking far to the north in the Atlantic. It became clear that very little was known about the physical properties of ocean waves and their propagation over long distances. A research team was set up at the Admiralty Research Laboratory in England to develop techniques to study the distribution of wave periods and heights. This work acted as a stimulus to developing more general oceanographic studies of wind and waves after the war. The research underpinned designs for the safety of coastal and offshore structures, especially those associated with oil and gas fields.

The gradual process of learning about storms and other extreme events has developed around the world. The challenge of flooding, which results from events ranging from severe thunderstorms to massive hurricanes, has been a major driving force for meteorological science. In the USA, a massive hurricane that wiped out much of Galveston, Texas in 1900 had a profound impact which led to the restructuring of the National Weather Service, with the emphasis on improved observations and communications.

Global patterns of drought and floods

In some instances it was the sheer magnitude of a calamity that had caused many authorities to rethink their approach to disaster mitigation. Yet, how the many aspects of climatic extremes around the world were interconnected would only become obvious during the latter part of the 20th century, although possible linkages were appreciated in reviews of the catastrophic events of 1877 and 1878. During this period drought was widespread across northern China, India, southern Africa, northeastern Brazil, Australia and the islands of the South Pacific. The Nile River was exceptionally low during the period, causing scarcity of food in Egypt and indicating severe drought in Ethiopia. In stark contrast, heavy rains and flooding lashed the Pacific Coast of South America and a rare tropical cyclone struck Tahiti in February 1878. Drought followed by debilitating floods caused famine and disease outbreaks in China and India that led to the massive scale of deaths, in the order of about 9–13 million people and 8 million people, respectively.

In 1888 a similar pattern of extremes affected many parts of the globe. Particularly severe impacts were reported in India, Ethiopia, northern Brazil and Australia. In India it is estimated that more than 1.5 million people died of hunger and disease, while in Ethiopia more than a third of the population may have perished. This event reinforced the already growing efforts to understand the patterns of tropical meteorology and develop forecasting techniques to anticipate dry years. However, it would not be until nearly a century later when the record-breaking episode of 1982–83 occurred, that the global implications of the El Niño in the equatorial Pacific became fully apparent. By the time of the next extreme event of 1997–98, the improved understanding of global impacts of the El Niño and real-time observations tracking its development were beginning to be used for mitigation.

The development of better meteorological services has depended on improved science, the availability of new technologies, and the stimulus of events whose connection with weather and climate was sometimes less than direct.

The loss of the 'unsinkable' liner the *Titanic* in April 1912 provided tragic evidence of the vital importance of making effective use of reliable information about prevailing iceberg conditions in the North Atlantic. The chart shows that April is the most hazardous month for icebergs. The map shows that the *Titanic* was within normal maximum range of iceberg sightings reached in April each year.

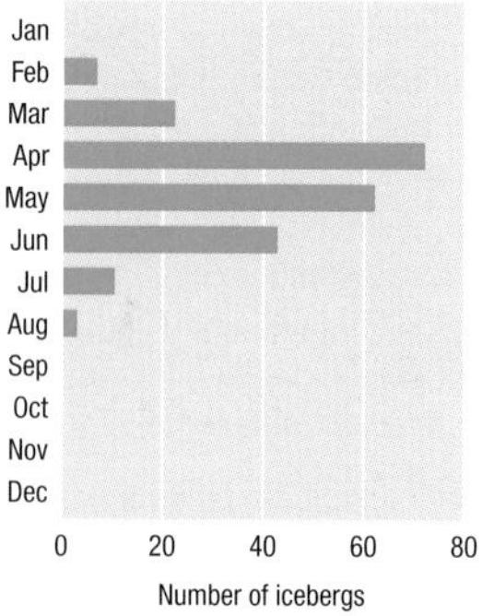

Many weather-related disasters led directly to public demands for improved services. In other cases, a realization that the scale of some disasters could have been reduced by application of existing knowledge and adaptation of emerging technologies had the same effect. The sinking of the *Titanic* transformed thinking about the use of radio communications to serve maritime meteorology. It is ironic that, while Wilhelm and Jacob Bjerknes were developing the classic model of extratropical depressions, the armies of Europe were bogged down in the mud, which contributed so much to the carnage of the Western Front. This experience convinced some European countries to develop a meteorological service as an integral part of their military capabilities.

Improved communications

The first meteorological services sought to exploit newly available measurement and communication systems to provide warnings of damaging storms. This approach relied on telegraphy, which enabled information about current weather conditions to be collected and used to provide some warning of advancing storms. In the United Kingdom, following the sinking of the steam clipper *Royal Charter* off the coast of Anglesey in October 1859 with the loss of 400 lives, the Meteorological Department of the Board of Trade started issuing storm warnings in February 1861. These were transmitted by telegraphy to coastal stations, where a system of 'cautionary signals' using cones and drums by day, and lights by night, was hoisted from a mast and yardarms, for shipping to see, when gales and storms were expected. Unfortunately, after a few years, the poor performance of these forecasts drew adverse criticism and they were abandoned in this form in 1867.

Losses at sea continued to have a major influence on the need for better weather science in the 20th century. Although not the consequence of severe weather conditions, the *Titanic* disaster in April 1912, with the loss of over 1500 lives, highlighted the importance of radio communications for informing ships of conditions at sea. The *Titanic* was equipped with radio and received reports from other ships about the extensive ice conditions south of Newfoundland during its fateful trip, but it failed to make effective use of this information. The scale of the disaster galvanized international action to establish more credible warning systems for ice conditions in the western North Atlantic. These have been built around ice-patrols maintained by the US Coast Guard since 1914, except during World War II. More generally, routine radio bulletins for shipping were introduced after World War I.

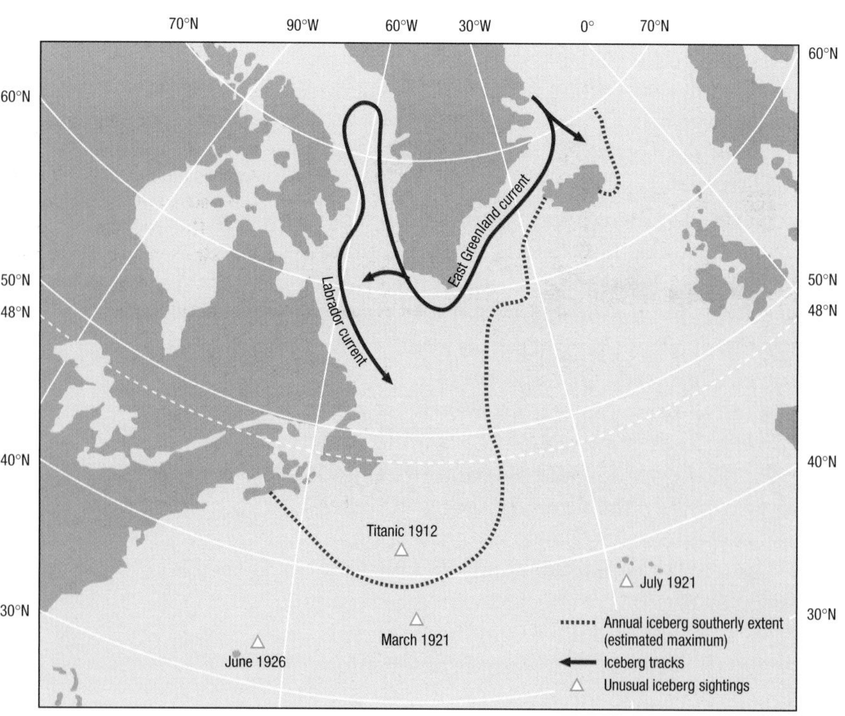

Forecasting the Indian monsoon

The famine in 1877, caused by the failure of the monsoon that year, led the Government of India to set up a system to forecast the monsoon. Over the next 50 years much progress was made in understanding the variety of factors throughout the tropics that could influence the Indian monsoon. It was not until the early 1980s and the major El Niño event, however, that the importance of much of this early work was fully appreciated. It was only then that it became generally accepted that changes in patterns of equatorial sea surface temperature could alter global atmospheric behaviour and cause droughts and floods, particularly around the tropics.

Understanding severe storms

The fact that the dynamics of everything from an isolated thunderstorm, through tropical storms to massive extratropical depressions more than 3000 km across are driven by the same physical processes has been a unifying influence in the development of meteorology. The input of energy in the form of both sensible heat and latent heat associated with condensation of moisture involves the same physical processes whatever the size of the storm. Consequently, meteorologists have been working on many aspects of cloud physics as part of the effort to understand and forecast the development of storms. This work varies from in-flight sampling of the properties of clouds to laboratory studies of individual ice crystals. A great deal of work on cloud physics throughout the middle part of the 20th century was aimed at efforts to modify the weather. Despite the importance of this basic research, it is somewhat ironic that many of the initial claims of successful weather modification proved illusory as more detailed statistically controlled trials failed to reproduce the earlier results.

Computer representations of the structure of hurricanes are used to improve understanding and make more accurate predictions of the likely landfall of these devastating storms. The graphic is a three-dimensional computer-generated visualization of Hurricane *Andrew* using actual data as it approached Miami in August 1992. The wind arrows follow the spiralling inflow near the surface, the rising air in the cumulonimbus clouds of the eye wall and the outflow in the high troposphere.

Remote sensing

The challenge of measuring conditions within intense storms, and the more general issue of measuring the dynamics of the weather from afar, has led to a wide variety of techniques for remote sensing of meteorological elements. The best-known method of measuring distant features of weather from the ground is radar. Scientists have also developed equipment using acoustics, radio waves and lasers to bounce signals off atmospheric features and clouds to detect their properties. In addition, they have developed various types of radiometer to detect the different forms of radiation emitted by the atmosphere and the amount of sunlight transmitted or reflected by the atmosphere at different wavelengths.

The same range of techniques is also being exploited in weather and research satellites. Measurements are made of the properties of the atmosphere, the land surfaces (including vegetation and snow characteristics) and the surface of the oceans (including sea level changes, wave height and direction, the strength of currents and the presence of sea ice). These techniques have provided new insights into the behaviour of weather systems and their climatic consequences. Only with better measurements is it possible to test and refine physical theories about how the climate system works. A good example is the use of a recently deployed geostationary satellite in 1969 to provide warning of the course of Hurricane *Camille*, the most intense storm to make landfall in the USA in the 20th century. This warning helped prevent significant loss of life and firmly established weather satellites as one of the most important weapons in the meteorological arsenal.

No adequate observing system existed during the major El Niño event of 1982–83. This limited scientific progress and led to the international Tropical Ocean Global Atmosphere (TOGA) project, which included a real-time observing system including about 70 moored buoys across the equatorial Pacific Ocean, to track the ocean and atmospheric changes in the tropical Pacific that occur with the onset of El Niño. The system has proved invaluable and remains operational today.

2.8 Northern middle latitude winter storms

Throughout the middle latitudes, winter storms in the form of low pressure systems have the potential to cause widespread property damage and death. At sea, wind and wind-generated waves are the principal cause of damage. On land, wind damage and flooding are the real scourges, whereas close to shore storm surges are a major threat.

The strength of middle latitude westerly winds and the frequency of storms in the winter half of the year exert a profound influence on the climate of Europe. Moreover, any change that might strengthen these winds, such as a consequence of global warming, would have major implications. Variations in the number and strength of these storms in recent decades have been watched closely for evidence of a link to global warming. Remarkably, three of the most devastating episodes of European storminess occurred near the end of the 20th century. The relatively rapid succession of these episodes reinforced concerns that global warming could lead to more weather of this type.

Asian cold waves

The East Asian winter is dominated by cold and relatively dry winds. The Siberian anticyclone is a persisting climatic feature that blows Arctic air over Siberia and northern China. Part of the flow sweeps out toward the North Pacific and part southward through China to the equatorial regions.

The winter monsoon throughout this region is characterized by successive outbursts of cold air, called cold waves, that produce sharp falls in temperature of more than 10°C and are accompanied by snow and, in the south, rain. Snowstorms can be especially violent over northern China with sub-zero temperatures and gales lasting many days. The snowstorms can be damaging to communities and are particularly disruptive to transport, including coastal shipping. Periods of frost following the cold outbreak last several days and are a major hazard over the south of China as they have far-reaching effects on agriculture, especially on plants and crops.

The frequency of cold waves varies greatly from year to year; as many as 10 per year or as few as one have been experienced. An active period is associated with higher pressure within the Siberian anticyclone and an intense low pressure system near the Aleutian Islands.

North American east coast storms

On average, a dozen major storms hit the east coast of North America each winter. Many of the most damaging winter storms involve heavy snowfall, most frequently affecting the Maritime Provinces of

Periods of storminess wreak havoc in Europe

The 'Great Storm' that hit southeast England on the night of 15–16 October 1987 came in the early hours of the morning, and so the death toll was relatively low, with about 20 people killed. The economic damage based on insurance claims, however, exceeded £1.2 billion. More memorable to many people was the damage to woodlands with over 15 million trees blown down in southeast England. Many of these trees were well over 100 years old and had seemed a permanent feature of the landscape, which made their destruction so much more distressing.

During a five-week period in January and February 1990, the North Atlantic bred nine intense storms that struck northern Europe with hurricane-force winds. The absence of any blocking anticyclone meant they penetrated unusually far into Europe. The first major storm struck on 25 January 1990 with winds up to 200 km/h. In northwestern Europe it killed nearly 100 people and caused several US$ billion damage. Only four days later, on 29 January, a second storm battered southwestern England, toppling trees and causing landslides and floods. A third major storm, which struck during the first week of February, was one of France's worst storms in decades. Finally, at the end of February winds up to 160 km/h levelled huge stands of trees in central Europe. In Germany, the number of downed trees represented twice the average annual harvest. This storm killed 63 people in addition to at least 25 forest workers who died in the clearing-up operations.

A series of severe storms in December 1999 brought the highest local wind speeds ever recorded in parts of Europe. Gusts reached 185 km/h on the island of Rømø, Denmark, on 3 December. Then at the end of the month two storms in close succession brought gusts of 173 km/h at Paris-Orly, France and 213 km/h at Feldberg (altitude 1498 m), Schwarzwald, southern Germany on 26 December; and 198 km/h at Saint Denise d'Oleron, France on 28 December. The storms killed over 100 people and resulted in damage estimated in US$ billions. Numerous buildings and vast tracts of forest were destroyed with hundreds of millions of trees blown down; transport and power were interrupted for days in some regions.

Trends in storminess

In spite of what looks like an upsurge in middle latitude storminess during the 20th century, when we look more closely at the evidence it is not all that convincing. In part, this is a result of limitations in the measurements. Long-term wind measurements are not representative of peak storm intensity. A more reliable source of information about long-term changes comes from pressure charts. Over the oceans the additional measurements of both wave heights and tides provide another means of indirectly evaluating changes in extratropical storm strength and frequency.

Analysis of pressure charts has produced a gale index for the United Kingdom since 1881, and estimates of gale-force winds for the German Bight region during the 20th century. Neither of these indices shows evidence of long-term trends. Century-long records of storminess along the east coast of the USA also do not show any long-term trends.

In contrast, an analysis of daily surface pressure maps over both the North Atlantic and North Pacific since 1899 suggests a large increase in intense extratropical cyclones (where the central pressure fell below 950 hPa) in both the Atlantic and Pacific basins since about 1970. This trend may be due, in part, to improved observation and detection methods including the use of satellite imagery.

The exceptionally heavy snow of March 1888 along the east coast of the USA paralysed transport services. It took several weeks to get things back to normal.

Canada. Farther south, once or twice each decade, an intense low pressure system runs up the east coast of the USA dumping anything from 30 to 75 cm of snow on the major cities of the region. These storms draw cold air down from Canada in their train, and the combination of heavy snow, strong winds and plummeting temperatures paralyses the worst-hit places. Perhaps the best known example was the 'Blizzard of '88' which struck New York in March 1888, with temperatures down to –20°C, and leaving up to 130 cm of snow and drifts up to 7 m deep. In January 1922, the storm that dumped some 60 cm of snow on Washington, DC, causing the collapse of the Knickerbocker Theatre and killing nearly 100 people, became known as the 'Knickerbocker Storm'.

Winter storms have continued to cause frequent major disruption in recent decades. In March 1993 the 'Blizzard of the Century' produced record-breaking snowfall in many places from Alabama to Massachusetts. Then in January 1996, a storm almost as intense brought record snowfall to the same places, caused over 100 deaths, and closed down transport, industry and commerce over much of the region for several days. Two more storms struck the area within the following 10 days. Even when storms involve relatively mild winds and rain along the coastal strip, inland they can produce immensely damaging ice storms.

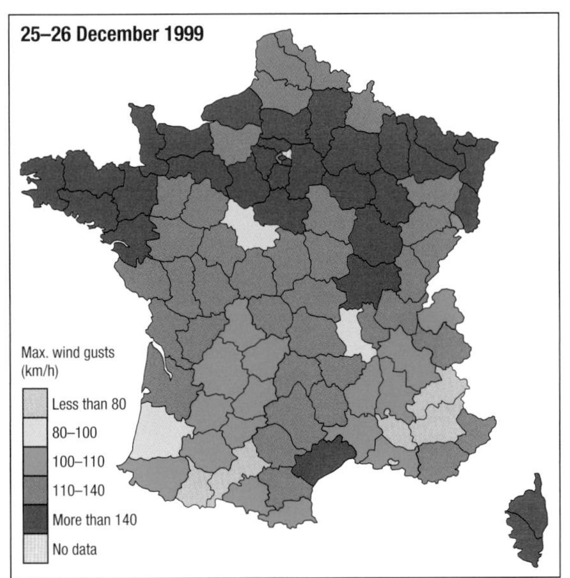

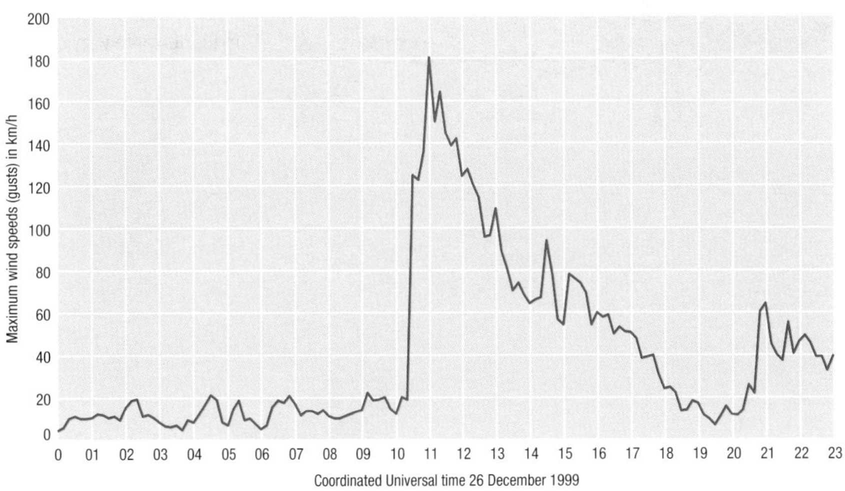

An intense low pressure system, named *Lothar*, roared across northern France into Switzerland and southwestern Germany in late December 1999. The map of strongest wind gusts on 25–26 December for France (left) and the wind trace for Brienz, Switzerland (right) showing the intensity of the peak gust on the morning of 26 December give some idea of why the storm caused such immense damage.

2.9 Cold winters and cool summers

Cold winters and cool summers have had a major impact on the economic and historical development of many middle latitude nations of the Northern Hemisphere during the 20th century. Abnormal amounts of snow and ice disrupt urban communities and crops fail to mature if the growing season is significantly shortened.

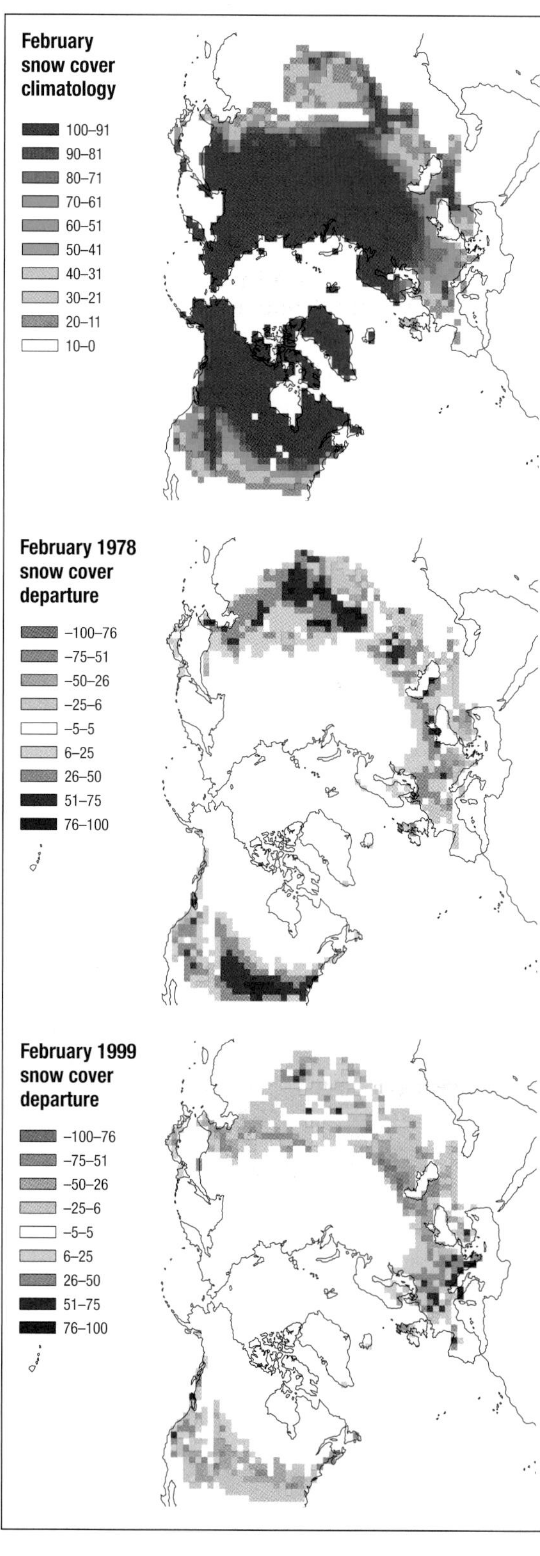

Satellites have been measuring winter snow cover across the Northern Hemisphere since the early 1970s. The map at the top shows the average February snow extent as a per cent of the area covered. Per cent coverages of snow vary in 10 per cent increments from 90–100 per cent (dark blue), through increasingly lighter shades of green to yellow to brown (10–19 per cent), to white (0–9 per cent). The year with most snow cover was February 1978 (middle) and the year of least snow cover was February 1999 (bottom). In these maps, positive anomalies range from dark blue (75–100 per cent) to light blue (5–25 per cent); anomalies between +5 and –5 per cent are shown as white; and negative anomalies range from dark orange (75–100 per cent) to tan (5–25 per cent). Anomalies are calculated from the 1967–99 average. The swathe of year-to-year variability is encompassed by the anomaly areas.

The economic and social impacts of extreme winters and summers illustrate how even modern societies are vulnerable to climatic variability. For example, during very cold winters the increased incidence of snow and ice can cause significant disruption, especially to the energy and transportation sectors. The increasing reliance on the automobile during the 20th century has raised people's mobility but has also raised the vulnerability of major cities to icy roads and snowstorms. Even a 2–3 cm snowfall can cause significant disruption, especially in cities where snowfalls are less frequent. By comparison, cool, cloudy summers may not provide sufficient warmth for the growth and ripening of crops, leading to local food shortages.

Mild winters

Throughout the 20th century there was very little overall trend in winter temperatures over northern Europe, and the frequent mild winters up to 1940 and since the mid-1980s were offset by more frequent colder winters in the years between. In the USA the trend for the contiguous 48 states shows an insignificant rise during the century. By contrast, for central and southern parts of European Russia there has been a warming trend in the cold half of the year. Ice now forms on the rivers 17–20 days later than at the beginning of the century and breaks up 7–11 days earlier. Winters in Canada, as well, have warmed overall, but the significant increase of 1.6°C over the period 1948–2000 (for the country as a whole) covers marked regional variations. Parts of western Canada have warmed as much as 3.9°C, while in the Atlantic Provinces winters cooled by 1.2°C.

The frequency of snow and the spatial extent of snow cover during a winter season provide a more effective measure than average temperatures because of the associated community disruption. Some estimates have been compiled for the whole century but, since the availability of satellite-derived snow-cover data in the early 1970s, more reliable data have been collected for the Northern Hemisphere. These data show that from 1971–72 onwards, decreases in snow extent have been large in spring, while in autumn and winter there has been no significant change. In spite of general warming in the climate, far northern parts of the world still must contend with cold winters.

Cool summers

In middle latitudes the interannual variation of summer weather is important for agriculture but the effects vary with crop type. Also, many different events can affect yield. For example, one-off extreme weather events, such as gales, storms and hail near harvest time, or severe frosts in late spring or early autumn, are especially destructive to otherwise healthy crops. The accumulated effects of build-up and drawdown of soil moisture act as a buffer in agricultural production for several months of the year. As a general rule, however, the combination of temperature and moisture during the growing season exerts the largest influence on yields.

In the grain-producing latitudes of the Northern Hemisphere, cooler, moister summers tend to produce good yields but prolonged heat and drought cause damage. This also applies to root crops and tubers (e.g. carrots and potatoes), which suffer most in drought conditions. However, persistently moist conditions are conducive to the spread of certain mildew and fungal spores that often require special measures for their control.

In northern Japan, by way of contrast, cool summers significantly reduce rice yields. There have been several very cool summers (June through August) in this region during the 20th century, notably 1902, 1913, 1941, 1954, 1983 and 1993, all of which caused disastrous crop damage. Overall, the first two decades featured cool summers with poor crops, which led to the need to import rice. Several cool summers in the 1930s and early 1940s caused harvest failures and social unrest. Since 1980, interannual variability of the temperature in the summer has increased with conspicuous cases of cool summers. Similar agricultural disasters have struck northeastern China. For example, the cool summers of 1972 and 1976 resulted in shortfalls in food production of 6.3 and 4.75 million tons respectively.

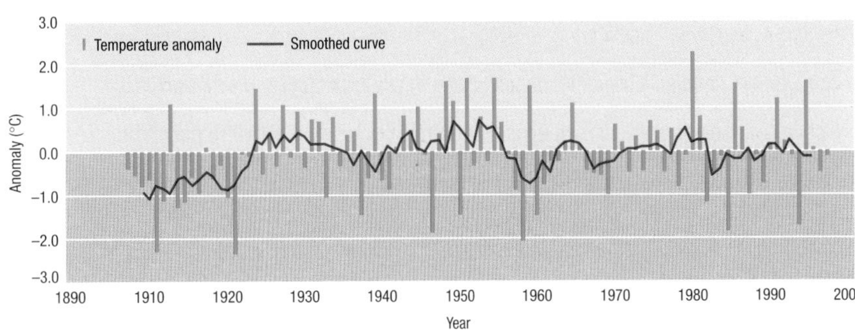

Over northern Japan during the 20th century there were considerable year-to-year fluctuations in the summer temperature. During the coolest years when negative temperature anomalies were greater than 1.5°C, poor rice crop harvests affected production.

Where is most vulnerable to cold?

In many populous parts of the Northern Hemisphere winter temperatures are often close to freezing. This means that during cold spells the temperature may drop below freezing or stay below freezing for relatively long periods. In countries within the temperate zones, there can be significant differences in severity of winters from one year to the next. Over many parts of England during the four coldest winters of the 20th century (1917, 1940, 1947 and 1963), the average daily temperature was below freezing for more than 40 days, whereas in six of the mildest winters there were no days in which the average temperature fell below this limit. Although the difference in the mean temperature between these two sets of extreme winters was only about 5°C, the disruptions caused by events of the colder winters were disproportionately large.

In general, major cities in the climatic belt that experience monthly average winter temperatures between –5°C and +5°C are much more likely to experience wildly differing amounts of snow and ice from year to year. In the Northern Hemisphere this belt covers a sizeable swathe across Europe, China, Japan and North America. Areas warmer than this, such as in countries bordering the Mediterranean and in the southern USA, rarely have below-freezing conditions for long enough (days or even weeks on end) for there to be a major problem. Nevertheless, short snap freezes in areas as far south as Florida in the USA, for example, can wreak havoc on the citrus-growing industry. In colder areas, such as far northern Europe and central and northern parts of Asia and North America, snow and ice are experienced sufficiently often that adequate preparations ensure most fluctuations have little impact. There are no large cities in the Southern Hemisphere that experience similar disruptive conditions during the austral winter. For the most part, unusually cold winters are a problem that is peculiar to communities living in the middle latitudes of the Northern Hemisphere.

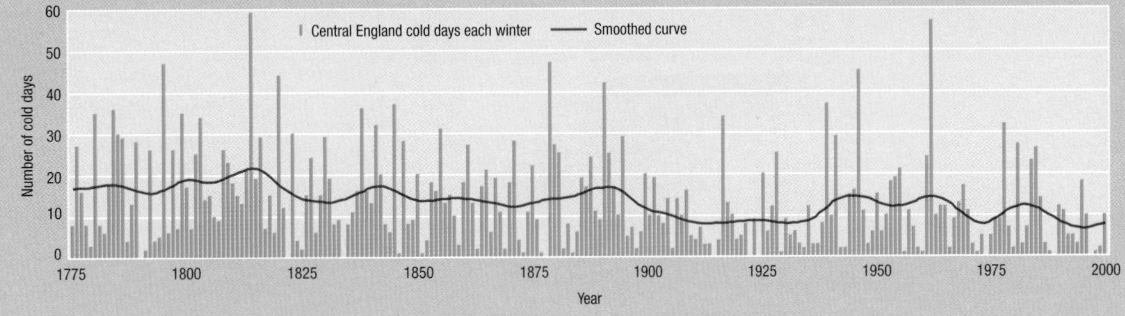

2.10 Drought and dust

The expansion of agriculture into the semi-arid grasslands of the world during the late 19th and early 20th centuries exposed the weaknesses of transferring agricultural practices without modification to different climate regimes. The crippling droughts of the 1930s Dust Bowl era in North America are perhaps the best known example of this bitter experience.

A dust storm looms over a homestead in Kansas in 1935. Such ominous sights were common in mid-western America in the Dust Bowl years.

Problems similar to those encountered from the expansion of grain production into marginal lands across the prairies of North America have been experienced in various forms on the steppes of Asia, the pampas of South America, the grasslands of Australia and in parts of Africa. In the former Soviet Union, the opening up of the Virgin Lands in the 1950s and 1960s repeated many of the mistakes of the Dust Bowl years in the USA. Between 1954 and 1958, more than 40 million hectares of new land (larger than three times the size of Great Britain) were put into cultivation. Initially, this bold expansion under favourable weather conditions was hailed as a great success. In the 1960s, however, erratic rainfall, bitter winters and searing summers led to dramatic falls in agricultural productivity. Although the Russian planners were initially inspired by the opening up of the Canadian prairie provinces in the 1930s, the rush to open up the Virgin Lands produced desiccation of topsoil, severe wind erosion and frequent dust storms.

Eventually, drawing on the bitter North American experience, soil conservation measures were introduced to over 20 million hectares of land. During the 1960s and 1970s yields rose. More recently, the combination of rising temperatures and declining precipitation has led to an increase in the incidence of drought, especially in the cold part of the year (October to April).

Drought indices

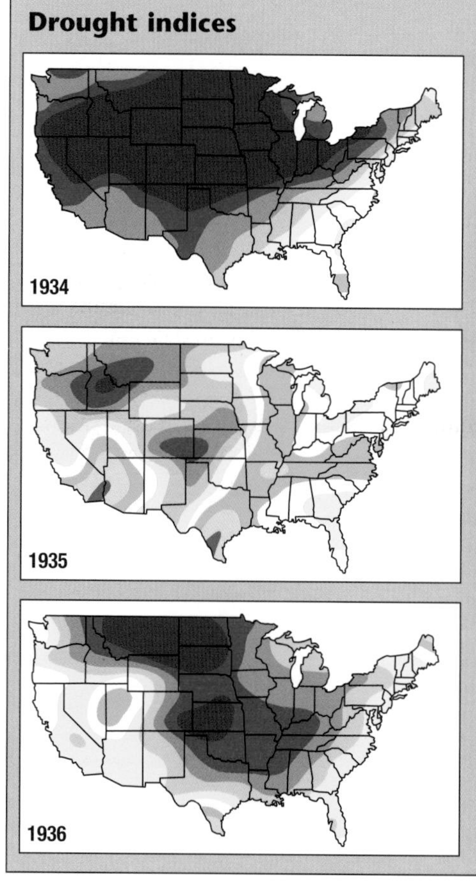

To provide a more accurate measure of drought in the USA, an index was developed by Wayne C. Palmer in the 1960s. Now used to produce a weekly analysis of the soil moisture conditions across the USA, this index (the Palmer Drought Severity Index) uses weekly precipitation statistics and mean temperature to compute evapotranspiration deficits or surpluses, which are compared with values in previous weeks and with climatological averages to derive a crop-moisture index. The index is usually expressed in a scale ranging from +4 or above (extremely moist) through zero (normal) to –4 or below (extreme drought). Calculated retrospectively, the maps for 1934 and 1936 show that the areas of drought were extensive, severe and prolonged, and more so than in the intervening year or in any subsequent year.

This index has also been applied to data from across Europe to provide estimates of the extent and severity of droughts since the end of the 19th century. A similar index is now also operated in the Russian Federation, and has been used to draw conclusions about changes in the frequency of occurrence of droughts over the past century. In Australia, drought conditions are defined in terms of accumulated rainfall deficiencies causing totals for a period to fall in the lowest fifth and tenth percentile ranges of historical rainfall.

Australian droughts

Over the years since Australia was first occupied by European settlers, grain production there has suffered dramatic ups and downs. Research during the 1980s revealed that these swings are driven principally by fluctuations in rainfall associated with the El Niño/Southern Oscillation (ENSO) phenomenon. The 1965–67 drought resulted in a 40 per cent drop in the wheat harvest, a loss of 20 million sheep, and a decrease in farm income of A$300–500 million. A chain reaction affected other industries, with heavy losses suffered by the railways and manufacturers of farm machinery. Water rationing had to be introduced in some urban and irrigation areas. During the 1982–83 El Niño event, the state of Victoria's 1982 cereal crop was the worst since 1944, when dislocation due to World War II and two years of severe dry conditions also associated with ENSO had caused very low yields.

Inappropriate forms of agriculture aggravated the drought problems. Even at the beginning of the century, dust storms were frequent. In south-eastern Australia the area of ground ploughed for wheat in 1900 was almost six times that ploughed in 1866, and much of that tilled soil, no longer held by moisture or grass, was blown away during what was to become known as the Federation Drought. The drought continued and on 21 November 1902 so much soil was blown from the Australian interior that Melbourne was coated in dust, and in the afternoon the Sun was almost hidden by dust-laden air. These scenes were to be repeated during the frequent and severe dust storms in the late 1930s and early 1940s and on 21 February 1983, near the end of the severe 1982–83 El Niño event.

Irrespective of improved economic and agricultural skills, the vulnerability of the land to drought has decreased only marginally. Over much of Australia loose sandy soils are exposed to wind erosion during drought. The floods that often follow can then be equally as damaging to the unprotected topsoil.

A dust storm sweeps right into the centre of Melbourne, Australia in February 1983, during a period when the country was experiencing extreme drought. This drought was linked to a record-breaking El Niño event prevailing in the equatorial Pacific.

A history of droughts

The challenges of dry-land agriculture were brought into sharp focus during the Dust Bowl years of the 1930s in North America. Early European settlers had found the extreme weather of the Great Plains of North America immensely demanding, and their traditional farming practices were often unsuited to the region. Summer rainfall is crucial for plant growth but the soil moisture reserve from accumulated winter snowfall is also important. A combination of bitterly cold, relatively dry winters and hot summers means crops have to tolerate the loss of a lot of moisture. Fluctuations of precipitation from year to year, however, are large. Moreover, the hottest summers are generally the drought years, so when the rains fail, agriculture is in double jeopardy as the crops wither rapidly in the blazing heat.

Agriculture in these harsh conditions has a history of setbacks. In both the 1890s and 1910s major droughts led to widespread crop failures and abandonment of farms across the Great Plains of North America to the west of the Mississippi; but nowhere near so intense as that of the 1930s. The summers of 1934 and 1936 were by far the hottest for the region during the 20th century, and 1936 was the driest by a large margin. Both summers were preceded by exceptionally dry springs. The analysis of past droughts in the USA can be extended back in time using tree-ring analysis. Particularly suited for studies of dendrochronology is the Earth's oldest living inhabitant, the Bristlecone Pine (*Pinus longaeva, Pinus aristata*). The average age is 1000 years, although some trees date over 4000 years. Each year the tree girth increases by a very thin ring of new growth. These annual rings are wider during wet years and narrower during dry seasons and this, combined with their age, makes counting and long-term climate analysis possible. Such analysis confirms the rarity of the Dust Bowl years, with no year back to 1700 matching 1934, although periods of bad years around the middle of the 18th century exhibited sustained dry conditions comparable to the 1930s.

2.11 Understanding tropical cyclones

The tropics spawn the most formidable storms on the planet. These monsters often sweep outward to higher latitudes and are an important part of the general circulation of the atmosphere in several regions in late summer and early autumn of each hemisphere.

Major storms in the tropics are known variously as tropical cyclones, hurricanes and typhoons depending on the region where they form. They all fall within the definition of a tropical cyclone, which is the generic term for a synoptic scale cyclonic circulation with organized convection and surface wind circulation originating over tropical waters.

These storms evolve from clusters of thunderstorms called tropical disturbances, sometimes associated with so-called easterly waves in the normal Trade wind pattern or the intertropical convergence zone (ITCZ), especially when it is located close to its maximum poleward shift in the summertime hemisphere. The majority of these disturbances develop no further, but some grow into major systems. A precondition for major storm development is a surface layer of warm ocean water, which explains why storms occur most frequently from late summer to early autumn. Tropical cyclones will only generally form in the 7–15° latitude belt in each hemisphere. Once the conditions are right for tropical cyclone development, a disturbance can grow rapidly, and tends to follow a characteristic pattern. As the pressure starts to fall rapidly at the centre of the disturbance, the winds rise in a tight band of some 30 to 60 km radius and the clouds form into a circular pattern about a central eye, becoming more symmetrical as the storm becomes stronger.

A satellite image of Hurricane *Mitch*, which devastated Honduras and Nicaragua in October 1998, shows the symmetry of the storm and the distinct eye at the centre.

As the storm grows, its speed and direction of movement are largely determined by the surrounding large-scale airflow. For example, a storm embedded in the Trade wind flow will move generally westward, usually in the belt 8° to 15° from the equator, and eventually start to curve towards higher latitudes. The maturing storm then expands while the central pressure stops falling. Some tropical cyclones can grow to a radius of more than 300 km before they start to decay. The declining phase is usually hastened by either passing over land or colder water.

At sea the greatest threats to safety are the high winds and waves. As the storm comes near to the shore, the waves pile up ahead of it forming a storm surge in the region of the greatest winds. Also, sea level is higher in the eye of the storm where the atmospheric pressure is lowest. This wall of water can do great damage when a cyclone strikes the shoreline, especially if it coincides with a high tide. For the Atlantic and Northeast Pacific basins the USA uses the Saffir–Simpson scale of estimated hurricane intensity to assess the potential for flooding and damage to property. Where storms develop in remote areas, and surface measurements are few and far between, their intensity is often estimated from satellite observations, using methods of pattern recognition.

The structure of a tropical cyclone

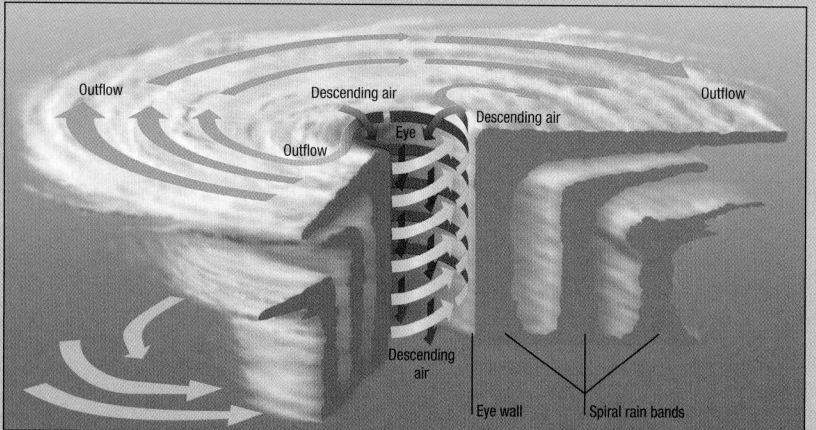

A tropical cyclone is driven by enormous quantities of warm moist air converging as a swirling mass in the layer near the ground. The air is lifted buoyantly in the convective clouds of the eye wall and outer spiral bands to release enormous amounts of energy as the moisture condenses and produces copious rainfall. As the rising air stabilizes at the top of the troposphere it is swept outward with the cloud pattern showing the characteristic near-circular canopy surrounding the eye that can be clearly seen in satellite images.

What drives tropical cyclones?

The most important ingredient needed to fuel a tropical cyclone is a copious amount of warm, moist and rapidly rising air. This condition is only really met when the underlying seawater temperature is above 27°C and evaporation and mixing allow the development of a deep overlying layer of warm, moist air. When tropical cyclones move over cooler oceans or over land they lose their energy source and tend to decay rapidly, although they may continue to bring very heavy rainfall. The increased surface friction over land further contributes to the decay. Changes in the temperature of the equatorial Pacific have a major influence on where tropical cyclones form. During an El Niño event, the incidence typically decreases in the Atlantic and in the far western Pacific and Australian regions, but increases in the central and eastern Pacific. Other effects can also exert a strong influence. For example, when the tropical North Atlantic Ocean is cooler than normal, enhanced low level easterlies (Trade winds) and strong upper tropospheric westerlies produce a wind shear that inhibits the formation of deep convection necessary for hurricane development. In contrast, warm conditions in the North Atlantic Ocean tend to reduce the amount of wind shear, and the deeper moist layer tends to stimulate incipient hurricanes into action.

The effect of winds higher in the atmosphere can also play a part in hurricane formation. In the lower stratosphere there is a periodic reversal of the winds at levels between around 20 and 30 km over equatorial regions. These winds go through a cycle every 27 months or so from being strongly easterlies to nearly as strong westerlies. This behaviour is known as the quasi-biennial oscillation (QBO) and appears to have some influence on hurricane formation in the Atlantic. The combination of being fuelled by the oceans from below, and steered and altered by high level winds, makes tropical cyclones a major forecasting challenge.

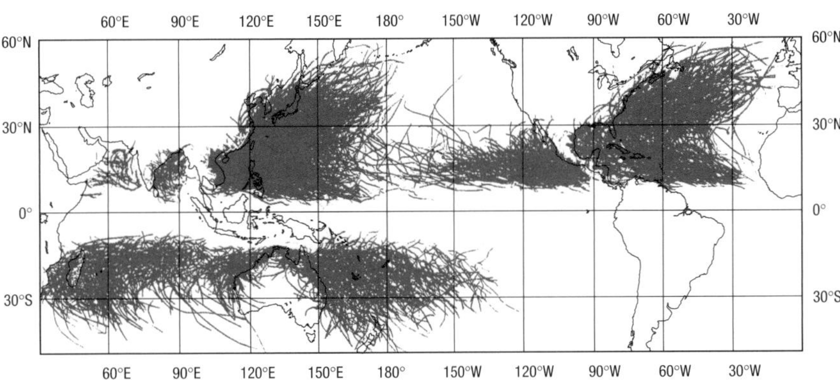

Mapping the paths of all the tropical storms in the period 1951 to 1995 shows where these storms are most likely to occur. The busiest region is the northwestern Pacific, followed by the tropical North Atlantic and the Caribbean, and the northeastern Pacific. In the Southern Hemisphere the storms are restricted to the southern Indian Ocean and the southwestern Pacific. The tropical South Atlantic and Southeast Pacific Oceans are not warm enough to generate these storms.

Classification of tropical cyclones

Region	Wind speed (knots)								
		17	28	34	48	64	90	116	120
Western North Pacific Ocean and South China Sea	Low pressure area	Tropical depression		Tropical storm	Severe tropical storm	Typhoon			
Atlantic and Northeast Pacific Oceans including Caribbean and Gulf of Mexico	Tropical disturbance	Tropical depression		Tropical storm		Hurricane			
North Indian Ocean including Bay of Bengal and Arabian Sea	Low pressure area	Depression	Deep depression	Cyclonic storm	Severe cyclonic storm	Very severe cyclonic storm			Super cyclonic storm
Southwest Indian Ocean	Tropical disturbance		Tropical depression	Moderate tropical storm	Severe tropical storm	Tropical cyclone	Intense tropical cyclone	Very intense tropical cyclone	
South Pacific Ocean and Southeast Indian Ocean	Tropical disturbance	Tropical depression		Tropical cyclone		Tropical cyclone (hurricane) Severe tropical cyclone			

The classification of tropical disturbances and cyclones varies regionally. The wind speed refers to the maximum sustained wind near the centre of the circulation.

Forecasting hurricane seasons

Monitoring and predicting of the day-to-day development and movement of tropical cyclones are essential for the provision of warning services. Computerized numerical weather prediction is a major tool in these operations. Seasonal forecasts of tropical cyclone activity being prepared in many parts of the world are a different matter. A good example is the statistics-based system for predicting hurricane activity in the North Atlantic and Caribbean that has been developed by William Gray at Colorado State University, USA. Forecast techniques are based on precursor atmospheric and oceanic signals observed (in historical data) to have predictive skill. Predictors include rainfall in the Sahel region of West Africa during the previous year, the phase of the stratospheric quasi-biennial oscillation (QBO), extended range estimates of El Niño/Southern Oscillation (ENSO) variability, the October–November and March strength of the Azores high pressure system and the configuration of broad-scale Atlantic sea surface temperature anomaly patterns. The hurricane season in the Atlantic starts in earnest at the beginning of August and runs to around the end of October. A series of forecasts are produced. The first is issued at the beginning of December for the coming year, followed by updates near the beginning of April, June and August. Each one uses a related but different set of predictors. Historically, these forecasts have generally done well. The relative success of the December predictions is particularly rewarding, as it suggests that useful warnings of active hurricane seasons can be provided in good time for many mitigating actions to be taken. The forecasts in June show considerable improvement but the August ones have been less skilful.

Tropical cyclone regions in the Southern Hemisphere (the Indian Ocean, Australian and Pacific islands regions) have all been shown to have some useful predictive relationship with ENSO variability.

2.12 Trends in tropical cyclone numbers

The search for knowledge of tropical storms has been driven by the death and destruction they cause. The scale of damage throughout the century has also provided the incentive to improve forecasts of their behaviour and hence improve our defences against their onslaught.

Relatively short observational records and changes in the methods of detection have yielded conclusions that have at times been mistakenly accepted as representing true variations in tropical cyclone activity. For the Atlantic basin, aircraft reconnaissance has helped to provide a nearly complete record dating back to the mid-1940s. Additionally, complete records of hurricane landfalls are available for the USA coast back to 1900. The Northwest Pacific basin also has valid records going back to at least the late 1950s. However, for the remaining basins, reliable estimates of open ocean tropical cyclones are limited to the satellite era beginning in the mid-1960s.

The analysis of trends in tropical storms must eliminate the bias in early measurements created by recording powerful storms but missing smaller ones far from land. In the Atlantic basin, from the mid-1940s to the late 1960s, the intensity of strong hurricanes was overestimated. After correcting for the true occurrence of intense hurricanes back to the mid-1940s, the data show that the number of tropical storms exhibits substantial yearly variability, but no significant long-term trend. The numbers of intense hurricanes nevertheless have varied dramatically from decade to decade. Active periods occurred from the late 1940s to the mid-1960s, quiet ones from the 1970s to the early 1990s, and then back to active conditions since 1995.

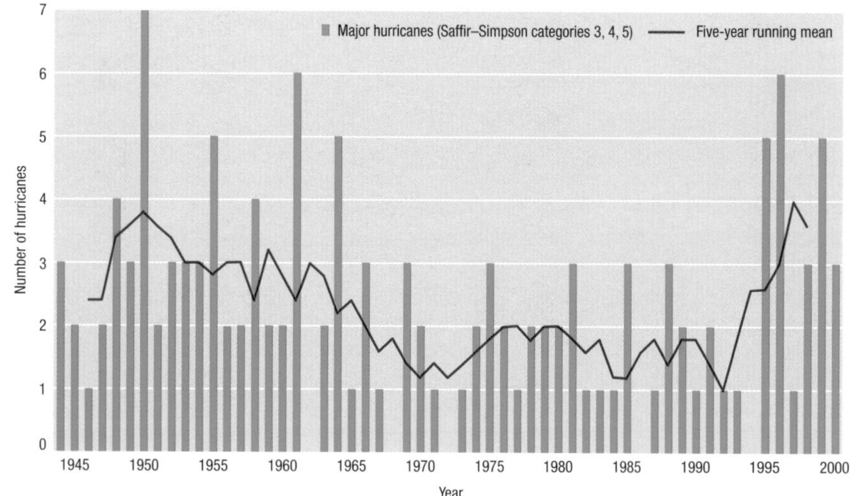

The annual number of major hurricanes (categories 3, 4 and 5) across the North Atlantic show considerable year-to-year variability. In some years there have been no hurricanes in those categories.

Changes in the classification of storms have also made it more difficult to detect trends. For instance, in the Australian region (90–160°E) forecasters no longer classify some weak (greater than 990 hPa central pressure) systems as 'cyclones'. This bias has had to be addressed and the revised record of moderate and strong tropical cyclones now reveals some decline in numbers in this region since the late 1980s. This decrease may be related to more frequent El Niño occurrences during the 1980s and 1990s.

In the Northeast Pacific the short record shows a significant upward trend in tropical cyclone frequency. There is no appreciable long-term trend to the annual number of tropical storms of cyclone strength observed in the North Indian, Southwest Indian and Southwest Pacific Oceans (east of 160°E). For the Northwest Pacific basin, both the frequency of typhoons and the annual number of tropical storms and typhoons have been variable. There was an increase from the early 1950s to the mid-1960s and the number then declined to a minimum in the 1970s. The number rose to a maximum at the beginning of the 1990s, thereafter declining with 1998 being the quietest year on record. Indeed, the decline in activity in the North Pacific as a whole has more than compensated for the rise in the Atlantic in recent years.

Hurricane hunters

Since 1944 the US Navy and Air Force have been flying reconnaissance missions to gather data and provide warnings of approaching typhoons and hurricanes. After the eastern coast of the USA was hit by three hurricanes in 1954, the National Hurricane Research Project was set up to conduct research into improving the understanding and forecasting of hurricanes. The advent of satellites in the early 1960s had a dramatic impact on reconnaissance work as it enabled aircraft to fly directly to tropical disturbances rather than mounting arduous 'fishing trips' looking for them. As a consequence, forecasters and researchers could watch the formation of a hurricane from the very start. Over the years the nature of the aircraft observations has shifted towards research into the properties of hurricanes.

Hurricanes and typhoons

The Galveston Hurricane of 1900

The storm that hit Galveston, Texas, in September 1900 (right) was, in human terms, the worst natural disaster in USA history. With no capacity to track the storm, by the time the danger became apparent it was too late to evacuate the city. A steamship torn from its moorings crashed through the three bridges that connected the mainland to the island on which the city stood, and the citizens were marooned. With nowhere in the city as much as 3 m above mean sea level and a storm surge of some 6 m, they had to seek shelter in the upper stories of their dwellings. The strong winds and crashing waves destroyed many buildings. Many people only survived by clinging to the debris of their houses but over 6000 died on the island, and another 2000 in surrounding areas. It is, however, a measure of the spirit of the survivors that they rebuilt their town, raised the height of the island by dredging huge amounts of material from the surrounding shallows, and built a massive protective sea wall. When a comparable hurricane struck in 1915 only eight people were killed.

The Great Hurricane of 1938

In September 1938 a monstrous hurricane devastated much of New England on the northeast coast of the USA. Forecasters completely failed to provide any warning because, at the time, they relied heavily on radio reports from ships. Although the storm had been watched closely for several days as it moved from the tropical Atlantic to west of the Bahamas, by the time it started tracking northwards off Florida it was so powerful that all shipping was giving it a wide berth. Then, within 24 hours, it roared at 70 km/h from near Jacksonville, Florida, to hit Long Island and New England with catastrophic consequences. In Providence, Rhode Island, the storm surge crested at 4.2 m above mean high water, and the downtown area was flooded to a depth of 3 m. Across New England some 250 million trees were blown down. Overall, more than 600 people were killed and damage exceeded US$ 400 million.

Typhoon *Cobra*

US Admiral Halsey's misreading of Typhoon *Cobra* in December 1944 is another example of the danger in not having a clear idea of the whereabouts of a tropical cyclone. The naval armada under his command on the US Navy battleship *New Jersey* received warning of the developing storm, but misinterpreted the wind patterns. This was because the system overtook a weak cold front and, as it intensified, it produced confusing winds ahead of it. As a result the fleet, in taking what was thought to be evasive action, steered directly into the path of the typhoon. Three frigates sank with the loss of over 800 men, and some 150 planes had to be ditched from the escort carriers (converted freighters). The damage to the task force was so great that it was unable to participate in the Allied invasion of the Philippines. The important meteorological consequence of this disaster was that the US Navy decided thereafter to establish aircraft reconnaissance flights to monitor the position and movement of tropical cyclones.

Hurricane *Camille*

Hurricane *Camille* hit the coast of the Gulf of Mexico just east of New Orleans in August 1969. It was the most intense hurricane to make landfall in the USA in the 20th century. It caused immense damage and killed nearly 300 people, many of them in the flash floods that swamped parts of the Appalachian Mountains where the decaying storm dumped more than 750 mm of rain in places (indeed floods remain, to this day, the major killer when hurricanes strike the USA). Although the loss of life was considerable, the figures were low compared to earlier great storms. For the first time, the US Weather Service had been able to use images of the hurricane, obtained from a geostationary satellite, to provide more accurate warnings of where the storm would make landfall and better target evacuation needs. The storm also added to weather folklore, as amongst those killed were 21 out of 22 people who stayed behind in an apartment block (below) to celebrate a 'hurricane party'.

2.13 Tornadoes and severe thunderstorms

Tornadoes produce the strongest winds observed on the surface of the Earth with speeds reaching 500 km/h. Understanding the physics and climatology of severe thunderstorms helps us predict them.

In Argentina, tornadoes are not uncommon, especially in the October to March half of the year. This car was destroyed after being lifted into the air during a violent outbreak on 10 January 1973 in San Justo, Santa Fe.

Tornadoes have killed people on every continent except Antarctica. They are most common in the south central part of North America. There, the combination of warm, moist, low level air from the Gulf of Mexico and cool, dry, upper air from the Rocky Mountains frequently provides the right conditions for generating severe thunderstorms. Tornadoes spawned from these thunderstorms most frequently strike between April and June, but can occur at any time of year. Tornadoes are also particularly damaging in parts of northern India, and in Bangladesh where there have been three tornadoes since 1989 that have killed at least 400 people each. Tornadoes are also quite frequently recorded in Japan, Australia, northern Argentina and parts of northern Europe, including the British Isles.

Tornadoes are often categorized as weak, strong, or violent. These correspond to F0/F1, F2/F3 and F4/F5 on the Fujita scale. Weak tornadoes account for around 80 to 90 per cent of tornado reports in the USA and the vast majority of reports from elsewhere in the world. They are most likely to go unreported since they are typically short-lived and cause little problem to people. Strong tornadoes make up about 10 to 20 per cent of the total and account for many of the deaths recorded each year. Outside of the Great Plains of North America they are very rare, perhaps numbering five or six a year. Although only 10 to 20 violent tornadoes are recorded each year in the USA, they cause a disproportionate amount of damage. An extreme example is the series of tornadoes that hit Oklahoma City in May 1999. One of these had a funnel 1.6 km across and its winds reached 508 km/h (this figure was estimated during the event using Doppler radar).

Causes of tornadoes

Despite an increasing amount of information about the nature of tornadoes, we still do not know a lot about predicting them well ahead of time, but we have learned when they are most likely to occur in association with severe thunderstorms. In particular, they are most often found where a storm is drawing in large amounts of moist, warm air at the surface and growing to form a supercell. The converging air provides the spin that characterizes a tornado. In addition, this rising, rotating air may interact with the horizontal wind field to produce more extensive rotation at low levels. Termed mesocyclones, these complex rotating storm systems can form a spinning column of air some 10 to 20 km across.

Damage

The violence of the winds associated with tornadoes and severe thunderstorms are such that they are capable of causing damage or destruction to everything in their paths. As populations increase and developments expand it is inevitable that there will be recurring damage, but loss of life can be minimized through warnings issued to

Tetsuya Theodore Fujita – Mr Tornado

In his youth Ted Fujita developed his passion for observing and analysing nature, with wide interests from astronomy to geology. As an assistant professor at the Kyushu Institute of Technology, he joined a special team to investigate the effects of the atomic bomb explosions over Nagasaki and Hiroshima in 1945. However, it was the remarkable analysis of thunder squalls on a mountaintop station in Kyushu in 1947 and the identification of downdrafts in their lower parts that was to set the course of his future work. In 1953 he was invited to join a group at the University of Chicago specializing in thunderstorm research. Thunderstorms, tornadoes and other related mesoscale storms became the focus of his life's work. Fujita developed the F-scale (or Fujita scale) to classify tornadoes according to their strength: weak (winds of 64 to 180 km/h), strong (winds of 181 to 331 km/h) and violent (wind speeds in excess of 332 km/h). His analysis of a disastrous aircraft accident at New York in 1975, initially attributed to pilot error, led to the discovery of the downburst (also called microburst) as a short-lived, very strong downdraft with strong outgoing radial wind at the ground. The power of Fujita's work came from his ability to mobilize all possible observations associated with the phenomenon he was studying.

vulnerable people and communities. Meteorologists look for and warn of the conditions that are conducive to severe thunderstorms, while satellite and radar systems are used to detect and locate the first signs of their development. Effective warnings and well-developed response plans mean that people can seek effective shelter.

Aircraft in the act of take-off or landing are particularly vulnerable to the strong low-level wind shears and microbursts generated by severe thunderstorms and there have been a number of accidents with loss of life as a consequence. Such winds have been recorded as gusts in excess of 160 km/h. Pilots and air-traffic controllers monitor thunderstorms in the vicinity of airports and, when necessary, operations are suspended until the danger has passed.

Thunderstorms can produce large hailstones exceeding 10 cm in diameter (a clenched fist) that can fall at over 60 km/h. A hailstorm on 22 April 1976 in Zejiang province, China was particularly damaging. Hailstones, generally as large as an egg, and some fist-size, fell on an area 2–3 km wide (7 km at the widest) and 13 km long. One hailstone weighed 1500 g (that is, a diameter approaching 14 cm). In that hailstorm, 160 people were wounded and eight people died, although the deaths were mainly attributed to the related weather, such as strong wind. Heavy rain associated with severe thunderstorms often leads to flash flooding.

In the USA, with the improved skill in forecasting and provision of early warning for major storms, it is lightning from thunderstorms that now kills more people than hurricanes, tornadoes and winter storms combined.

Trends

Severe thunderstorms have associated with them phenomena such as tornadoes, lightning, hailstorms, wind, dust storms, water spouts, downpours, macrobursts and microbursts, all belonging to a group defined as small-scale severe weather phenomena. They occur widely but are often short-lived and local in extent, so it is difficult to study them and establish their climate patterns. It is also very difficult to determine how many are missed, particularly in less populated rural and mountainous areas. The annual number of tornadoes of all severities observed in the USA increased nearly tenfold since the 1920s. Although data suggest that the number may have continued to rise in recent decades, most of the increase comes from more efficient reporting of smaller twisters. The number reported in the strong category was almost identical in the 1920s and the 1990s. The number of reported tornadoes has thus risen largely because of the increased ability to detect them using radar and satellites and because of the 'storm spotter' networks established in many countries. A similar pattern of increasing numbers is seen in the records from Argentina, where much effort has gone into collecting reports of tornadoes. The number of tornado-related deaths in the USA has declined from 300 per year in the 1920s to 50 per year in recent decades because of better warnings.

Other measures of the trend in severe thunderstorms are few and far between and relate principally to hail damage. Worldwide losses to agriculture from severe thunderstorms in a typical year are more than US$ 200 million. Individual hailstorms have also caused great damage to cities. Storms in Sydney, Australia in 1999, Dallas–Fort Worth, USA in 1995 and Munich, Germany in 1984 caused damage of more than US$ 500 million each. More general measures of hail frequency show a decrease during the 20th century over most of the USA except for an increase across the High Plains, where most of the USA crop-hail damages occur. These trends correspond well with observations of thunder activity and crop-hail insurance losses. In south Moravia, Czech Republic, decreases in thunderstorms, hailstorms and heavy rain have been observed during the period 1946–95.

Our increased ability to observe these short-lived, small-scale phenomena is contributing to the compilation of stable, credible climatologies that in future years should give rise to better warning systems.

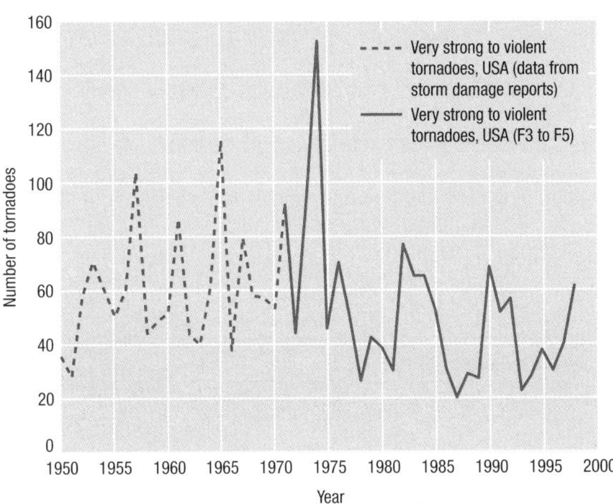

The annual number of tornadoes in the USA that were estimated to be F3 or above on the Fujita scale (wind speeds of 255 km/h or greater) since 1950 suggests, if anything, a slight decrease in the number of such severe tornadoes towards the end of the 20th century.

The US National Oceanic and Atmospheric Administration's National Severe Storms Laboratory scientists witnessed this deadly tornado near Oklahoma City on 3 May 1999 as they collected valuable radar and atmospheric data for the VORTEX-99 project. The trail of damage caused by this and several other massive tornadoes, some F5 intensity, exceeded US$ 1 billion.

2.14 Monsoons

The term 'monsoon', of Arabic origins, refers to a steady seasonal wind, and became widely associated with the Indian subcontinent and the onset of the main rainfall season. In fact, monsoon systems are a major feature of the general circulation of the atmosphere in subtropical latitudes of most regions of the world.

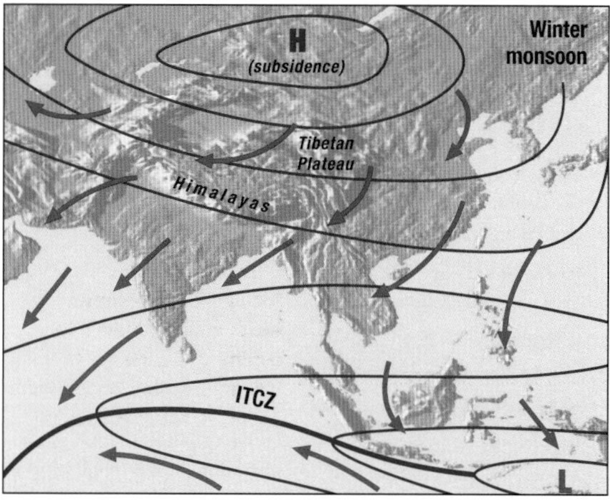

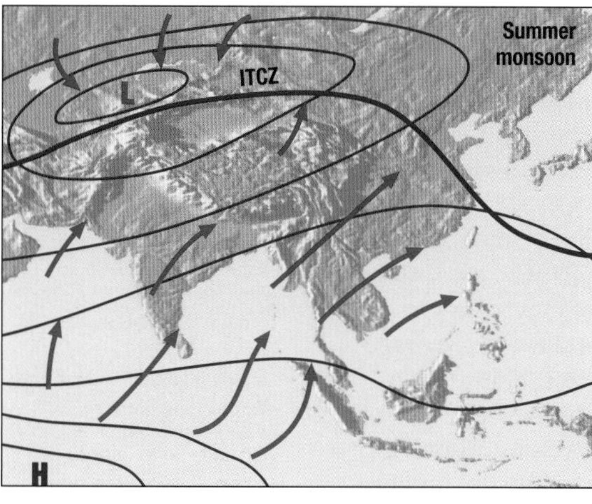

In winter when Eurasia is cold, dry northeasterly winds flow over much of the Indian subcontinent. In summer, with the warming of the Tibetan Plateau, warm moist southwesterly winds are drawn over the Indian subcontinent. This air rises and cools as it approaches the Himalayas, resulting in clouds and often copious rain.

Sir Edmund Halley first proposed the basic explanation for the Asian monsoon in 1686. He suggested that as Asia warms up in the summer it has the effect of drawing huge quantities of moisture from the tropical Indian and Pacific Oceans. In winter the reverse occurs with the winds blowing from the cold continent to the warm sea. This simple explanation is broadly correct, but does not explain why the summer monsoon is so much more regular over India. Here the controlling influence is the Himalayas and the Tibetan plateau. As the latter heats, an upper atmosphere high pressure system forms and the jet stream, which guides weather systems, switches from south of the Himalayas to north of Tibet. This is a big shift and moisture-laden tropical air is drawn north over India, producing heavy rainfall in the region. Elsewhere around the globe there are no similar controls over the location of jet streams and the summer monsoon rains are more variable.

Variations from year to year in the Indian monsoon

In any one year the movement of the monsoon up across India takes place in fits and starts. Torrential storms are followed by days of sunny weather. More importantly, the timing of the start and end of the monsoon varies considerably from year to year. Late arrival in June and July, or early departure in August and September, can ruin crops.

Attempts to explain variations in the timing of the monsoon have a long history. Following the failure of the monsoon rain and resultant famine in 1877, the Government of India called upon H.F. Blandford, the Chief Reporter of the India Meteorological Department, to prepare monsoon forecasts. He concluded that "the varying extent and thickness of Himalayan snows exercise a great and prolonged influence on the climate conditions and weather of the plains of north-west India". This relationship suggested that deep snow cover during the spring would lead to a dry summer, whilst a thinner cover was associated with wet seasons. Initial forecasts between 1882 and 1885 encouraged Blandford to start operational forecasts covering the whole of India and Burma in 1886. Since then, predicting the summer monsoon has

The progress of the summer monsoon across the Indian subcontinent every year is largely predictable. Starting in the far south in late May, it sweeps with a great scything motion north and west to reach Pakistan and Punjab in July. The withdrawal of the monsoon follows a similar pattern in the September to December period.

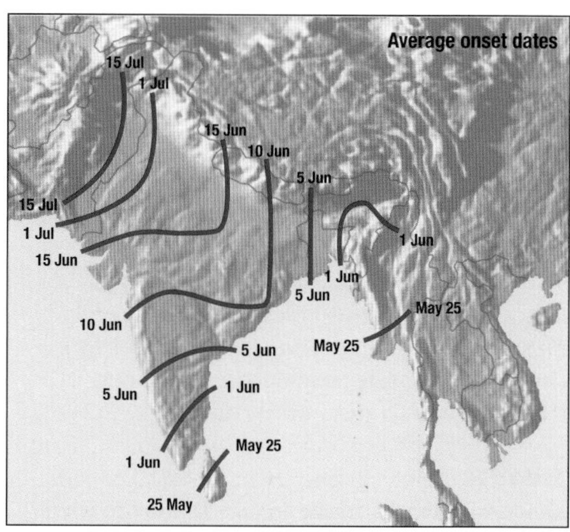

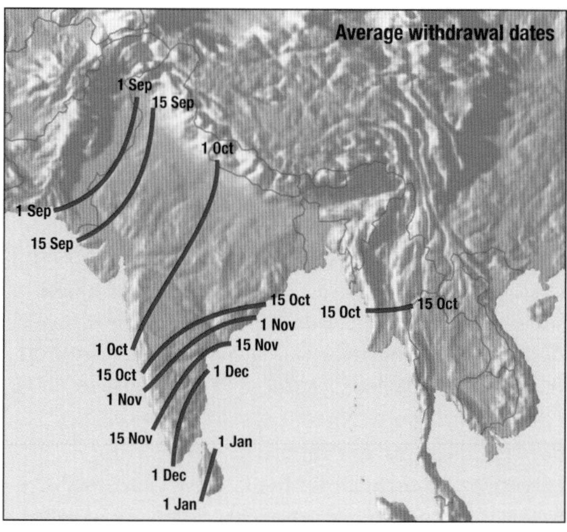

been an important task for the India Meteorological Department.

After 1895, additional factors across India and over the Indian Ocean and Australia were also considered. Sir Gilbert Walker developed this wider perspective by studying global scale oscillations. This established the importance of the Southern Oscillation to the development of the summer monsoon, and put the forecasting process on a more formal footing. Thereafter, from the 1920s to the early 1980s, the forecasting methods made little progress. Although new factors were identified and some of the established relationships fell from favour (the connection between the monsoon and Himalayan snow cover appeared to shift and was dropped from forecasts in the 1950s), the general performance remained static.

The forecasts have improved since the early 1980s. This is the result of a growing understanding of the empirical relationships between indicators around the world and the subsequent monsoon. One reason for these advances has been the rising quality of the data. Recent satellite observations have also revived interest in Himalayan snow cover as a predictor. They show that the relationship first identified by Blandford is a useful guide, but that the extent of the all-Eurasian winter snow cover is a better indicator, given the geographically uneven and variable nature of snow cover over the Himalayas, Tibet and Siberia.

The East Asian monsoon

Over China the monsoon rains first appear in May in the south. The monsoon rains slowly move northward and are associated with the typical flooding of the major river systems (Yellow River (Hwang Ho) and others) in the April/May time frame. There are local names for the monsoon rains as they move into southern China and Japan – the mei-yu (China) and bai-u (Japan) are also known as the plum rains. By July the rains have made their way northwards into Japan and the Korean peninsula.

As over India, the progression from south to north of the East Asian monsoon during the summer is influenced by the position of the jet stream but large variations occur from year to year. The pattern of variability has also shifted from north to south. Since the mid-1960s the incidence of drought in the north has increased sharply, while flooding has almost ceased to be a problem. In the central region the reverse has happened with serious flooding becoming more frequent since 1980, whereas drought has been a rare occurrence. In the south flooding increased in the 1990s.

Australian monsoon

The summer monsoon over northern Australia is the natural complement of its namesake in Asia. As the overhead sun progresses south of the equator, the desert interior of Australia heats and draws humid equatorial air towards the northern coast of the continent. From December to March this can bring bouts of heavy thundery rain to the tropical north of the country. The local timing of maximum summer thunderstorm activity is influenced by the seasonal patterns of neighbouring sea surface temperature. Maximum sea surface temperatures over the Timor Sea occur in December and over the Coral Sea in January and February. The peak thunderstorm activity is earlier over the northwest than over the northeast of Australia.

When the summer monsoon fails, drought becomes widespread in India and famine stalks the land. Villagers in Ghodra district in eastern Gujurat queue for water during the drought at the end of the 20th century.

There is no mountain massif equivalent to Tibet locking the position of the jet stream and so the occurrence of rainfall events is highly irregular. Also, the El Niño/Southern Oscillation is a significant factor in year-to-year variability; years with low rainfall often coincide with El Niño event years.

North American monsoon

Over northern Mexico and parts of southwestern USA there is a pronounced maximum in the summer rainfall. This has become known as the North American monsoon, because the way in which the mountains in the region heat up is analogous to the way the larger Tibetan Plateau controls the monsoon over the Indian subcontinent. The steady surface southwesterly airflow created by the upper atmosphere high pressure draws tropical moisture and produces a peak in the rainfall during July and August. The interannual variability of the North American monsoon is not well understood and is the subject of major research.

South American monsoon

As in North America the annual movement of the overhead sun leads to a peak in summer rainfall in parts of South America. Here it is the formation of an upper atmosphere high over the Bolivian Andes that exerts influence over rainfall patterns, drawing warm moist air in from the tropical Atlantic. The heaviest rainfall occurs across the southwestern half of Brazil and parts of Bolivia and northern Argentina.

Monsoon waters are vital to life in many parts of the world. This girl calmly keeps her pet safe and dry as she traverses a flooded street in Dhaka. A heavy monsoon season, though, can bring severe flooding to many parts of the Indian subcontinent, leaving hundreds dead and countless suffering.

2.15 El Niño/Southern Oscillation (ENSO)

The strongest natural fluctuation of climate on interannual timescales is the El Niño/Southern Oscillation (ENSO) phenomenon. ENSO originates in the tropical Pacific but affects climate conditions over many parts of the world.

Across the equatorial Pacific the atmosphere and the ocean are coupled through the exchange of heat, moisture and momentum at the interface. The easterly Trade winds drive warm surface water westwards and both deepen the thermocline and feed energy to the convection and rising air over Asia and the islands of Indonesia. The upper air returns and sinks over the eastern Pacific, thus completing a cycle known as the Walker Circulation. During an El Niño event (left) the Trade winds weaken, the focus of convection and rainfall shifts eastwards, sea surface temperatures become warmer in the equatorial eastern Pacific and the thermocline becomes less tilted. During La Niña conditions (right) the Trade winds strengthen, convection is firmly anchored over Asia and the islands of Indonesia, and sea surface temperatures cool across the central and eastern Pacific. The strengthened Trade winds deepen the thermocline in the west but cause stronger upwelling of deeper cold water and a rise of the thermocline towards the surface in the east. Normal conditions (middle) represent the intermediate condition between the El Niño and La Niña states.

Dramatic changes occur from year to year beyond the march of the seasons. Next year's summer is hardly ever the same as last year's summer; in some places one winter will have a drought, while the next will bring floods. Elsewhere tropical storms tend to cluster in some years and not in others. Given that the geographical structure of the Earth does not change from year to year and the Sun delivers almost exactly the same amount of energy year in, year out, these fluctuations may seem odd. Much of the progress during the last 100 years in understanding them has centred on how the atmosphere and the oceans interact in the longer term. Foremost in this process was the discovery of the atmospheric Southern Oscillation and later how it interacts with the tropical oceans.

What are El Niño, La Niña and the Southern Oscillation?

The Southern Oscillation was first described and named in the early part of the 20th century by Sir Gilbert Walker. It is now usually defined in terms of the monthly or seasonal fluctuations in the air pressure anomaly difference between Tahiti and Darwin (the Southern Oscillation Index (SOI)). The large-scale surface air pressure pattern see-saws between one extreme, with above normal sea-level pressure over Indonesia and northern Australia and below normal pressure over much of the eastern Pacific (negative SOI), and vice versa.

For centuries fishing people of the coastal communities of northern Peru and Ecuador have used the term El Niño to describe an annual warming of the offshore waters during December (from the Spanish, the time of El Niño – the boy or Christ Child). El Niño is now used to describe extensive warming of the ocean surface across the eastern and central equatorial Pacific lasting three or more seasons. When this region switches to below normal temperatures, it is called La Niña.

The Southern Oscillation and El Niño are closely linked with each other and are collectively known as the El Niño/Southern Oscillation phenomenon. ENSO swings between warm (El Niño), with negative SOI, and cold (La Niña) conditions, with positive SOI. Over the 20th century there were, using one definition for these events, 23 El Niño events and 22 La Niña events.

What causes ENSO?

In first describing the large-scale zonal air circulation across the Pacific (the Walker Circulation), Walker noted in particular the easterly Trade winds picking up moisture across the Pacific to feed the monsoon rains of Australasia and India. In the high atmosphere, the air that has risen in the convective storms returns eastward and subsides over the eastern Pacific, thus completing this circulation. Walker linked the changing strength of the circulation (the Southern Oscillation) with the behaviour of the Indian monsoon.

Modern thinking about ENSO rests on a hypothesis first put forward by Jacob Bjerknes in the mid-1960s. He noted that in normal conditions, the persistent tropical Trade winds 'push' the ocean's surface water westward causing upwelling of cold subsurface water off the coast of Peru. During an El Niño event, the appearance of positive sea surface temperature anomalies over the eastern equatorial Pacific Ocean is accompanied by falls of atmospheric pressure and a reduction of the normal

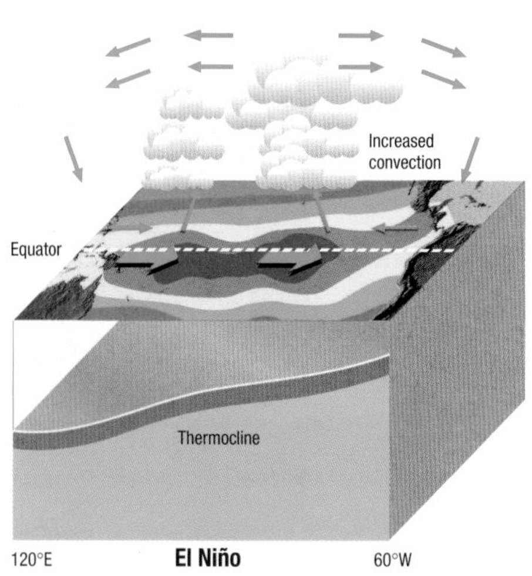

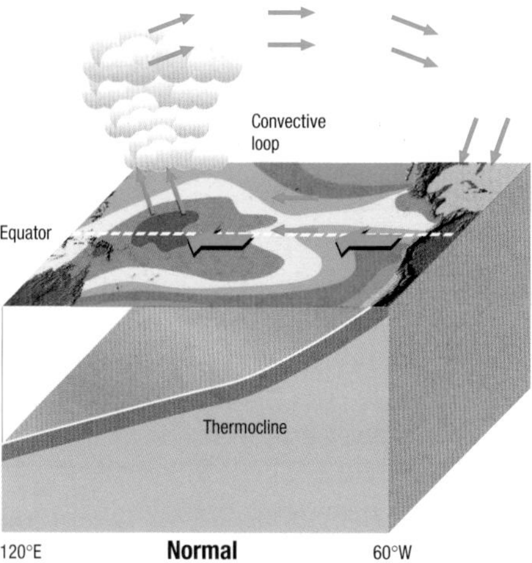

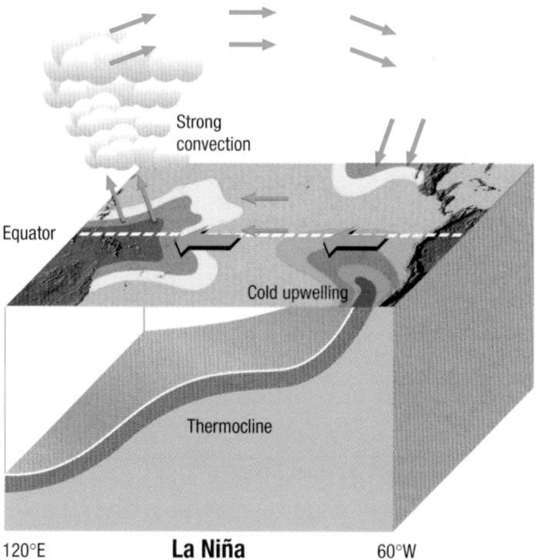

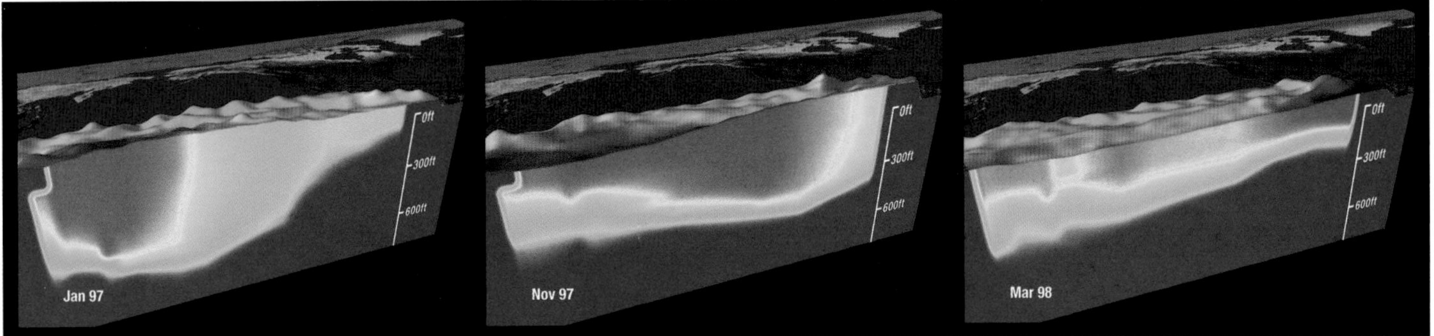

A combination of ocean temperature measurements across the equatorial Pacific from buoys, satellites and other sources can provide a detailed picture of ENSO events. Here the development of the El Niño during 1997–98 is clearly seen as the pool of deeper warm water shifts from west to east.

sea-level pressure gradient that drives the Trade winds. The Trade winds are weakened and the upwelling of cold water off the coast of Peru is reduced, thus reinforcing the initial positive temperature anomaly.

The net effect of these interactions gives the appearance of large quantities of warm water slowly sloshing back and forth across the equatorial Pacific and a large east–west oscillation in the heat supply to the atmosphere from the Pacific Ocean. At the peak of an El Niño event, the tropical Pacific Ocean overall is warmer than normal and the global near-surface air temperature warms up as the ocean gives up heat to the atmosphere. The enormous heat capacity of the ocean compared to the atmosphere makes it slow to respond to forcing (seasons to years) from the fast-changing atmospheric system (days to weeks). This huge difference in response times leads to self-sustaining oscillations in the tropical Pacific which peak every 2–7 years producing the El Niño/La Niña phenomena.

An interesting feature of El Niño and La Niña episodes is that there are decades when one or the other has been more frequent and intense. Early and late in the 20th century El Niño events were quite active. La Niña events were less frequent during the late 1970s and 1980s.

Ringing around the world

ENSO is a primary reason for climate anomalies that may last a season or more in many parts of the world. Over the tropical Pacific, the weakening and even reversal of the normal pattern of rising air over Asia and sinking air over the eastern Pacific during an El Niño event alters the circulation of neighbouring tropical regions, especially across the Indian Ocean to Africa and over South America to the Atlantic Ocean. The shift in tropical convection, and thus the dominant heating locations of the atmosphere, triggers long waves or pulses in the middle and high latitude westerly winds. These large-scale waves or pulses in the atmosphere change the locations of the jet streams and middle latitude storm tracks, altering weather patterns far afield.

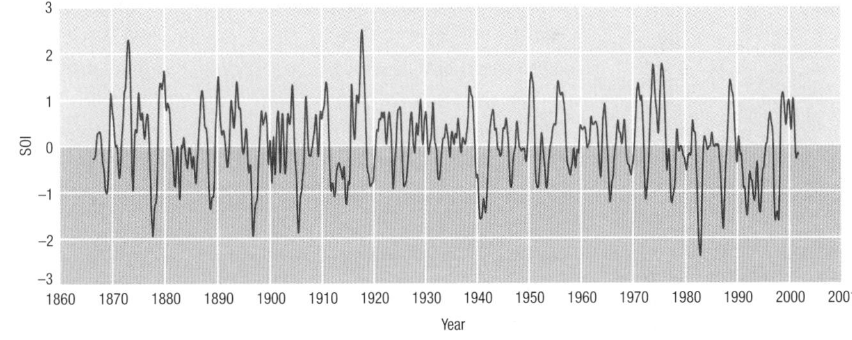

The Southern Oscillation Index (SOI) has been recorded since the late 19th century. There was a period towards the end of the 20th century when the SOI tended to be negative more than usual, linked to a period of more frequent El Niño events.

Sir Gilbert Walker (1868–1958)

Walker was trained as a mathematician and became the second Director General of Observatories in India in 1904. He returned to England in 1924 to become the Professor of Meteorology at Imperial College, London. In 1909 he noted that the intensity of the Indian monsoon was related to pressure patterns across the Pacific. He went on to identify three slowly varying large-scale atmospheric ('see-saw') patterns. These pressure swings were the Southern Oscillation between the tropical Pacific and Australasia, the North Atlantic Oscillation between the Azores and Iceland, and the North Pacific Oscillation between Hawaii and southern Alaska. Walker recognized that, at the time, the Southern Oscillation and other related patterns could not lead to real forecasts and described them as 'seasonal foreshadowing' which indicated 'a vaguer prediction'. They have, however, re-emerged as important factors in understanding longer-term fluctuations of regional climate as our knowledge of atmosphere–ocean interactions has grown in recent decades through observation and research.

2.16 Tropical modes of variability

Equatorial regions receive more heat from the Sun than the middle and high latitudes. They behave, therefore, like the heat engine for the climate system. Any tropical variability is bound therefore to have repercussions out into the higher latitudes. Identifying systematic modes of variability beyond the El Niño/Southern Oscillation (ENSO) may provide additional forecast potential.

The El Niño/Southern Oscillation dominates tropical climate variability but there is emerging evidence of similar atmosphere–ocean coupling elsewhere in the tropics. Recent research suggests the possibility of both the Atlantic and Indian Oceans behaving in similar ways. Neither of these oceans' tropical sea surface temperature (SST) patterns rank alongside the Pacific ENSO in either strength or global reach. The tropical Pacific Ocean is more than twice as wide as either of them and there is more scope for larger areas of anomaly and bigger geographic displacement. Accordingly, the Pacific packs a much bigger punch, and the modes of equatorial variability of the other ocean basins have less dramatic consequences on the global climate. Nevertheless, they are an important part of the story.

Indian Ocean SST patterns

Unlike the Pacific Ocean, the surface wind flow over the tropical Indian Ocean does not have a persisting easterly component. Strong coastal seawater upwelling off the Horn of Africa during the summer monsoon provides seasonal cooling of sea surface temperatures and reinforces the surface pressure patterns that drive the monsoon winds of India. Recent analysis of historical ocean and atmospheric data suggests that every few years or so there is an east to west oscillation of warm waters similar to the El Niño and La Niña events of the Pacific. For instance, when the temperature in the western Indian Ocean is much above normal, as it was in 1997, it tends to produce unexpectedly high rainfall in East Africa, and dryer conditions over Indonesia and northern Australia. This behaviour involves both large-scale shifts in precipitation and SST patterns. These patterns of variability are linked to the ENSO although the coupling is not well understood. They may, however, help to explain why the strength of the monsoon over India and rainfall over East Africa have, over the years, responded in sometimes unexpected ways to El Niño and La Niña events in the Pacific.

Madden–Julian (Intra-seasonal) Oscillations

Not all climatically important atmosphere–ocean couplings relate to interannual timescales. Certain shorter-term variations in the atmosphere appear to play a critical part in major oscillations. A good example of this type of behaviour was discovered by Roland Madden and Paul Julian in 1971–72. They reported that surface and radiosonde data from various equatorial sites, and satellite images, showed the roughly periodic nature of tropical convection. Over the Indian and Pacific Oceans, in particular, tropical convection was often active for about a week or so and then there was a relatively quiet period of around a month. The active regions of convection travel from west to east close to the equator with a period of somewhere between 30 and 70 days. A significant percentage of the rainfall in the tropics occurs during the active convection phases of this cycle, as does the formation of many tropical cyclones. This cycle is

Atlantic sea surface temperature patterns

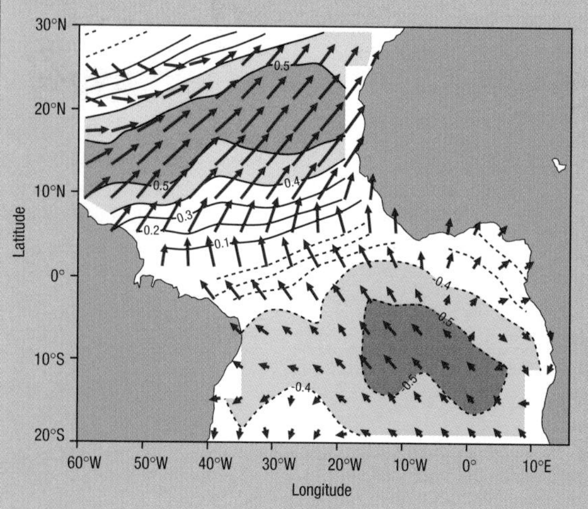

The Atlantic Ocean has the characteristics of two large ocean basins linked at the equator, rather than of a single large basin. This difference means that it is likely to exhibit somewhat different interannual behaviour to the Pacific and Indian Oceans. Studies suggest that the dominant mode of interannual and longer variability in the tropical Atlantic is a north–south dipole pattern of anomalous sea surface temperature and wind stress with focii at 15°N and 15°S about which the SSTs oscillate every 10 to 20 years. The variations of the sea surface temperature patterns are in phase with the strengthening and weakening Trade winds. The links between these fluctuations and ENSO and their importance in the climate of the tropics remain a puzzle. However, it is recognized that anomalies of sea surface temperature in the Gulf of Guinea do have an influence on rainfall in the Sahel.

now known as the Madden–Julian Oscillation (MJO) or Intra-seasonal Oscillation.

The MJO is at its strongest from September to May, when the focus of convection on the western limb of the Walker Circulation has shifted from the Asian continent to the maritime region of Indonesia and the South-West Pacific. Moreover, if one of these bouts of activity, with its strong convection and anomalous westerly winds, hits the western Pacific just as an El Niño event is ready to hatch, it can stimulate its rapid development. This sequence, reinforced by a long-lived tropical cyclone named *Justin* in the Coral Sea, is what appears to have happened in early 1997.

The trigger factors for the MJO are not fully understood and they are largely chaotic like other aspects of atmospheric weather behaviour. The fact that they do have a roughly periodic nature, however, suggests underlying dynamics of a more ordered phenomenon. Unfortunately, forecasts made on the basis that, if the last two or three cycles were 45 days apart, it is reasonable to predict the next one will come along in about 45 days, have not been very successful. However, once an active area is identified, satellite imagery can be used to predict the short-term future movement of the active regions. The MJO itself can be strong in some years and almost absent in others, thus making the forecasting of its behaviour that more difficult.

The Met Office (UK) and Météo-France use ODAS buoys in the North Atlantic to measure atmospheric and ocean conditions. Such automatic buoys are an essential research tool, and their data have enabled scientists to obtain a much better picture of the large-scale properties of the coupling between the atmosphere and the oceans. Data from these buoys will be incorporated in GODAE with other observations to produce better initial inputs for climate predictions and analyses for validation of climate simulations.

South Pacific Convergence Zone

In the vast tropical South-West Pacific, there is an interesting region where the generally warm sea surface temperatures produce lower surface air pressure, and where converging, rising air produces cloud and rainfall. Its position and the associated rainfall have a major impact on the peoples of many Pacific Island countries in the region. Known as the South Pacific Convergence Zone, it runs diagonally southeast from the Solomon Islands to Samoa and beyond. While it usually shifts little during the year, its position is linked to ENSO variations. During El Niño events it is displaced east, and during La Niña events west, of its mean position. Over the longer term the zone has shown a striking eastward displacement since 1977, compared with the period 1948–76, again reflecting the more prevalent occurrences of El Niño events during the later period.

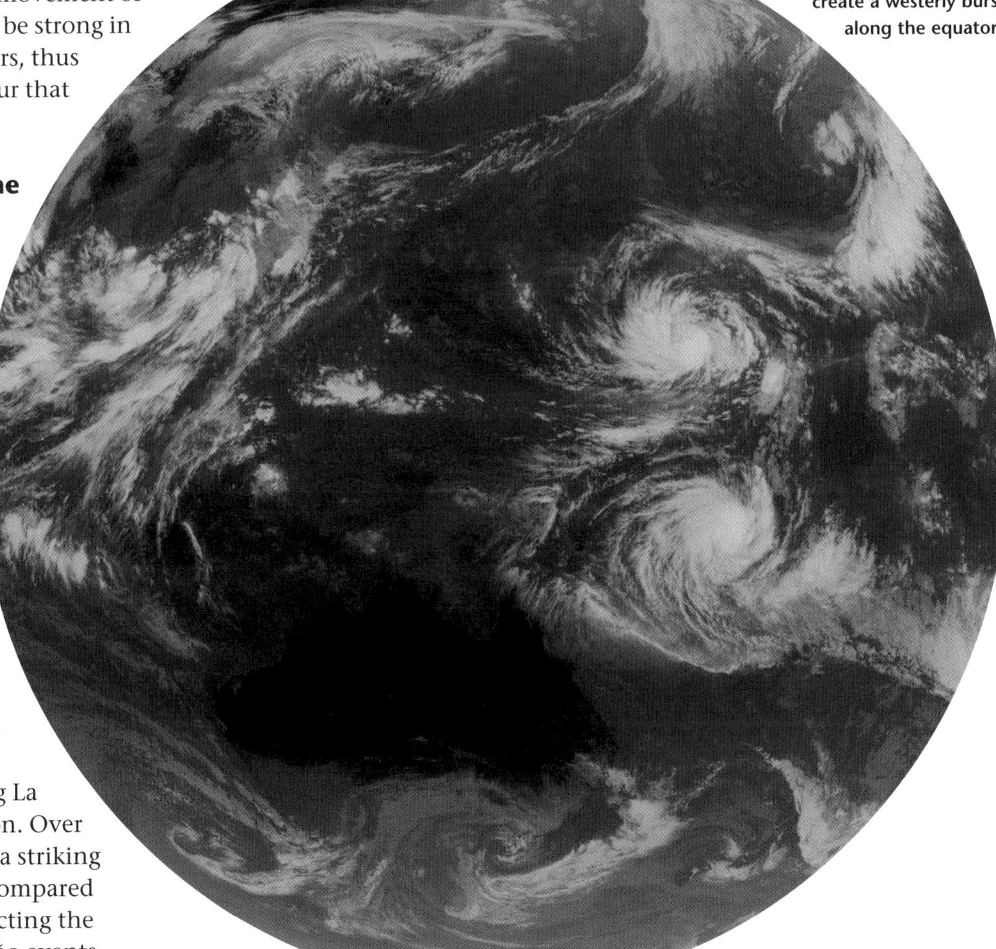

The occurrence of twin tropical cyclones, such as *Lola* (north of the equator) and *Namu* (south) on 18 May 1986, is thought to be one of the triggering mechanisms for a warm ENSO event. Such pairings create a westerly burst along the equator.

2.17 Middle latitude circulation anomalies – blocking

Persistent spells of abnormal weather in middle latitudes are closely linked to quasi-stationary patterns of circulation in the atmosphere. These patterns are known as blocking, and are central to understanding many features of extremes in our climate.

The circulation of the atmosphere in the middle latitudes of the Northern Hemisphere has a propensity to get stuck in a given pattern for weeks or even months at a time, producing extreme weather in many places, a phenomenon climatologists call 'blocking'.

The key to these long-lasting weather patterns is found in the dynamics of the westerly wind circulation of the middle latitudes. The thickness of the atmosphere between the 1000 hPa and 500 hPa surfaces (approximately sea level and 6 km altitude) is proportional to the mean temperature of the layer – lower thickness values correspond to cold air and high thickness values correspond to warm air. Contour maps of the layer thickness clearly show the cold air over the Polar regions and waves indicate outbreaks of cold air towards the equator.

A contour map of the 500 hPa surface has lower height over the cold Polar regions and the winds generally follow the contours from west to east forming a vortex around the poles in both hemispheres. The shape of the contours is linked to the outbreaks of cold polar air but is also influenced strongly by the underlying topography and land and sea as heat sources. The mean 500 hPa map of the Northern Hemisphere winter is of an asymmetric vortex with a primary centre over the eastern Canadian Arctic and a secondary one near eastern Siberia. This asymmetry in the climatic normal pattern is primarily due to the deformation of the flow by major mountain ranges, notably the Tibetan Plateau and the Rocky Mountains. In summer the pattern is similar but much less pronounced due to the warmer air over the Polar regions. The Southern Hemisphere, being largely covered by water in the middle latitudes, exhibits a more symmetric mean pattern with little variation from winter to summer.

The winter of 1962–63

The winter of 1962–3 was a 'classic' example of blocking. In the upper atmosphere the circulation at 500 hPa (upper right) meandered much more than normal (lower right). At low levels during mid-December through January an intense, high pressure system persisted over the northeastern Atlantic and channelled cold Arctic air across Europe. Over Asia, the cold Siberian high was also stronger than normal while persistent highs over North America brought icy blasts down across the northeast. In contrast, Alaska, western Greenland and parts of central Asia experienced unusually warm conditions.

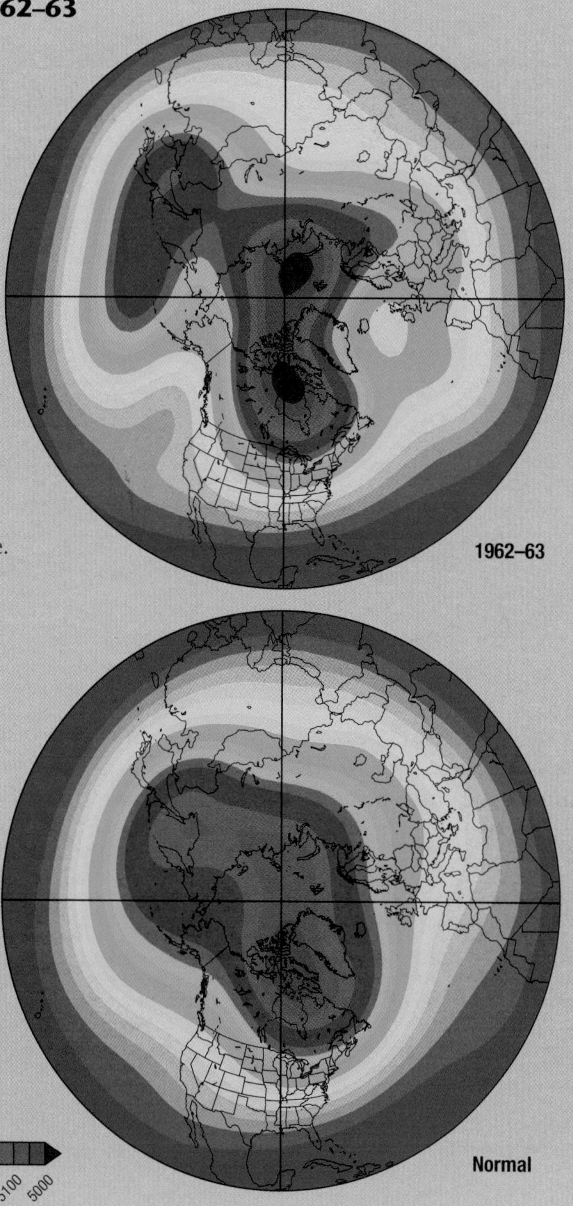

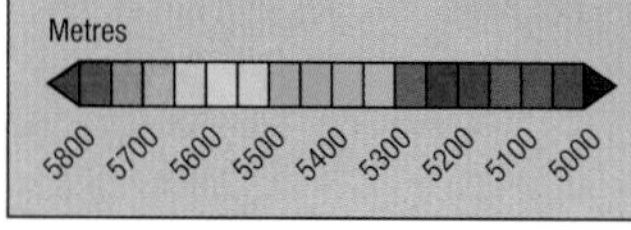

The jet streams

In both hemispheres the waves in the westerly flow are described as troughs of low pressure and ridges of high pressure. At any time the number of troughs and ridges may vary as will their position and amplitude. These patterns are known as long waves (or Rossby waves – after the Swedish meteorologist Carl-Gustaf Rossby who first provided a physical explanation of their origin). The strongest winds of the upper atmosphere westerly flow are the subtropical jet streams concentrated in a narrow region on the poleward side of the Hadley cell, often situated near 30 degrees of latitude and at an altitude between 9 and 12 km. However, in middle latitudes additional jet streams are often formed as the horizontal temperature gradients tighten during outbreaks of cold polar air. The cores of these polar front jet streams are associated with principal troughs of the Rossby long waves. The surface weather systems are steered by the upper atmosphere circulation and the jet streams influence their intensity, often causing rapid development of weather systems in middle latitudes.

During January 1963 bitterly cold air sweeping out of Siberia brought blizzards to western Japan. The destruction of this railway bridge in Hokkaido was the result of the huge snowfalls.

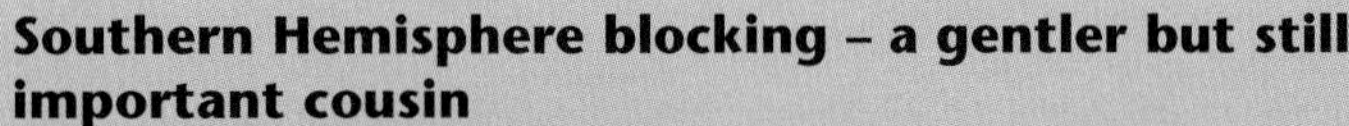

Southern Hemisphere blocking – a gentler but still important cousin

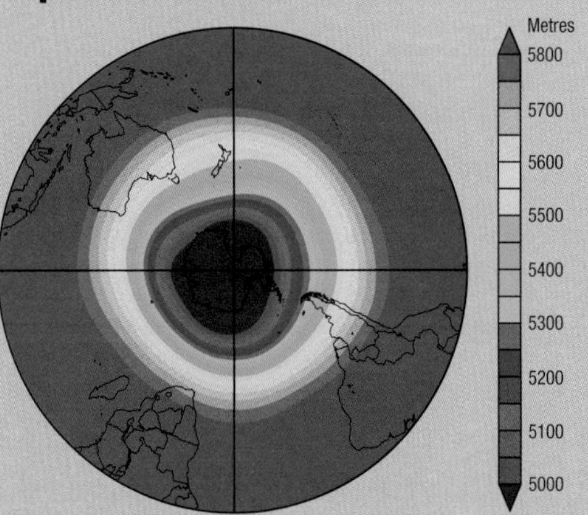

In the Southern Hemisphere the westerlies are far more symmetric because of the weaker land-sea contrast and just one substantial middle latitude mountain range, the Andes.

Blocking episodes are, at first glance, insipid by contrast with their Northern Hemisphere counterparts. Moderate strength blocking episodes occur over the southern parts of the Pacific Ocean and are an important factor in producing seasonal rainfall anomalies over eastern Australia and New Zealand. They occur relatively frequently but instead of being locked in place, they drift slowly eastward, with a new blocking pattern tending to form again upstream. Another favoured location, but far removed from people, is a high pressure ridge bulging out from Antarctica that produces almost as frequent blocking as the Northern Hemisphere counterparts.

Blocking patterns

Blocking occurs when the seasonal circulation pattern is radically different from normal and is frequently associated with extreme weather around the globe. These blocking patterns are particularly common in the Northern Hemisphere, largely because land/ocean boundaries and the major mountain ranges help to amplify and anchor the anomalies. The circulation anomalies occur irregularly, and can last a month or two, but are more pronounced in winter when the circulation is strongest. Ridges and troughs can become accentuated, adopt different positions and even split up into cellular blocking patterns, obstructing the normal west-to-east motion of weather systems. The semi-stationary intense high pressure systems, which are an essential feature of a blocking pattern, are known as blocking anticyclones. The normal westerly flow is replaced by locally persistent easterly (east to west) or meridional (north to south or south to north) flows producing alternating regions of warm and cold weather around the globe, as in the winter of 1962–63.

Incidence of blocking

Blocking can occur throughout the year but, as already noted, is more common during the cooler months (i.e. September to May in the Northern Hemisphere). The amount of blocking varies from year to year and the reasons are not well understood. In the Northern Hemisphere over the eastern Atlantic, during the cooler seasons, blocking is more common when there are La Niña conditions in the tropical Pacific. Elsewhere there is no clear linkage to El Niño/Southern Oscillation events.

Predicting blocking

Changes in the incidence of extreme seasons, like the winter of 1962–63, are a major factor in variations of the climate of middle latitudes. It is thus important to know what defines their occurrence relative to more symmetric circulation patterns. Although in the Northern Hemisphere the distribution of land masses and the major mountain ranges tells us where they are likely to occur, it does not tell when they will occur. Moreover, since blocking anticyclones are regions of relatively warm air aloft, once they have become established, the persistent northward transport of warm air by weather systems moving along a set track may reinforce the phenomenon. However, this does not explain why the number of waves around the globe may range from three to six or why they can vary from only small ripples on a strong circumpolar vortex to exaggerated meandering with isolated cells. One of the most important goals of improved weather forecasting and climate modelling is to handle these elusive systems more effectively.

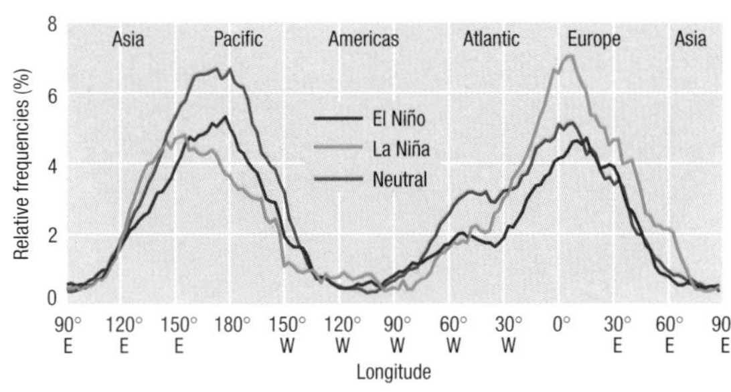

This plot shows for the Northern Hemisphere the frequency of mid-latitude atmospheric blocking as a function of longitude for the interval 1958–98 for El Niño, La Niña and neutral periods. Note the tendency towards more frequent blocking over Western Europe (around 0°) during La Niña episodes.

2.18 Atmospheric oscillations of the extratropics and high latitudes

The discovery during the 1970s and 1980s of the global significance of El Niño/Southern Oscillation (ENSO) variability reawakened interest in other long-term 'oscillations'. Some of these, over the North Atlantic and North Pacific, were identified in the 1920s and 1930s, but have assumed new importance in the context of global teleconnections and with recent and growing concern about climate change.

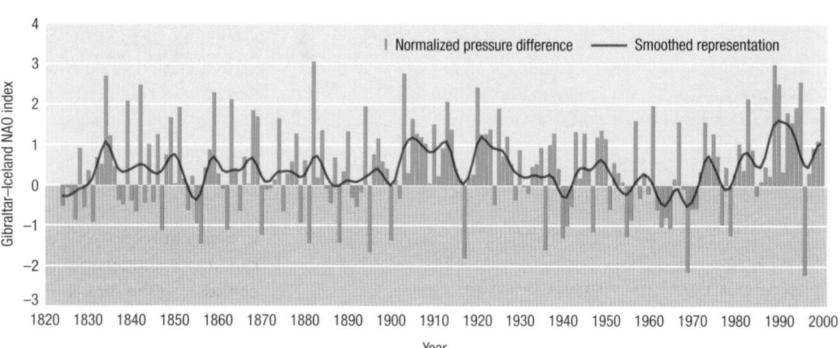

The sea level pressure patterns in the North Atlantic define the North Atlantic Oscillation (NAO) that is associated with substantial variations in the strength of the westerly winds over the North Atlantic and Europe from year to year and decade to decade. Between 1900 and 1940 and since 1980, when the index was positive, the westerly winds were strong, but during the intervening years the westerly winds were much weaker.

During the early part of the 20th century attempts to explain major fluctuations in seasonal weather from year to year had to rely on looking for recurring patterns in the global circulation. These involved identifying statistical links between pressure, temperature and precipitation patterns over large areas of the globe. The Southern Oscillation of the tropical Indian and Pacific Ocean regions is the best known of these links, but others were identified for the North Atlantic and North Pacific regions. More recent studies, particularly those associated with the international Tropical Ocean Global Atmosphere (TOGA) research project, have shown that these 'oscillations' occur more widely. Some are triggered from the tropics and many have important implications for understanding natural climatic variability, especially on interdecadal timescales.

The North Atlantic Oscillation

An important mode of variability in the extratropics of the Northern Hemisphere is the North Atlantic Oscillation (NAO). First analysed by Sir Gilbert Walker, the NAO is a measure of the surface westerlies across the Atlantic. Positive values of the NAO index indicate stronger-than-average westerlies over the middle latitudes with low pressure anomalies in the Icelandic region and high pressure anomalies across the subtropical Atlantic. The positive phase is associated with cold winters over the northwest Atlantic and mild winters over Europe, as well as wet conditions from Iceland through Scandinavia and drier winter conditions over southern Europe. A negative index indicates weaker westerlies, a more meandering circulation pattern, often with blocking anticyclones occurring over Iceland or Scandinavia, and colder winters over northern Europe.

Over the last two decades of the 20th century the pattern of wintertime atmospheric circulation variability over the North Atlantic apparently shifted in an unprecedented manner. A sharp change in the index began around 1980, since when the NAO has tended to remain in a highly positive phase. The recent upward trend in the NAO accounts for much of the regional surface warming over Europe and Asia, as well as the cooling over the northwest Atlantic.

There are also clear indications that the thermohaline circulation of the North Atlantic Ocean varies significantly with the variations in the circulation patterns of the overlying atmosphere. These changes are linked not only to the NAO but also to wider circulation patterns involving both the North Pacific and ENSO, and including variations on decadal timescales.

Changes over the North Pacific

Studies have identified that year-to-year variability over the Pacific is dominated by ENSO but the influence of lower frequency sea surface temperature variability is also present and extends more widely. The low frequency variability is

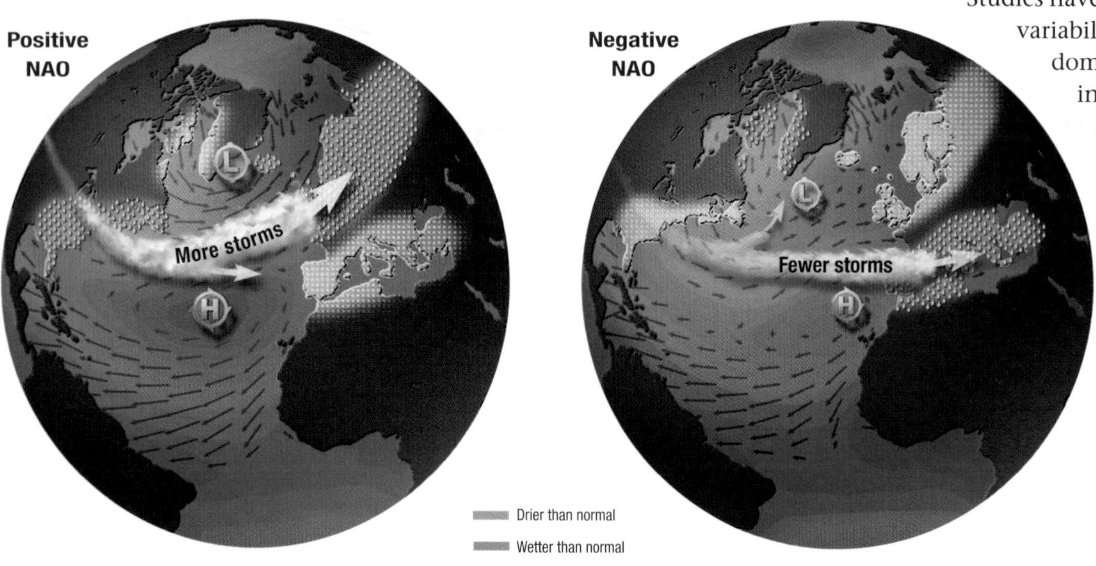

The NAO fluctuates between positive and negative phases, and surface westerly winds correspondingly alternate between stronger (positive) and weaker (negative) phases. This change in circulation is associated with the surface temperature and precipitation anomalies shown.

prominent over the North Pacific where, in wintertime, its behaviour more closely resembles changes in the intensity of the semi-permanent low pressure system near the Aleutian Islands (the Aleutian Low). This low has become deeper than normal since about 1976, especially during the winter half of each year (November to March), and the westerly winds across the central North Pacific have strengthened. Consequently, over much of the last two decades of the 20th century there were related increases in surface air temperature and sea surface temperature over much of western North America and the eastern North Pacific respectively, but decreases in sea surface temperatures over the central North Pacific.

In addition, there appear to be more lengthy interdecadal fluctuations in atmospheric pressure variability (the Interdecadal Pacific Oscillation (IPO)). These are associated with cooler than average sea surface temperatures occurring over the central North Pacific when the Aleutian low pressure is below average, and vice versa. These changes extend over the entire Pacific Basin and exhibited three major phases during the 20th century. The IPO had positive phases (southeastern tropical Pacific warm) from 1922 to 1946 and 1978 to 1998, and a negative phase between 1947 and 1976. The two phases of the IPO appear to modulate year-to-year ENSO precipitation variability over Australia, and bi-decadal climate shifts in New Zealand. The positive phase enhances the prevailing west to southwest atmospheric circulation in the region, and the negative phase weakens this circulation.

Teleconnections and extratropical low frequency variability

The part played by ENSO remains central to a better understanding of extratropical variability on decadal timescales. The 1985–94 TOGA project led to the identification of various large-scale circulation anomalies that are linked to ENSO events. The most prominent of these is the Pacific North America pattern. During a warm ENSO event enhanced convection over the abnormally warm sea surface temperatures of the eastern equatorial Pacific, particularly during the Northern Hemisphere winter, locally strengthens the Hadley cell in each hemisphere. This triggers a sequence of long amplitude waves in the westerly wind flow of the upper atmosphere (Rossby waves). Each El Niño event will trigger a similar characteristic long-wave response that propagates into the middle and higher latitudes.

The Pacific North America pattern is the manifestation of the forced Rossby waves and influences rainfall and temperature patterns over North and Central America. For example, enhanced subsidence of the Hadley cell expands the area of dry winter conditions over northern Mexico and into the Texas region of the USA. Also, the southward shift and strengthening of the jet stream brings stronger winter storms to southern California and enhanced precipitation to southeastern states of the USA.

The climate of the polar regions is linked by a series of complex connections to events in the tropics. Changes at high latitudes can only be understood by considering the global climate as a whole.

Variability of the extratropical Southern Hemisphere circulation on timescales of a few months to a few decades is also linked to how tropical convection is modulated by the Madden–Julian Oscillation (see 2.16) and by ENSO. A corresponding Pacific South America pattern has also been identified that brings more frequent winter storms and rain to central Chile and east of the Andes over the Paraná and Paraguay River Basins. The cause of decadal-scale variations is the subject of ongoing research. Nevertheless, it is known that variability in blocking frequency across the far south Pacific is tied to sea surface temperature anomalies and variations in the sea ice extent around Antarctica.

In the Southern Ocean sea surface temperature anomalies tend to move eastward in the general flow of the Antarctic Circumpolar Current, suggesting a coupling in the ocean–atmosphere system of the region. This has become known as the Antarctic Circumpolar Wave. Four huge pools of alternating above and below normal temperature water appear to circulate every nine years or so, suggested to be linked with changes in rainfall patterns over the southern continents. It is too early to say whether the influence of this pattern extends farther afield, or whether it is a permanent feature of the climate of the Southern Hemisphere.

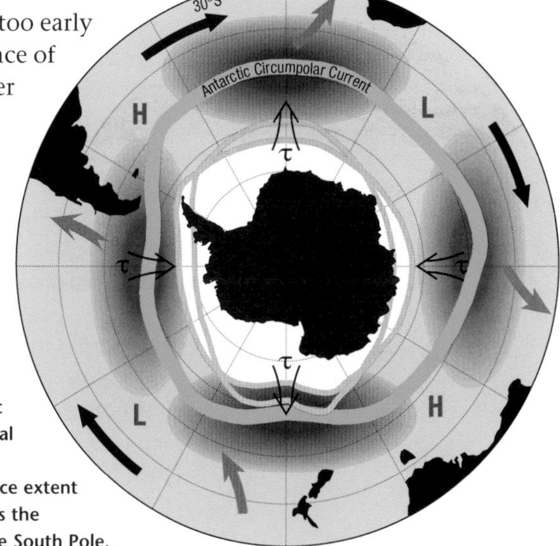

Satellite observations of variations in long-term temperature anomalies in the Southern Ocean reveal a phenomenon called the Antarctic Circumpolar Wave. The meridional forcing by the winds is critical in stretching and compressing the ice extent around Antarctica. Over the years the anomalies swirl slowly around the South Pole.

2.19 Drought in Africa

The causes of drought in Africa are a complex set of interactions extending throughout the tropics. The consequences are much easier to define: adequate rainfall produces healthy economies, drought brings famine and death.

The dominant features of the climate system over Africa are the zone of rising air with embedded convective clouds through the equatorial regions and the adjacent Hadley cells in each hemisphere with descending dry air between 10 and 20 degrees latitude. The zone of convective activity and heavy rains – the intertropical convergence zone (ITCZ) – moves north and south with the seasons and reaches furthest from the equator during late summer. The Hadley cells are strongest during their respective winter season, causing rainfall over the subtropics to be a minimum at this time.

Generally clear skies associated with subsiding air over the arid and semi-arid regions of North Africa and the Middle East stand out clearly in this Meteosat image. To the south of the deserts the cloudiness of the intertropical convergence zone (ITCZ), from time to time, brings rainfall to the Sahel.

The fluctuations of rainfall across Africa have provided some of the clearest examples of how tropical ocean–atmosphere variability can lead to widespread and damaging consequences over years and even decades. The prolonged period of below-average rainfall in the Sahel since the late 1960s has been one of the most marked large-scale climatic changes anywhere in the world during the 20th century. The human suffering from drought in Africa has motivated much research to understand the global climatic processes at work and develop seasonal forecasts of rainfall. Promising results are being achieved.

Sub-Saharan Africa

The Sahel is a semi-arid region that stretches in a narrow band across the southern fringes of the Sahara from Mauritania to the Red Sea. The annual motion of the ITCZ defines the climatology and the region has two main seasons: the rainy season between June and September and the dry season between October and May. The main characteristic of the dry season is the dustiness that occurs with the prolonged period without rain. Rainfall amounts vary from 200 to 800 mm a year from the north to south across the band, and are restricted to the rainy season. These amounts also fluctuate appreciably from year to year and are linked to variations in the northward movement and intensity of the ITCZ.

Two basic features in the lower atmosphere define how far the rainfall extends northwards. Over West Africa the intense summer (June–September) heat creates a thermally induced low pressure cell across the Sahara with its axis located between latitudes 18°N and 22°N. At the same time, over the South Atlantic, the Southern Hemisphere winter subtropical high pressure system intensifies and extends towards the equator. The resulting pressure gradient between these two systems induces a low-level southwesterly airflow (southwest monsoon) that penetrates northwards. At the surface, this warm, moist air can reach as far inland as the southern fringes of the Sahara, reaching latitudes 20°N to 22°N in August. The rainfall occurs in the moist air, mostly as organized bands of thunderstorms called squall lines. The vertical depth of this moist airstream decreases as it moves inland and so does the amount of rainfall. The amount in any year depends on how strong and sustained the push from the south is, and this varies with factors such as tropical sea surface temperatures in the Atlantic and less understood changes in the large-scale atmospheric circulations.

The dustiness of the atmosphere during extended drought periods becomes all-pervasive during the dry season in the Sahel.

Equatorial West Africa

Droughts in equatorial West Africa are less severe than elsewhere on the continent. Since 1990 the region has become moderately wetter than in the 1980s, with the last five years of the 1990s being comparable with the last five years of the 1960s. However, only a couple of years in the 1990s reached the long-term average rainfall of the 20th century.

Southern Africa

A summer monsoonal circulation also affects Southern Africa and occasional tropical cyclones from the Indian Ocean. Most of the precipitation is again related to the presence and strength of the ITCZ, which in the east penetrates to around 15°S in the southern summer. In addition, in the more southerly parts of the continent there are occasional rain events resulting from interactions between tropical moisture and middle latitude weather systems moving from the South Atlantic.

Rainfall over Southern Africa shows little evidence of long-term trends, and seasonal variations are more clearly linked to El Niño/ Southern Oscillation phases, especially along the southeastern quadrant of the continent. During La Niña events, moist surface winds are drawn from the Indian Ocean over southeastern Africa and produce plentiful rainfall. During El Niño events, the region of rising air shifts offshore closer to Madagascar, and descending dry air over southern Africa is more likely.

Dustiness

Dustiness, sometimes defined as the number of days when the horizontal visibility is lower than 5 km, is linked to the lack of rainfall during the dry season and to the deficiency of rainfall during the rainy season (June to October) over the Sahara and sub-Saharan Africa. Changes in dustiness have closely mirrored the incidence of drought during the 1980s (see 3.11). In the dry season (November to May) there was a significant increase in dustiness over the second half of the century, suggesting overall land degradation and encroachment of the deserts.

The scale of the impact has become clear from both satellite observations and surface measurements far afield. Enhanced satellite images have recorded huge clouds of distinctive sandy coloured Saharan dust being blown from North Africa westwards across the tropical Atlantic and reaching the Caribbean. In Barbados, measurements showed that these clouds sometimes reduce the amount of sunlight reaching the surface.

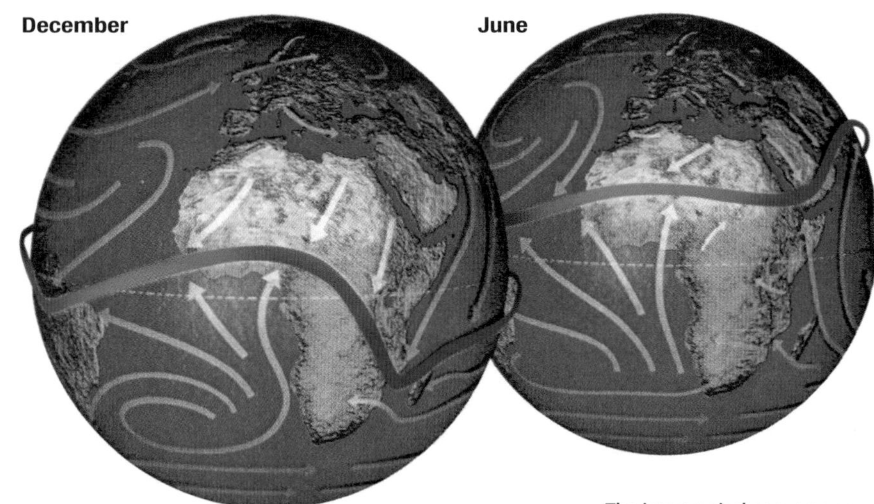

The intertropical convergence zone (ITCZ) separates the near-surface airflow of the hemispheres. In western Africa the ITCZ is always north of the equator but intensity of the associated convective activity changes with seasons. By contrast, over eastern Africa there is a pronounced seasonal shift between hemispheres in the mean position.

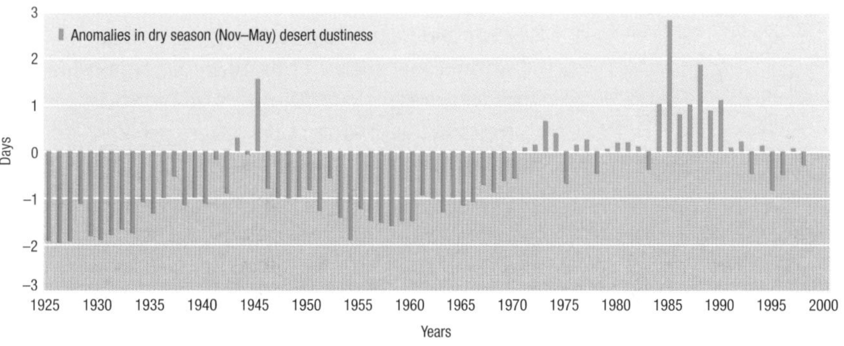

The general rise in dustiness in Niger since early in the 20th century is related to the decline in rainfall in that region over the same period.

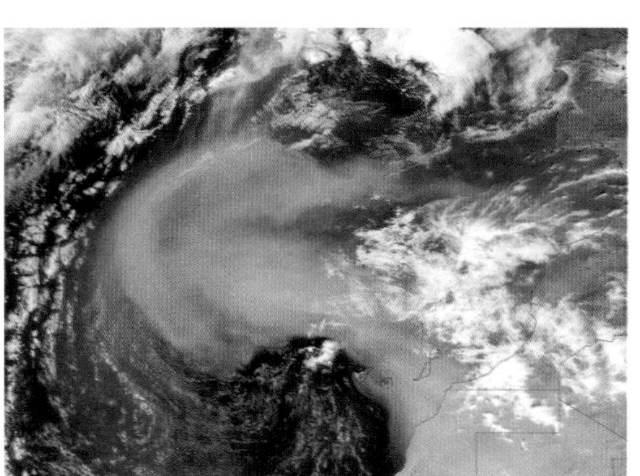

Satellite images, two days apart in February 2000, provide a graphic illustration of the amount of clouds of dust blowing out over the Atlantic from the Sahara.

2.20 Land surfaces and the atmosphere

Almost all life on land lives close to the ground, and so interactions between the atmosphere and the land surface, including its vegetation cover, define much of the natural world around us.

Boundary layer studies need frequent observations at many levels to measure temperature and moisture gradients and turbulent motions that are important for understanding land-biosphere exchanges with the atmosphere. Specially instrumented towers, such as this one, are set up in forested and open locations for recording these measurements.

Within the atmospheric boundary layer (the first few tens of metres above the ground) there are many complex physical processes at work. Understanding these processes is an essential part of improving our knowledge of climate, developing better climate forecasting models, estimating the impact of human activities on climate, and understanding how a changing climate might affect us.

We already know a lot about these boundary layer processes, including how those involving trees and plants affect the local climate. On a hot day, evapotranspiration coupled with the shading effect makes it cooler within a canopy of leafy trees than where the soil or grass is exposed to direct sunlight. In winter, ground frost develops first out on exposed grass rather than under trees. Such features of our lives are common knowledge, but the processes present formidable challenges to anyone trying to predict the detailed behaviour of the climate. Understanding the processes in the boundary layer is crucial to our ability to assess the likely effects of broadscale agriculture and tropical deforestation on climate, and whether human activities might be leading to an expansion of the world's deserts.

Better measurements

Better measurement of boundary layer and land-surface processes is the key to understanding how changes in vegetation cover might affect the climate. In recent years several large-scale international experiments have examined surface processes across the prairies of the Midwestern USA, the forests of Manitoba and Saskatchewan in Canada, over the Pyrenées between France and Spain and in a number of other climate regimes around the world. These experiments are designed to obtain a better understanding of how the atmosphere interacts with the Earth's surface. Using a wide variety of surface equipment plus aircraft and satellite observations, these studies have provided insights into land surface boundary layer processes, and have confirmed the critical role that surface conditions have on the amount of energy exchanged with the atmosphere.

Getting the effects of the surface right can appreciably improve the accuracy of computer models of the earth–atmosphere system. The European Centre for Medium Range Weather Forecasts (ECMWF) found that in the early development of its weather forecasting model, its predictions for winter night-time minima were often 10°C too low over much of Europe. By including improved representations of soil moisture and the time taken for it to freeze, errors have been significantly reduced.

Soil moisture and evapotranspiration rates also exert a major influence throughout the lower atmosphere. Weather forecasting models had difficulty predicting the location and intensity of the unprecedented rainfall (and consequent record-breaking floods along the Mississippi) over the midwest of the USA during the summer of 1993. Subsequent computer model studies showed that the forecasts of rainfall could be improved by incorporating more realistic measurements of soil moisture in the model from the outset.

Snow-covered northern forests reflect much less sunlight than open snow.

Developing improved models

Computer models are used to produce weather forecasts, to predict seasonal climate anomalies and to generate scenarios of climate change. Until recently, the representation of the land surface in computer models of weather and climate was quite inadequate (see above). Deficiencies included poor representation of water and energy storage and release by the land; inadequate representation of river discharge to the ocean; and for climate change studies, an inability to represent the responses of vegetation structures to climate change and the neglecting of long-term change in land carbon storage. Significant scientific progress is being made on all of these fronts. However, it is only in the first two that progress benefited the coupled models used to estimate climate sensitivity. Most coupled models now employ some representation of how vegetation controls evaporation and most can estimate river runoff for input to the ocean component of the model.

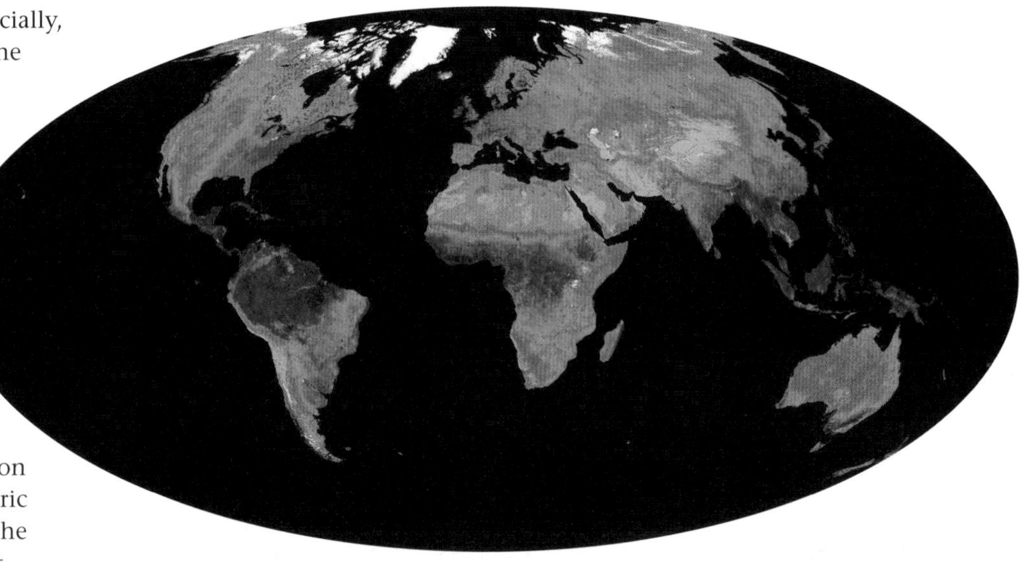

Progress with climate models, especially, depends on better measurements of the processes involved and their representation by the models (also called parameterization). Without these and further advances it will not be possible to estimate the extent to which regional differences in land surface response might translate into different climate sensitivities to radiative forcing. At the global scale, these limitations have not yet been shown to be significant. However, for the simulation of land surface climate and atmospheric climate change at the regional level, the sensitivities may be highly significant.

Freshwater runoff and local rainfall affect the salinity distribution of the oceans and together are an important part of the development of the latest climate models. Within the models, simulated runoff is collected over geographically realistic river basins and mixed into the ocean at the appropriate river mouths. Although this routing is performed instantaneously in some models, the trend is toward incorporation of a significant time lag of a month or more, from runoff production to river–ocean discharge.

The feedback process whereby climate-induced changes in vegetation affect the climate system, which further affects vegetation, potentially has large climatic implications. So far, however, it has proven difficult to incorporate this feedback process adequately in the coupled-model experiments used to estimate climate sensitivity. Also, the amount of carbon that is either extracted from the soil or stored in it by decaying vegetation is another source of considerable uncertainty.

Snow, with its high reflectivity, is an important component of the land surface. Current climate models have some capability in simulating the seasonal cycle of snow extent but tend to underestimate interannual variability. These weaknesses limit confidence in the details of changes, particularly at middle and high latitudes, simulated by current climate models.

Human activities

Land-use changes have led to changes in the amount of sunlight reflected from the ground (the surface albedo). The scale of these changes is estimated to be about one-fifth of the forcing on the global climate due to changes in emissions of greenhouse gases. About half of the land use changes are estimated to have occurred during the industrial era, much of it due to replacement of forests by agricultural cropping and grazing lands over Eurasia and North America. The largest effect of deforestation is estimated to be at high latitudes where the albedo of snow-covered land, previously forested, has increased. This is because snow on trees reflects only about half of the sunlight falling on it, whereas snow-covered open ground reflects about two thirds. Also, the snow lies on the open ground for longer than it hangs on the trees and hence the higher albedo persists longer. Overall, the increased albedo over Eurasian and North American agricultural regions has had a cooling effect. Other significant changes in the land surface resulting from human activities include tropical deforestation which changes evapotranspiration rates, desertification which increases surface albedo, and the general effects of agriculture on soil moisture characteristics. All of these processes need to be included in climate models, but for climate change studies there are few reliable records of past changes in land use.

One way to build up a better picture of the effects of past changes is to combine surface records of changing land use with satellite measurements of the properties of vegetation cover. Such analyses show that forest clearing for agriculture and irrigated farming in arid and semi-arid lands are two major sources of climatically important land cover changes. The two effects tend, however, to cancel out, because irrigated agriculture increases solar energy absorption and the amount of moisture evaporated into the atmosphere, whereas forest clearing decreases these two processes. Continuous accurate satellite measurements and ongoing surface observations are needed to establish whether either or both of these changes are having appreciable impacts on the climate.

Satellite measurements can quantify the concentrations of green leaf vegetation around the world, allowing scientists to monitor major fluctuations in vegetation and develop understanding of how they affect, and are affected by, regional climate trends. In this MODIS Enhanced Vegetation Index (EVI) map, very low values of EVI (white and brown areas) correspond to barren areas of rock, sand or snow. Moderate values (light greens) represent shrub and grassland, while high values indicate temperate and tropical rainforests (dark greens).

2.21 Climate variations with altitude

How the climate has varied at altitude is of importance both in understanding global climate variability and in how activities in mountain regions will be affected by any future changes.

Mountains create their own weather. Here, in the setting sun, Mount Hood in Oregon, USA, is causing the air to rise and form graceful lenticular clouds over its peak. Measurements made high in the mountains provide valuable insights into the behaviour of the atmosphere and the climate in general.

It has long been known that the temperature of the air decreases with increasing altitude. Snow-covered peaks show clearly how much colder it is at high levels. Other features of the climate vary appreciably with altitude (e.g. precipitation and cloudiness), and so understanding the climate of upland areas is an important aspect of meteorology. Of particular interest is how the climate at altitude has changed in the last 100 years.

Vertical temperature trends

The broad picture for temperature trends since 1958 for the globe at various atmospheric levels are:

- a substantial warming at the surface and in the boundary layer air;
- a broadly similar warming throughout the lower and middle troposphere, but with a more rapid warming until the late 1970s and less rapid since then; and
- significant cooling in the stratosphere.

Since 1979, microwave radiometers on weather satellites have been used to measure temperature trends in the lower troposphere (centred on an altitude of around 3 km) and in the stratosphere. These show rather less warming in the troposphere than otherwise observed at the surface, but there is a clear cooling trend in the stratosphere. The stratospheric observations also clearly show the warming effect of the two major volcanic eruptions in the last 20 years (El Chicon in 1982 and Mount Pinatubo in 1991).

The surface warming and stratospheric cooling are a signature pattern of the enhanced greenhouse effect. This makes observations in mountainous regions particularly important since they provide temperature readings above the normal level of the boundary layer air. Unfortunately, there are relatively few lengthy records in many remote mountain areas.

There are a number of explanations for the different temperature trends observed from the surface to the stratosphere. For example, the global cooling that follows large volcanic eruptions is greater in the troposphere than the surface. By way of contrast, the clearest impact of volcanic eruptions is seen in the warming of the stratosphere. Understanding the processes in the boundary layer is crucial to our ability to assess the likely effects of broadscale agriculture and tropical deforestation on climate, and whether human activities might be leading to an expansion of the world's deserts. In addition, the variability of tropospheric temperature increases with altitude so that the temperature effects of warm and cool phases of the El Niño/ Southern Oscillation are exaggerated at higher levels relative to the surface. Another possibility for the remaining differences is related to the gradual decrease in concentrations of stratospheric ozone, since the upper troposphere would be expected to cool due to stratospheric ozone depletion.

Recent global warming is also reflected in data from mountainous areas or the tropics. The freezing level in these areas has risen 100 m since the mid-1970s, with a consequent retreat in the tropical glaciers of South America, East Africa, Irian Jaya and Tibet.

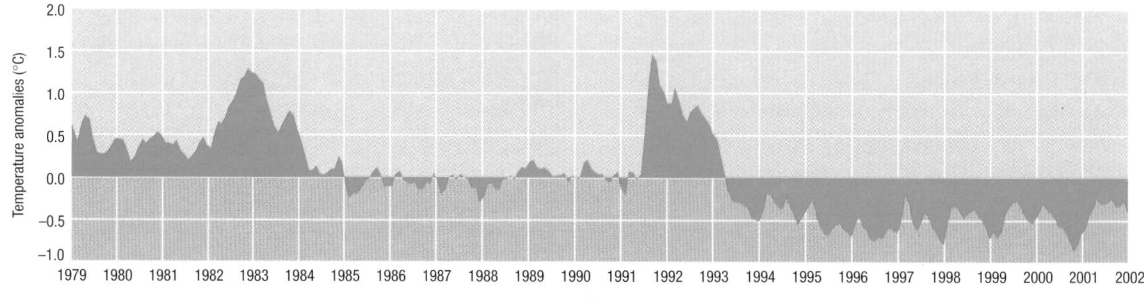

Satellite measurements of the temperature in the stratosphere show a marked cooling trend since 1979. They also show how the eruptions of El Chicon (Mexico) in 1982 and Mount Pinatubo (Philippines) in 1991 cause substantial temporary warming.

The retreat of glaciers

The 20th century glacier retreats of most regions is consistent with a mean warming in alpine regions of 0.6–1.0°C. Data on glacier recession come in two forms: (i) mass balance observations and (ii) measurement of extent, notably in length of a glacier. The specific mass balance is defined as the net annual gain or loss of ice mass at the glacier surface. It is expressed in metres, averaged over the entire area of the glacier. Glacier mass balance is closely coupled to the flow of energy to and from the surface averaged over the summer half year and therefore to temperature, and the radiation balance. Warm, dry, sunny summers reduce the mass dramatically, whereas cool, cloudy summers coupled with heavy snowfalls in winter add to the mass.

Continuous observations for the period 1980–95 for 32 glaciers in North America, Eurasia and Africa show continued or even accelerating losses. This conclusion is, however, strongly influenced by the preponderance of Alpine and Scandinavian glaciers in the sample, so the general significance of the accelerated trend remains uncertain. Inland glaciers appear to have decreased steadily in mass. The mass balance of glaciers in humid areas is not so clear-cut – for instance, recent increased winter precipitation in southwest Norway has outweighed any loss due to a warming trend.

The other method of assessing glacier retreat is through estimates of changes in glacier length. Records showing the worldwide retreat of glaciers, in some places, go back many centuries. Interpreting these is difficult, however, as the complex chain of dynamical processes linking glacier mass balance and length changes has only been calculated for a few individual glaciers. When this work is extended, it will enable scientists to convert glacier length measurements to a consistent set of mass balance time series. In turn, it may then be possible to relate these series in a consistent way to regional and global temperature changes representative of the altitude of the glaciers.

The Swiss Alps

What has happened in the Swiss Alps highlights the challenges of interpreting trends. This region experienced a remarkably strong warming during the 20th century, with near-surface temperatures rising at twice the global average. Night-time temperatures have risen more rapidly than daytime values, especially at lower levels where increased cloudiness has reduced the amount of sunshine while at the same time reducing the loss of heat from the surface overnight. There is also worrying evidence of an increase in extreme rainfall events, especially in summer. Perhaps the most evocative instance of the effects of the recent warmth in the European Alps was the finding of the Oeztal 'ice man', who had remained entombed for over 5000 years until warm late-summer weather in 1991 melted the ice around him. This discovery suggests that the current warming has led to glacier recession reaching levels not seen for several thousand years.

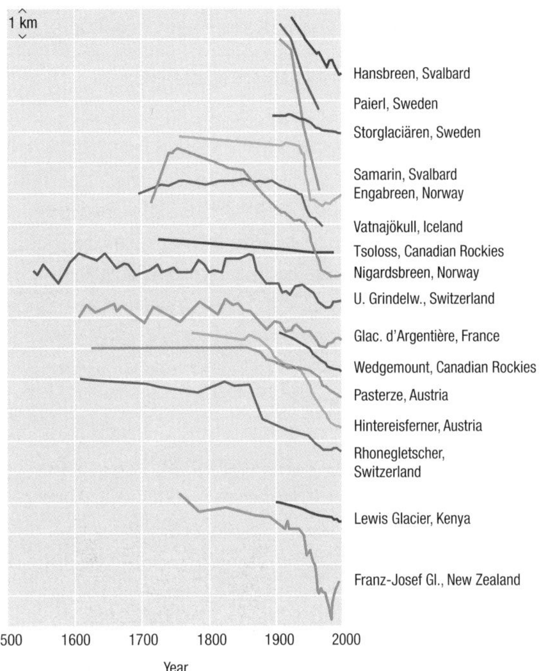

Measurements of the extent of glaciers around the world show a consistent pattern of rapid retreat (the downward slope of the graphs) in the second half of the 20th century.

Two photographs of the Athabasca (middle distance) and Dome Glaciers in the Columbia Icefield, Alberta, Canada, show the dramatic retreat between 1906 and 1998.

2.22 Variations in the cryosphere

Measurements of changes in the extent of snow cover and sea ice at high latitudes provide a sensitive measure of climate change.

Snow and ice are good reflectors of sunlight and so the cryosphere, the totality of snow, ice and permafrost above and below the land and oceans, is an important component of the climate system. Variations from year to year and on longer timescales are important factors in the energy budgets of the Polar regions, and for climate change.

Snow cover

Until the advent of weather satellites it was not possible to obtain hemisphere measurements of the extent of snow cover and its variation over the years. Since the 1960s, we have been able to obtain some valuable information from satellite imagery. The mean annual Northern Hemisphere snow cover extent is 25.3 million km^2, with 14.7 million km^2 over Eurasia and 10.6 million km^2 over North America (including Greenland).

Snow cover was more extensive in the first half of the satellite record. Between 1972 and 1985, annual means of snow extent fluctuated around a mean of 25.9 million km^2. An abrupt transition occurred in 1986 and 1987, and since then the mean annual extent has fallen to 24.2 million km^2. What appeared to be a gradual rebound from low extents in the late 1980s and early 1990s, peaked in 1995, at least temporarily. More recently, values have fallen back to those previously observed between 1988 and 1994.

Fluctuations of North American snow extent between 1900 and 2000 have been analysed by calibrating historical surface station observations with satellite measurements. This record suggests spring snow extent peaked in the 1960s and has since fallen to its lowest values; winter snow extent peaked between the 1970s and 1980s; and autumn snow extent continued to increase into the late 1990s. These changes indicate a shift in seasonality of the North American snow season. Moreover, since the mid-1970s, February snow cover extent across the continent showed the highest interannual variability of the 20th century.

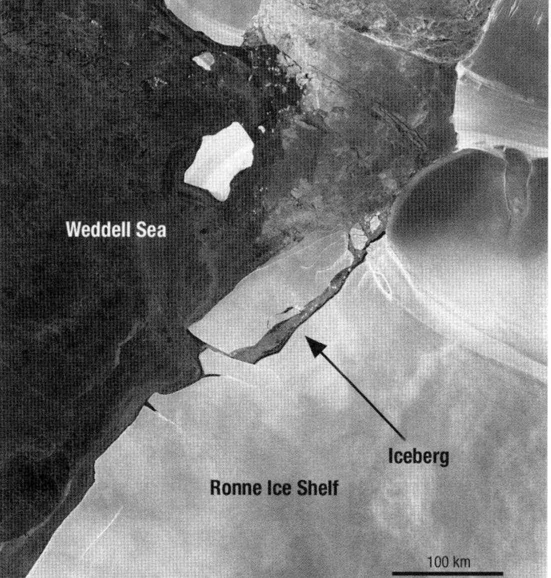

The formation of icebergs at the margins of glaciers and ice shelves is a natural process. The Canadian Space Agency RADARSAT satellite image shows a massive iceberg breaking off the Ronne Ice Shelf in Antarctica. With an area of some 5800 km^2, such a huge iceberg may drift for several years in the Southern Ocean before it finally breaks up and melts away.

Trends in sea-ice extent and thickness

Regions where sea ice forms are frequently covered with clouds and it was not until microwave radiometers, which can 'see' through clouds, were mounted on weather satellites in the late 1970s that reliable measurements of the extent of ice could be made on a regular basis. Arctic data over the past two decades show a decrease of 2.9 (± 0.4) per cent per decade in sea-ice extent. This decrease is strongest in the eastern part of the hemisphere and has been most apparent in summer. Earlier surface data are not complete, with the largest gaps in the autumn and winter, but they can be used to make some observations about trends in the 20th century. The summer decrease that is largely responsible for the overall downward trend during the satellite era is present in the record during the

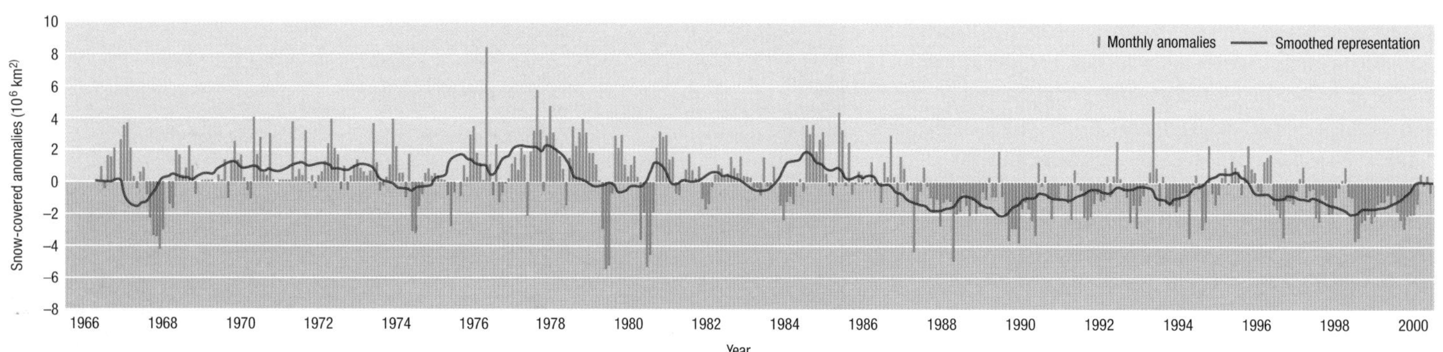

Snow cover in the Northern Hemisphere is a sensitive indicator of climate change. Snow cover declined considerably after the mid-1980s.

entire second half of the 20th century. This decrease represents about 15 per cent of the estimated average summer extent in the first half of the 20th century. Springtime values show a somewhat weaker negative trend over the same period with a total reduction of about 8 per cent, but there is only a slight and uncertain downward trend in autumn and winter since about 1970.

Analysis of data obtained from submarines on the thickness of the permanent ice cover has indicated the same sort of reduction. A comparison of a series of measurements made from the late 1950s to the mid-1970s with a set made in the mid-1990s showed that the mean ice thickness had decreased by about 40 per cent at the end of the melt season (September). The decrease was greater in the central and eastern Arctic than in the Beaufort and Chukchi Seas.

The overall late 20th century decrease of Arctic ice extent, at first sight, is consistent with the contemporary pattern of high-latitude temperature change, which included a warming over most of the sub-arctic land areas. Some of this pattern of warming has been attributed to recent trends in the North Atlantic Oscillation (NAO). The extended positive phase of the NAO during the 1980s and 1990s, corresponding to reduced sea-level pressure in the Arctic, is broadly compatible with the observed warming and associated ice retreat over the eastern Arctic. This connection with the NAO must, however, be treated with caution as there is no evidence of the same response during the period of positive NAO years in the 1910s and 1920s.

Sea-ice extent in the Antarctic shows a weak increase of 1.3 ($\pm$0.2) per cent per decade since the late 1970s. Prior to this there was a marked reduction in the early 1970s. Records are not sufficient to construct a longer time series for the Antarctic, but whaling ship logs suggest significantly greater ice extent in the Southern Ocean during the 1930s and 1940s than during recent decades.

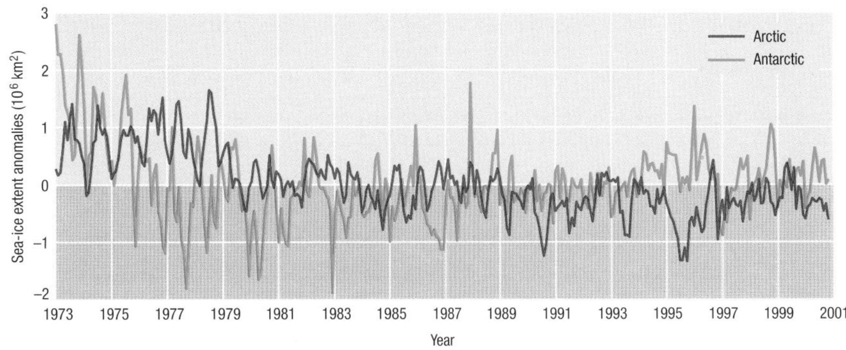

Satellites have been used to obtain reliable measurements of the extent of sea ice in the Arctic and Antarctic since the 1970s. There has been a significant reduction in extent over northern polar regions but around Antarctica there has, if anything, been a slight increase in extent.

Sea ice

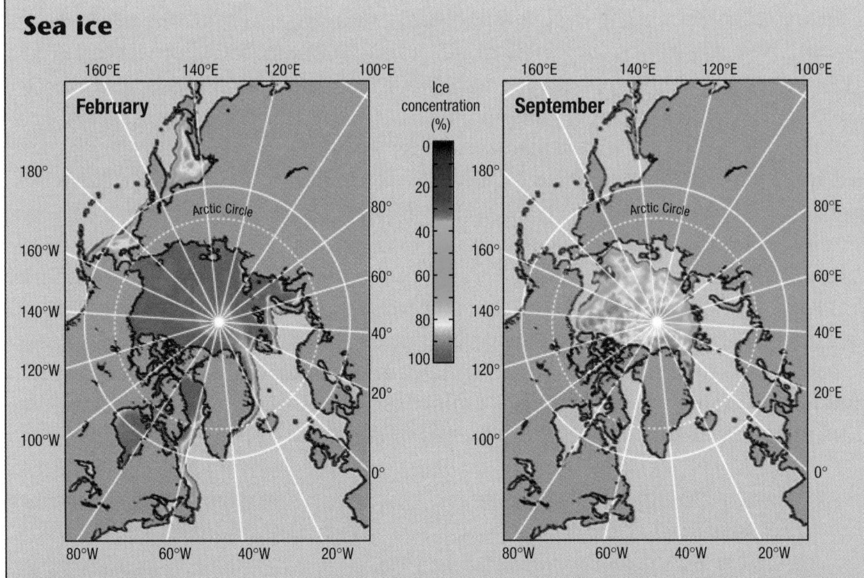

In the Northern Hemisphere, the extent of pack ice varies from a minimum of some 9 million km^2 in late summer, when it is confined to the central part of the Arctic basin, to a maximum of 16 million km^2 in late spring. The annual average extent of the ice in the Arctic is about 13 million km^2, within which there are about 2.5 million km^2 of open water.

The maximum extent of the pack ice fluctuates from year to year. These variations are closely linked to the strength of the westerly zonal wind flow during the winter half of the year, especially in the Atlantic sector. When the zonal flow is strong, low pressure systems regularly push through the Norwegian Sea and into the Barents Sea and keep the pack ice back. At the same time, this stream of depressions continually pulls Arctic air down the Davis Strait and pushes pack ice farther out beyond Newfoundland.

In Antarctic regions the pattern is much simpler. Although the average extent of the sea ice is a little less than in the Arctic, being some 12 million km^2, it expands and contracts much more. This creates a gigantic annual pulse ranging from a maximum extent of nearly 20 million km^2 each August to a minimum of only 3 to 4 million km^2 each February. As with the Arctic, within this ice boundary there are some 3 million km^2 of open water in the form of 'leads' and, occasionally, much larger areas of open water known as 'polynyas'.

The cycle of ice formation releases salt, which contributes directly to Antarctic deep-water production through large-scale sinking of surface ocean waters and is a substantial factor in the Great Ocean Conveyor Belt. Variations in the amount of ice formed from year to year and on longer timescales affects the strength of this circulation. In addition, the formation of polynyas up to 350 by 1000 km can have a substantial impact on both the climate of the region and the formation of deep water.

2.23 The effect of volcanoes on climate

It has long been recognized that volcanoes have the capacity to alter the climate. In recent decades, studies of large eruptions have provided the evidence to estimate just how big their impacts are.

The massive eruption of Mount Pinatubo in June 1991 injected large amounts of dust and sulphur dioxide into the upper atmosphere and its effect on global climate could be detected for several years.

The impact of volcanoes on the climate has intrigued scientists, at least since Benjamin Franklin suggested that the bitter winter of 1783–84 in northern Europe was caused by the dust cloud produced by the huge eruption of Laki in Iceland in July 1783. Explosive eruptions can inject vast amounts of dust and, more significantly, sulphur dioxide into the upper atmosphere where this gas is converted into sulphuric acid aerosols. In the stratosphere, at altitudes of 15 to 30 km, there is very little mass exchange with the turbulent troposphere below. The minute aerosol particles remain suspended for up to several years and are spread round the entire globe forming a veil. The particles absorb sunlight and locally heat the stratosphere but at lower levels cause compensating cooling as less solar radiation reaches the Earth's surface. The worldwide brilliant sunsets which followed the eruption of Krakatau, Indonesia in 1883 added to speculation about the effects of volcanoes on the climate.

Eruptions which inject large quantities of sulphur compounds into the stratosphere in equatorial regions have the most impact on climate. Mount Pelée (Martinique), Soufrière (St Vincent) and particularly Santa Maria (Guatemala) that all erupted in 1902 were the only significant equatorial volcanoes to erupt during the first half of the 20th century. Then in 1963, the major eruption of Agung on Bali occurred, a few years before the publication by Hubert Lamb of his research describing the first evidence that major eruptions were followed for periods of a year or two by lower surface temperatures in the Northern Hemisphere. After large explosive tropical eruptions, the Southern Hemisphere shows a cooling (somewhat smaller than the Northern Hemisphere) in the three years following the eruptions, but the spatial patterns of the responses have been less well studied than in the Northern Hemisphere. The fact that climatically significant eruptions have, in recent centuries, occurred roughly every decade, and sometimes come in more concerted bunches, means that they are a significant factor in understanding both interannual climatic variability and climate change.

During the last two decades of the 20th century two important eruptions, El Chicon (Mexico) in 1982, and Mount Pinatubo (Philippines) in 1991, provided the opportunity to make much more detailed measurements of the composition of volcanic dust veils and their effects on the climate. In particular, Mount Pinatubo appears to have injected the greatest amount of sulphur compounds into the stratosphere in the 20th century. This eruption produced an extensive dust veil, which provided the opportunity to acquire considerably more information on how volcanoes affect climate. It was also used to check how well the computer models of the global climate handle the presence of dust high in the atmosphere.

Hubert Lamb (1913–97)

Hubert Lamb's great contribution to climate studies was to identify the wealth of information about changes in the climate in historical records and early observations. After joining the UK Meteorological Office in 1936, he had a varied career until he joined the Climatology Branch in 1954. Here he discovered extensive archives that enabled him to make a series of groundbreaking assessments about climate change over recent centuries. These included establishing the value of sea surface temperature measurements, and, perhaps most importantly, building up an estimate of the dust veil index of major volcanic eruptions extending back to AD 1500. His contribution to climate change as a scientific discipline was recognized in his being made the first Director of the Climatic Research Unit at the University of East Anglia in 1972. He is also renowned for his major contributions to climatology in his two-volume book *Climate, Present, Past and Future.*

Volcanoes and the global temperature record

Analyses of the global temperature record since the mid-19th century confirm the general assumption that major volcanic eruptions lead to an overall cooling at the surface. The Mount Pinatubo eruption in June 1991 generated significant cooling for several years after the event. This cooling lasted a similar period and was similar in magnitude to, or somewhat greater than, that which followed the earlier equatorial eruptions in 1883, 1902, 1963 and 1982.

The spatial patterns of temperature and circulation anomalies associated with major eruptions are more complicated. The most direct effect is on summer temperatures in middle latitudes, where the dust veil blocks out some solar radiation and reduces temperatures. Somewhat surprisingly, however, a warming was observed over the continents of the Northern Hemisphere at higher latitudes in the first winter after the Mount Pinatubo eruption. The precise details behind this response are hard to pin down but, broadly speaking, what appears to have happened is that the westerly winds over the North Atlantic were pushed to higher latitudes, a pattern associated with the positive phase of the North Atlantic Oscillation. The initial warming response was then followed by a general cooling of the Northern Hemisphere, but limited change in atmospheric circulation, during the next two years.

The eruption of Mount Pinatubo caused quite a strong cooling of the global surface temperature (about 0.2°C) and in the troposphere (perhaps 0.4°C) from late 1991 to 1994. Satellite measurements also showed that its maximum impact on the radiative balance of the troposphere was about 3–4 W/m^2. It is likely that other major eruptions like Krakatau in 1883 reduced surface temperature for one to a few years. However, major effects from volcanoes lasting many decades seem less likely because the volcanic materials and their by-products, unlike CO_2, slowly settle out and do not survive for decades in the atmosphere.

Modelling the impact of Mount Pinatubo

Mount Pinatubo provided a good test of computer models of the climate, as its cooling effect through the reduction in the amount of solar radiation reaching the Earth's surface was comparable to the warming effect equivalent to a doubling of CO_2 in the atmosphere. The predicted changes due to the eruption show models can do well in predicting the global consequences of specific climatic perturbations. This gives us greater confidence in using these models to determine the likely effects of human activities on the global climate.

Major volcanic eruptions are fabled for producing spectacular sunsets, as this example after the eruption of Mount Pinatubo shows.

Mount Pinatubo Stratospheric Aerosol (SAGE II)

10 Apr 1991 to 13 May 1991

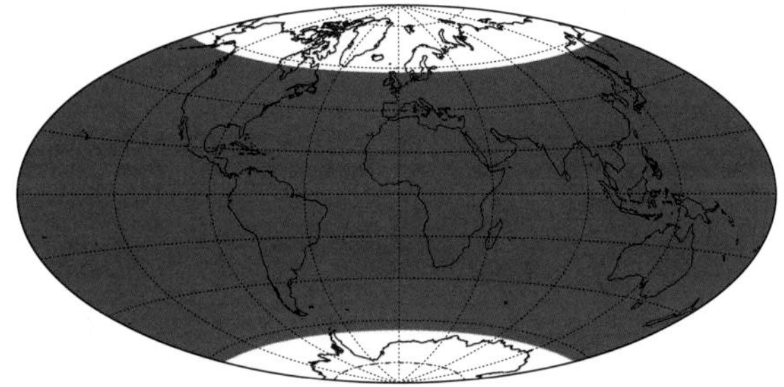

15 Jun 1991 to 25 Jul 1991

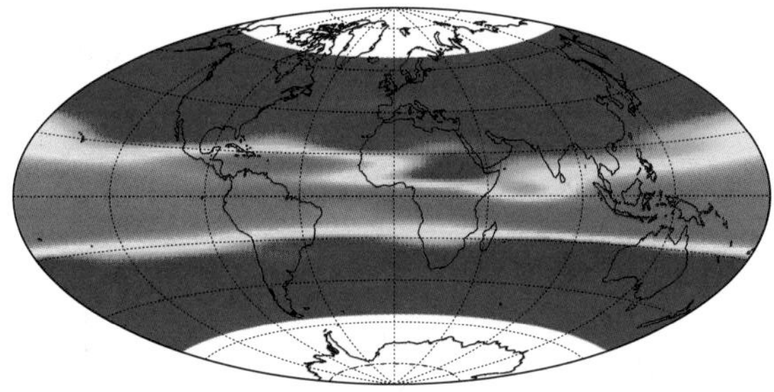

23 Aug 1991 to 30 Sep 1991

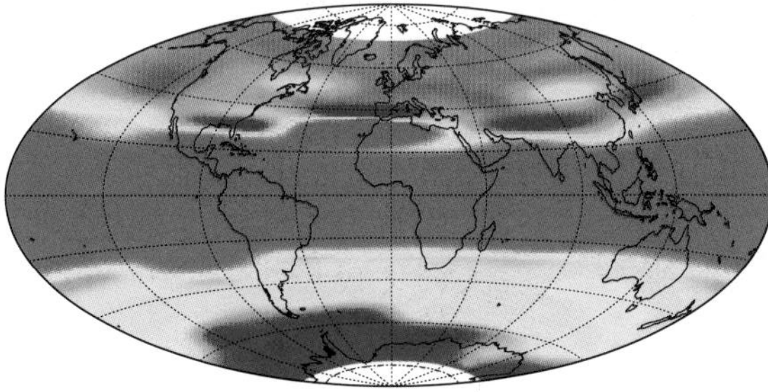

5 Dec 1991 to 16 Jan 1994

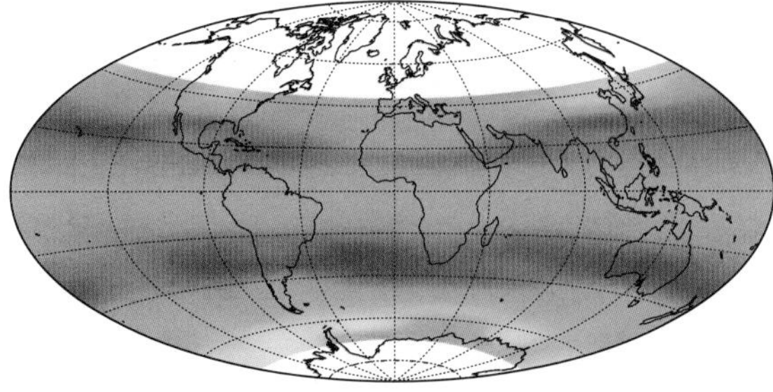

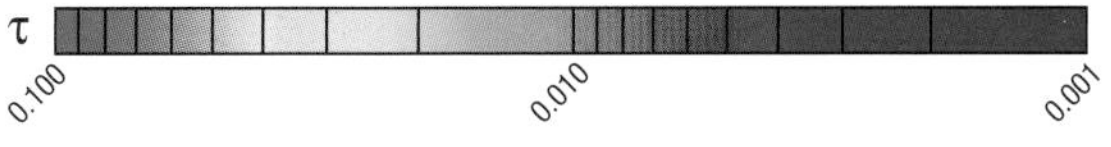

The Stratospheric Aerosol and Gas Experiment II (SAGE II) aboard the Earth Radiation Budget Satellite measures stratospheric optical depth (τ) for a wavelength of 1020 nm, and provided a detailed profile of the evolution and dispersion of the Pinatubo aerosol over several years. The values of τ range from less than 10^{-3} (dark blue) to greater than 10^{-1} (red). The maps show that prior to the eruption of Mount Pinatubo, the optical depths around the world were low. Within weeks of the eruption in mid-June 1991, the dust cloud it created in the stratosphere had been transported around the world. Even in 1994, optical depths had not returned to pre-Pinatubo levels.

2.24 Solar influences

Of all the possible extraterrestrial influences on the Earth's climate, solar activity (sunspots) has been the most promoted. Throughout the 20th century a scientific debate raged as to whether the influences were real. Recently, the evidence for a solar influence on climate has again grown.

The output of the Sun varies on all timescales. The best-known variation in solar behaviour is the regular fluctuation in the number of sunspots, which show up as small dark regions on the solar disk, and affect the energy output of the Sun. Other aspects of solar activity include changes in the solar magnetic field, which influence the number of cosmic rays entering the Earth's atmosphere from deep space, and variations in the amount of ultraviolet radiation from the Sun that may lead to photochemical changes in the upper atmosphere. All these variations have the potential to induce fluctuations in the climate.

Does the Sun's energy output vary enough to affect the climate?

The fundamental issue in deciding whether the Sun could be responsible for a significant part of the observed fluctuations in the climate is whether its output varies enough to produce the type of changes we see. At the simplest level this would be a matter of showing that the amount of energy from the Sun varied significantly during the 11-year sunspot cycle. Ground-based efforts during the first half of the century to show that there were appreciable changes in the output were plagued by problems in correcting for the effects of atmospheric absorption.

It was only in 1980, with the launching of specialized satellite instruments, that it was possible to measure accurately the changes in energy radiated by the Sun. Orbiting high above the atmosphere, the effects of atmospheric turbulence and clouds were eliminated and the output of the Sun was monitored with sufficient accuracy. Observations during the last 20 years show a modulation of about 1.5 W/m^2 in the solar output received by the Earth over the 11-year solar cycle. This is equivalent to about 0.1 per cent of the average incoming solar radiation (1370 W/m^2). These changes cannot, however, be explained in terms of sunspots alone. Sunspots are areas of lower temperature and an increase in their number might be expected to coincide with reduced solar output. On the contrary, the energy output from the Sun peaks when the sunspot number is high.

To explain behaviour of the energy output of the Sun we need to consider other aspects of solar activity. In particular, the output appears to be a balance between increases due to the development of bright areas, known as faculae, at times of high solar activity and the decrease resulting from the upsurge in sunspot number. Overall, the heating effect of the faculae outweighs the cooling effect of the sunspots. Various estimates have been made of the longer-term fluctuations in solar energy output over the last two or three centuries using various proxy records to infer changes in solar activity. These suggest that the output may have varied by as much as two to three times the fluctuations observed by satellite instruments during the last 20 years.

The possibility that the Sun's energy output may have varied more appreciably in the past could explain the marked parallel between these changes and estimates of the Earth's surface temperature over much of the past four centuries. The changes could be sufficient to explain as much as half of the climate change observed from the 18th century until the early 20th century, but they cannot explain the warming that has occurred

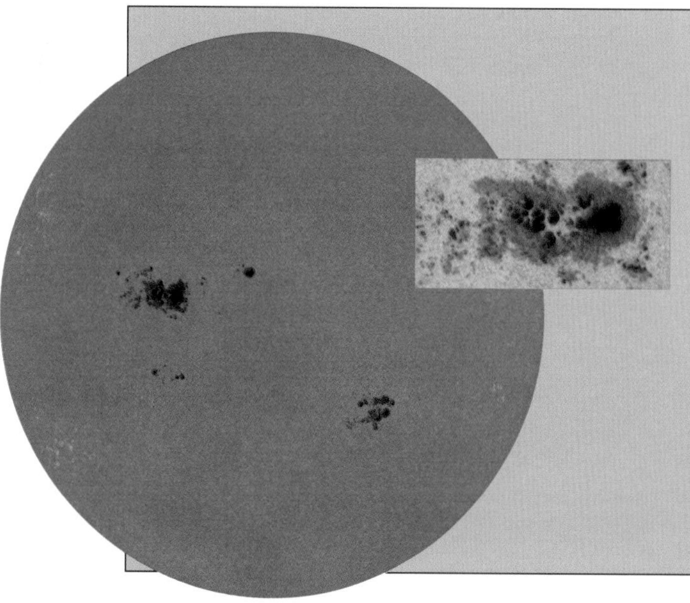

Sunspots

Sunspots feature in early Chinese records and again in early 17th century observations by the Italian scientist Galileo. By the mid-19th century it was realized that the number of sunspots went through a well-defined cycle of about 11 years in length. Ever since, there have been attempts to show that this cycle influences the climate, and solar-terrestrial physics has been an active research topic. These studies have concentrated not only on the 11-year cycle, but also longer periodicities. In particular, there is a 22-year cycle that may be linked to the fact that the magnetic field associated with sunspots reverses between alternate 11-year cycles. Longer cycles in the peak number of sunspots occur every 80 to 90 years and around 200 years.

during the second half of the 20th century. Furthermore, the satellite measurements indicate that the levels of solar radiation were approximately the same during the sunspot minima in 1986 and 1996. Changes in solar energy output therefore would have appeared to have played no part in the approximately 0.15°C global surface warming that occurred during that period.

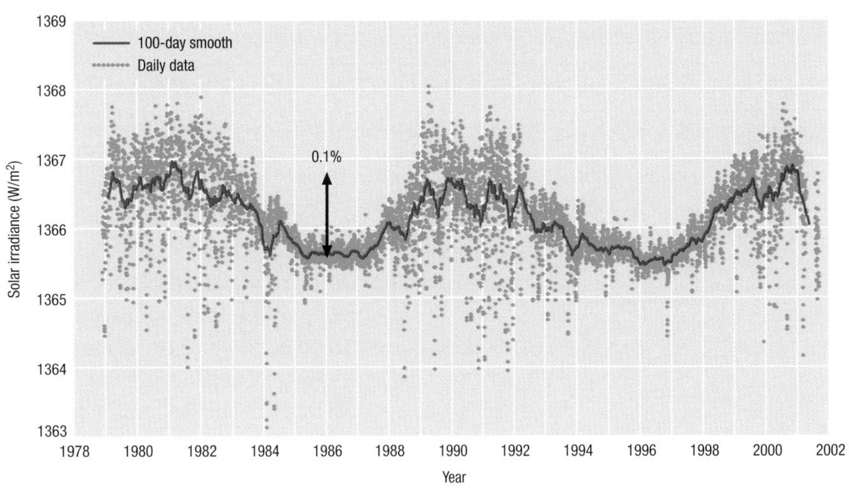

Measurements by a series of satellites since 1978 have confirmed that the energy output from the Sun rises and falls in line with the 11-year sunspot cycle.

Possible amplification mechanisms

If changes in the absolute energy output of the Sun were too small to cause directly the observed global climate change in recent centuries, what other possible explanations could there be? There is no shortage of theories as to how other features of the Sun's variability might be responsible for climate change. The majority of theories are linked to changes in the Sun's magnetic field that are capable of affecting events on Earth. For example, changes in the solar magnetic field mean that galactic cosmic rays are less able to penetrate the Earth's atmosphere when the Sun is more active causing the cosmic ray penetration to be inversely related to solar activity. This might affect the efficiency of cloud formation and hence the solar cycle could modulate cloudiness through variations in the number of cosmic rays reaching the Earth. At present, however, there is insufficient evidence to support this theory.

Changes in the intensity of solar radiation at ultraviolet wavelengths are proportionately greater than for the longer wavelengths, thus the effects on the amount of ozone in the Earth's upper atmosphere could be correspondingly greater. Ozone concentrations measured by space-based instruments since 1978 exhibit fluctuations of 1.5 to 2 per cent, apparently in response to changing solar ultraviolet radiation during the 11-year solar cycle. This might have a significant consequence on the circulation of the stratosphere. It is possible that various links in the behaviour of these induced variations could propagate downwards to affect surface temperature. Simulations using a climate model provide some support for this proposal and suggest that the decadal solar-ozone cycle, which is comparable in amplitude to the overall long-term ozone decrease in the past two decades, may influence tropospheric circulation patterns.

One problem in confirming or disproving the proposed ozone-related mechanisms connection between solar activity and climate variability is the lack of long-term, high stability, space-based solar ultraviolet monitoring. Another is the need to establish the extent to which volcanic activity has been responsible for recent fluctuations in stratospheric ozone. The nearly coincidental occurrence of volcanic eruptions in recent decades (1982, 1991) with peaks of solar activity (1980, 1990) is a complicating factor. A great deal of work is clearly needed to establish the response of ozone to solar variability and on how it might affect climate.

The number of sunspots has been recorded since the beginning of the 18th century. This record shows that there is a dominant 11-year cycle and several less obvious cycles with longer periods.

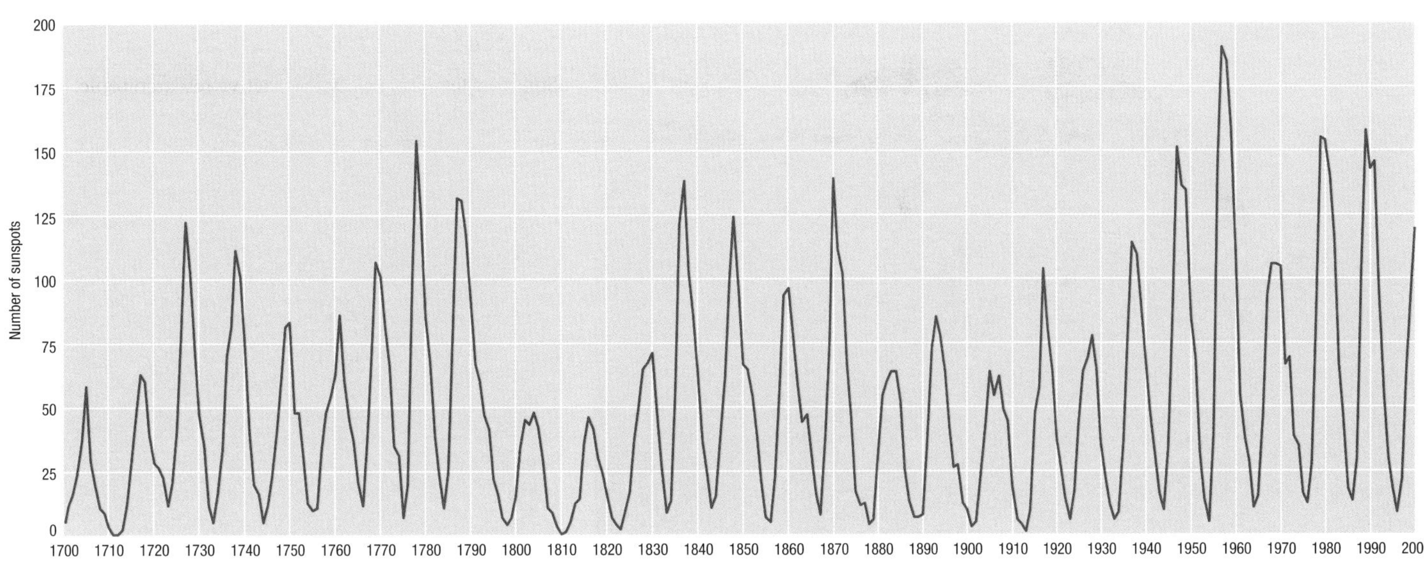

2.25 Human influences on climate: (a) Urbanization

All urban areas alter their local climate, and the bigger the city, the bigger the impacts. Many aspects of the local weather reflect the differences between conditions in cities and the surrounding rural areas.

The worldwide movement of people into the cities has caused profound changes in the local climate as the cities have expanded to accommodate the influx. At the beginning of the 20th century some 250 million people lived in cities – only about 15 per cent of the world's population. At the end of the century, nearly half of the 6000 million people on the planet lived in urban areas, with this proportion likely to increase appreciably. The fact that cities generate many aspects of their own local climate is clearly of direct relevance to the people who live in them.

Urban pollution

As well as being a major environmental hazard, the build-up of pollutants in cities has a significant effect on the local climate. The most dangerous conditions occur during calm sunny conditions. Higher levels of pollutants such as sulphur dioxide, oxides of nitrogen and particulates, notably from internal combustion engines, are a permanent feature of many urban areas. In the presence of strong sunlight these pollutants react chemically to form photochemical smog. This is exacerbated by the higher temperature within the so-called urban heat island. Increased particulate levels reduce the amount of sunlight appreciably and, combined with the additional updraft from the heat island, tend to increase the amount of cloudiness over cities and downwind. There is also evidence of these effects leading to more precipitation and, in particular, heavy showers and thunderstorms. Many cities have been located in geographical areas that add to the pollution problem by having poor natural airflow through them. Los Angeles,

Mexico City, like many major cities, experiences serious levels of photochemical smog when conditions are calm and sunny. This smog is made worse by the urban heat island which means that it can be much warmer in the centre of the city than in the surrounding countryside.

The urban heat island

A series of physical processes is at work in rendering the climate of cities different from that in surrounding rural areas. The removal of trees and vegetation and the presence of buildings alter the surface roughness and windflow patterns. The brick and concrete canyons formed by buildings in addition to the asphalt of roads and parking lots increase the amount of heat absorbed from the Sun, which is conducted more efficiently into these materials. The result is more heat stored, and its release at night slows down the rate of cooling of the surface, just like a giant storage heater.

The impervious nature of so many surfaces in cities also accelerates the runoff of any rain, reducing potential evaporation and its associated cooling. This effect is exacerbated by the removal of vegetation and trees which reduces the amount of water vapour transpired into the atmosphere. Furthermore, the concentration of buildings with their high energy use, either by burning fossil fuels or consuming electricity, further warms up the area, as do vehicles, which also release large quantities of pollutants.

The overall impact of these effects makes cities substantially warmer than their surroundings, as shown in this example from Mexico City. On calm clear nights the difference can exceed 10°C for the largest cities. These temperature differences largely disappear when wind speeds rise above about 24–36 km/h. Other meteorological variables change by smaller amounts, with rainfall slightly increasing, especially from convective storms, and wind speeds decreasing appreciably, although eddy effects caused by high buildings can lead to unpleasantly gusty wind conditions in urban canyons unless care has been given to their aerodynamic design. The more rapid runoff of rain increases the risk of damaging flash floods.

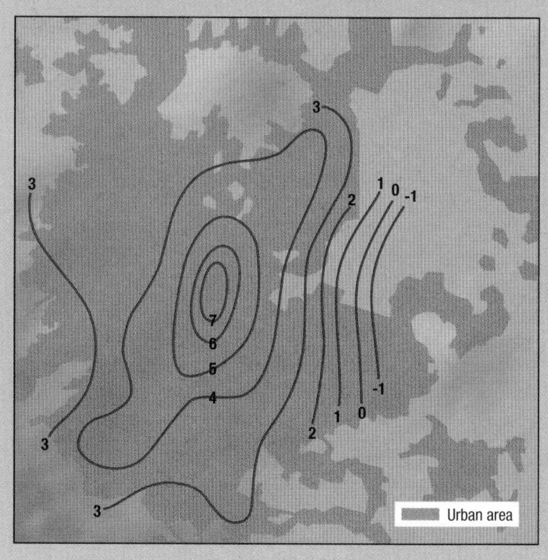

California is one of the more notorious cities with this problem.

Improving life in the cities

Some of the first, and most effective environmental legislation, concerned efforts to reduce the level of air pollution in cities. Following the 'Great Smog' in London in December 1952, in which the attributed mortality was over 4000 people, comprehensive legislation was introduced in 1956 to control the emission of smoke from coal fires (the Clean Air Act). The effects of this legislation were dramatic, with the amount of winter sunshine observed in central London rising by over 70 per cent between the late 1950s and the 1980s. Over the last two decades, much the same occurred in the Los Angeles basin where the introduction of exhaust controls on automobiles dramatically reduced the incidence of photochemical smog.

These results provide an indication of the strategies that will be needed to address the threat of higher temperatures and worsening air pollution likely to occur as a result of further increases on urbanization and with global warming. Many health problems are now associated with summer heatwaves, which are amplified by the effects of the urban heat island in the large cities of warmer temperate regions (e.g. New Delhi, Shanghai, New York and Tokyo). As daytime temperature rises above 35°C, the attributable mortality in many middle latitude cities rises sharply, especially if the temperatures remain high during the night. When combined with local pollution, the overall effect of prolonged hot weather tends to make some cities even more unhealthy places.

The solutions will lie in reducing vehicle emissions, promoting cleaner public transport and reducing domestic power consumption. Pressures for local action will reinforce the case for reducing the release of greenhouse gases that are contributing to global warming, and implementation will contribute to beneficial outcomes. Strategies designed to minimize the intensity of the urban heat island include the planting of more trees to create shade in suburban areas. There are also major gains to be had through reducing the absorptivity of the surfaces of buildings (e.g. by painting roofs white), parking lots and roads. Black surfaces, such as the black asphalt commonly used for parking lots, can become up to 40°C hotter in the sun than more reflective surfaces. Using more reflective surfaces wherever possible in major cities may therefore be a highly cost-effective solution to reducing the intensity of the urban heat island.

The magnificent Taj Mahal near Agra in India is today less often seen in direct sunlight because of air pollution from the nearby city. This pollution is also causing damage to its delicate carvings.

Helmut Landsberg (1906–85)

Helmut Erich Landsberg, Professor Emeritus of the Institute of Physical Sciences and Technology, University of Maryland (USA), was one of the world's leading climatologists. Born and educated in Germany, Professor Landsberg moved to the USA in 1934. In addition to his university career as research professor, and 20 years in the service of the US Government, he had a distinguished international career. In 1979 he was honoured with the World Meteorological Organization's IMO prize in recognition of his outstanding contributions to climatology.

Professor Landsberg was keenly interested in many of the sciences, and his work in bioclimatology was outstanding. He strove to raise awareness of the human influences on climate, how man might induce climate change and how subsequent changes might influence our climate and hence our living conditions. His knowledge of condensation nuclei, dust and pollution and regional climates was applied with great enthusiasm to studies of urban climates. Landsberg used his insights into urban habitation and micrometeorological conditions in his seminal study (1979) of the impact of the local climate on the building of the planned community of Columbia, Maryland (between Baltimore and Washington). This provided the clearest measure of how, even at the early stages of development, the urban heat island and other features of urbanization quickly develop.

2.26 Human influences on climate: (b) Greenhouse gases

The clearest example of an anthropogenic impact on the global climate is the change caused by the emission of radiatively active gases from human activities. Most attention has been devoted to the rise in carbon dioxide (CO_2) from the burning of fossil fuels, but a number of other emissions are also of potential importance to climate.

Carbon dioxide and other greenhouse gas emissions from power stations and motor vehicles are major contributors to the overall influence of human activities on the Earth's climate.

The build-up of greenhouse gases in the atmosphere during the 20th century was a direct consequence of growth in the use of energy with expansion of the global economy. Over this period industrial activity grew 40-fold, and the emissions of gases like carbon dioxide (CO_2) and sulphur dioxide (SO_2) grew 10-fold. The increase in emissions has also resulted in an increase in the atmospheric concentrations of other trace gases that contribute to the enhanced greenhouse effect; and to the presence of some new ones.

Which emissions matter?

The emissions pertinent to human activities are those of the radiatively active gases, usually referred to as greenhouse gases. The most closely studied of these is CO_2. Its atmospheric concentration has been monitored continuously since 1958 at Mauna Loa in Hawaii and, for nearly as long, at a variety of other sites, including the South Pole. When combined with ice-core data the conclusion is that the amount of carbon dioxide in the air increased from approximately 280 parts per million by volume (ppmv) at the beginning of the century to about 367 ppmv at the end of 1999.

The amount of CO_2 varies within each year as the result of the annual cycles of photosynthesis and oxidation. Photosynthesis, in which plants take up carbon dioxide from the atmosphere and release oxygen, dominates during the warmer and wetter parts of the year; oxidation, including biomass burning and the decay of plant material that release carbon dioxide, occurs all the time but dominates during the colder and drier parts of the year. Overall, then, CO_2 in the atmosphere decreases during the growing season and increases during the rest of the year. As the seasons in the Northern and Southern Hemispheres are inversed, CO_2 in the atmosphere increases in the north while it decreases in the south, and vice versa. The magnitude of this cycle is strongest nearer the poles and approaches zero towards the equator, where it reverses sign. The cycle is more pronounced in the Northern Hemisphere, which has relatively more land mass and terrestrial vegetation than the Southern Hemisphere, which is dominated more by oceans.

Of the other gases, methane (CH_4), which is formed by anaerobic decomposition of organic matter (notably in rice paddy fields and during digestion by ruminants as well as a considerable amount being emitted directly from coal mines, gas plants and pipelines), has shown a marked rise in atmospheric concentration. The level of methane had risen from a pre-industrial level of around 700 parts per billion by volume (ppbv) to about 1730 ppbv by the end of the 1990s. Other important greenhouse gases that are increasing in the atmosphere are oxides of nitrogen, notably nitrous oxide (N_2O) and halocarbons (these include the chlorofluorocarbons (CFCs) and other chlorine and bromine containing compounds).

The observatory at the top of Mauna Loa in Hawaii has maintained a continuous record of the rising level of carbon dioxide in the atmosphere since 1958. This record has been central in raising awareness throughout the world of the potential impact of rising levels of greenhouse gases on the climate.

Satellite observations can be used to monitor the changing biological activity of the oceans, as this is one of the important factors that causes carbon dioxide concentrations in the atmosphere to vary from year to year. Here, the amount of biological activity in the tropical Pacific, and the uptake of carbon dioxide, shows increased production from normal years (left) to during the La Niña event of 1998 (right) in response to cold upwelling nutrient-rich waters.

Changing rates of emissions

Although the build-up of greenhouse gases in recent decades seems inexorable, closer examination of the Mauna Loa CO_2 data shows that the rate of rise has fluctuated appreciably. Rapid growth in the late 1980s was followed by a marked slowdown in the early 1990s and then more rapid growth in the late 1990s. El Niño events have been marked by high rates of increase in atmospheric CO_2 concentration compared with surrounding years. The variation in the rate of increase in the globally averaged atmospheric concentration of CO_2 from year to year is as much as 3 to 4 billion tons of carbon per year. This variability cannot be accounted for by fossil fuel emissions.

Fluctuations in the rate of build-up in CO_2 seem to reflect climatic variability (e.g. El Niño/Southern Oscillation) and the associated uptake of carbon in the biosphere. The explanation is complicated, however, as it involves two opposing effects. The first is that during the warm (El Niño) episodes, the warm water covering the equatorial Pacific prevents cold upwelling CO_2-rich waters in the eastern tropical Pacific reaching the surface and releasing CO_2 to the atmosphere. Other ocean basins may also play a part in this process. In addition, the changed precipitation patterns and droughts over some of the most productive forest regions, such as the Amazon rain forest, reduce growth and photosynthetic CO_2 uptake. In contrast, increased incidence of forest fires during these episodes can increase CO_2 production.

Conversely, the biosphere tends to use more CO_2 during cold (La Niña) episodes both on land, owing to the strengthened Asian monsoon, and across the equatorial Pacific through increased photosynthetic activity. On balance, CO_2 tends to build up less rapidly in the atmosphere during La Niña events. In the longer term, the amount of the gas in the atmosphere will build up broadly in line with the emissions from human activities.

As for methane, the rate of build-up in the atmosphere has also fluctuated dramatically in recent years. It slowed from an annual increase of 15 ppbv in 1980 to about 10 ppbv in 1990 and then in the 1990s slowed even more appreciably, although fluctuating considerably from year to year. This means that the concentration of CH_4 may slowly approach a limiting value of around 1800 ppbv in the next few decades. As for CFCs, which represent potentially important greenhouse gases, international action to protect the ozone layer has already led to stabilization in concentrations of these pollutants.

Altering the radiation balance of the atmosphere

The overall effect of the build-up of greenhouse gases is to alter the radiative balance of the atmosphere. The net effect of this change is to produce an effective warming of the surface and the lower atmosphere because these gases absorb some of the Earth's outgoing heat radiation and reradiate it back towards the surface. The overall effect of the build-up of greenhouse gases in the troposphere is often defined as radiative forcing. The scale of this effect from 1850 to the end of the 20th century amounted to about 2.5 W/m^2; CO_2 contributed some 60 per cent of this figure and CH_4 about 25 per cent, and N_2O and halocarbons the remainder. The radiative forcing that would arise from the equivalent of doubling of CO_2 from pre-industrial levels is estimated to be 4 W/m^2.

Observations from around the world show a consistent pattern of rising carbon dioxide levels in the atmosphere. They also show how the annual cycle of plant growth and decay modulates the levels at high latitudes in the Northern Hemisphere (for example Barrow).

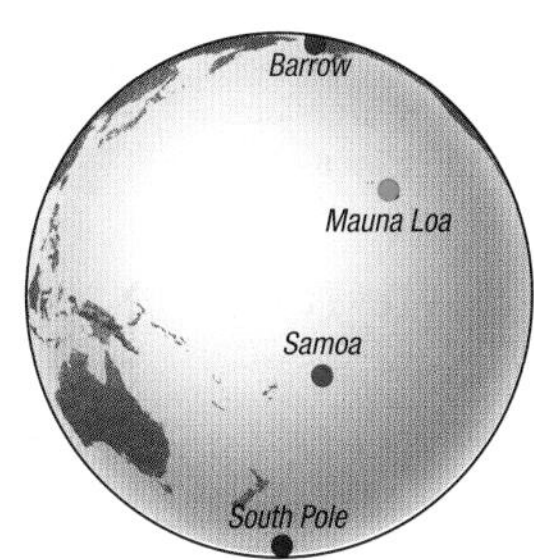

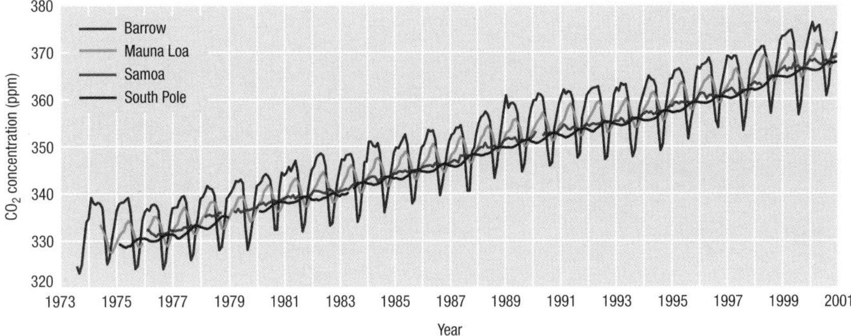

2.27 Human influences on climate: (c) Particulates in the atmosphere

Sometimes termed the 'human volcano', the injection of dust and liquid particles (aerosols) into the atmosphere by agricultural and industrial activities is one of the least understood aspects of the human impact on climate.

Atmospheric aerosols are able to alter climate in two important ways. First, they scatter and absorb solar and infrared radiation and, second, they may change the microphysical and chemical properties of clouds and possibly their lifetime and extent. The scattering of solar radiation acts to cool the planet, while absorption of solar radiation by aerosols warms the air directly instead of allowing sunlight to be absorbed by the surface of the Earth. Aerosols also absorb and scatter outgoing terrestrial radiation that further warms the atmosphere around them. The direct effect of the presence of aerosols is, therefore, to alter the vertical temperature profile of the atmosphere in a way that tends to reduce the rate of cooling with altitude (the lapse rate), which will have the effect of reducing the overall rate of upward vertical motion in the atmosphere.

Different types of clouds

Aerosols and cloud droplets come in a wide range of sizes; therefore the extent to which they scatter and absorb solar and terrestrial radiation will vary substantially depending on the type of clouds they form. The emission of terrestrial radiation from warm liquid-water clouds at low altitudes is not very different from surface emissions and so the strong scattering of solar radiation is the dominating effect. For the thin ice clouds at high altitudes the reduced emission of terrestrial radiation dominates over the weak backscatter of solar radiation. Overall, increases in low clouds would act to mainly cool the planet while increases in high clouds would tend to warm it.

Aerosols have an additional role to play on clouds by acting as condensation nuclei for cloud droplets. Increases in aerosol numbers will tend to increase the number and reduce the size of cloud droplets, and may also increase the number of ice particles. As an increased number of smaller cloud droplets greatly increases cloud brightness and hence the scattering of solar radiation back to space (while having little effect on their interaction with terrestrial radiation), the effect of increased concentrations of aerosols in the lower troposphere is to cool the planet. The overall effect of increases in ice particles may be even more complex.

Mechanized agricultural practices contribute considerable quantities of dust to the atmosphere.

Regulations are designed to minimize direct emissions of pollutants into the atmosphere but often condensation nuclei in the exhaust plume from cooling towers, such as at this paper mill in Finland, can affect the cloud-formation properties of the atmosphere.

The human volcano

The human contribution to the amount of aerosols in the atmosphere takes on many forms. Dust is a by-product of agriculture. Biomass burning produces a combination of organic droplets and soot particles. Industrial processes produce a wide variety of aerosols depending on what is being burned or produced in the manufacturing process. In addition, the exhaust emissions from transport (principally road vehicles, but increasingly aircraft) generate a rich cocktail of pollutants that are either aerosols from the outset, or are converted by chemical reactions in the atmosphere to form aerosols. All these effluents may have consequences on the climate.

Making long-term measurements of this 'human volcano' is particularly difficult, because so many of the effects are local or regional. One satellite experiment has produced a 20-year global record of a qualitative aerosol index. When calibrated with ground-based equipment, the record shows, for example, a significant increase in the amount of biomass burning smoke during the 1990s from African savannah regions.

Sulphate particulates

One of the most intensively researched aspects of particulate formation is the creation of sulphate aerosols by the combustion of fossil fuels and their subsequent life history in the atmosphere. Fossil fuels that contain sulphur compounds produce sulphur dioxide when they burn. Much of this is converted in the atmosphere to sulphuric acid which then combines with other compounds to form sulphates. These exist as tiny particulates that are efficient reflectors of sunlight, and it is estimated that the additional amount formed by the industrial nations of the world has had an appreciable cooling effect through the scattering of sunlight back into space.

The direct impact of sulphate particulates on the climate has been the subject of much computer-modelling work. The indirect effect that these particles have on the formation of clouds, by increasing the number of droplets (see above), has only recently been included in the models. Until we have a better measure of how possibly sulphate particulates are influencing the nature and extent of clouds, there is a possibility that we have been underestimating one of the more important effects that human activities have on the climate.

In the Northern Hemisphere the concentrations of condensation nuclei are about three times that of the Southern Hemisphere. This higher concentration is estimated to result in radiation forcing that is only about 50 per cent higher for the Northern Hemisphere. In other words, a small addition to the condensation nuclei concentrations in regions remote from human activity could have a much larger effect on cloud formation and hence on radiative forcing of the climate than the same addition for regions with an existing significant influence from anthropogenic sources.

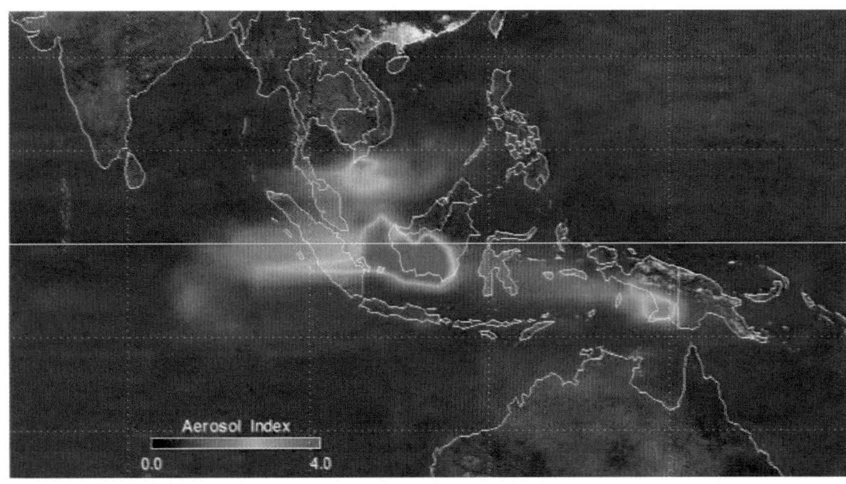

In September 1997, fires in many parts of drought-stricken Indonesia created a blanket of smoke in the region as well as smog (tropospheric low-level ozone), which rapidly spread out over Malaysia and the Indian Ocean. The unhealthy pollution was tracked by NASA's Earth Probe Total Ozone Mapping Spectrometer (TOMS), and was exacerbated by the El Niño-dominated atmospheric conditions.

Desert dust

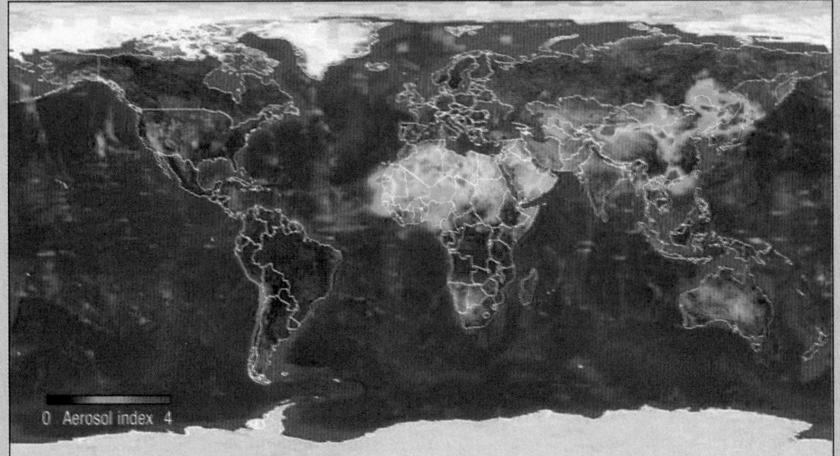

Dust from the deserts of the world contributes a major part of the particulate load in the atmosphere. There is uncertainty about the part played by human activities in this process as compared to the role played by natural climatic variability. Ground-based observations of dust in the Sahel region show that during the summer the dust levels change according to drought severity, whereas during the rest of the year they have been generally increasing (see 2.19). Another measure is to use satellite images to track Saharan dust clouds out across the tropical Atlantic and up into the Mediterranean. An analysis of images from the geostationary Meteosat satellite between 1982 and 1994 showed that the amount of Saharan dust spreading into the tropical Atlantic and up into the Mediterranean was closely linked to the strength of the North Atlantic Oscillation.

What is now clear is that desert dust has a substantial influence on the properties of the atmosphere. These influences seem greatest where dust from arid regions streams out over the oceans, since dust clouds will have a much higher reflectivity than the underlying ocean, and hence, will have a cooling effect. The areas where this cooling is likely to be greatest are over the tropical Atlantic to the west of Africa, and east of China, both regions of high aerosol loading as shown in this National Aeronautics and Space Administration (NASA) April 2001 TOMS analysis.

2.28 Human influences on climate: (d) Deforestation and

Removing and altering the vegetation cover of the land has had different impacts on the climate. These activities have been going on since the dawn of human history, but in recent decades the destruction of the tropical rain forests and changes in the extent of the world's deserts, in particular, have become major environmental issues.

On either side of a narrow dirt road, the Amazonian jungle is being cleared for settlement and farms. The river, a tributary of the Amazon, is coloured brown by the large amount of sediment it carries, possibly the result of deforestation upstream. This IKONOS satellite image was produced as part of the Large Scale Biosphere-Atmosphere Experiment, in which scientists are studying the environment of Amazonia, the role of Amazonia in global climate and the effects of human-caused change in the region.

Although burning vegetation to flush out game and to generate new grass may have had an impact on forest cover for many thousands of years, especially in the subtropical areas of the world, significant deforestation only really began around 10 000 years ago with the advent of agriculture in the Middle East. Since then large parts of the Northern Hemisphere land areas have had their vegetation cover altered by agricultural expansion and by the exploitation of forests for construction materials and fuel. In Europe the forest cover was reduced from about 80 per cent cover in AD 900 to only 20 per cent in AD 1400. In the eastern USA three-quarters of the forest cover was removed in the 18th and 19th centuries. At higher latitudes the most important consequence for the climate has been a change in the albedo, especially in winter.

Current deforestation

More than one-fifth of the world's tropical forests has been cleared since 1960. Tropical deforestation increased from 11.8 million hectares per year in the 1970s to 15.4 million hectares in the 1980s. Forest clearing and burning currently account for anywhere between 7 and 30 per cent of the annual atmospheric carbon emissions from human activities. The reduced forest cover also cuts down the overall photosynthesis and the amount of CO_2 that is extracted from the air and stored (sequestered) in the biosphere, thereby further increasing the rate of build-up of CO_2 in the atmosphere.

In the tropics the rapid exploitation of the rainforests is a major environmental concern, but the consequences on the climate are far more complex than might be expected. Replacement of tropical forest by degraded pasture reduces both evapotranspiration and absorption of incoming thermal radiation, as more sunlight is reflected by the surface; surface temperature still increases since cloudiness is reduced. This is what might be expected to happen on the basis of local analysis. There are, however, large uncertainties about the effects of large-scale deforestation on the hydrological cycle, especially over the Amazon. Some numerical studies point to a reduction of the amount of moisture drawn into the region, while others tend to show an increase in this inflow. This lack of agreement relates particularly to the rainy season and leads to major uncertainties in the computed amount of cloudiness, rainfall and surface runoff. These differences reflect our poor understanding of the interaction between convection and land surface processes. In southeast Asia, the question of moisture flow is less important because the rain forests lie closer to the oceans.

The indirect consequences of deforestation spread far and wide. In hilly areas increased runoff carries away topsoil which, in turn, silts up the rivers, causing deforestation to have a twofold impact on flooding problems. It both accelerates the rate at which water reaches the rivers and leads to an impeding of the river flow in the floodplain.

Desertification and land degradation

Desertification may be defined as the degradation of drylands leading to the permanent loss of living plants. It is due mainly to a combination of climatic variability and unsustainable human activities. The most commonly cited forms of unsustainable land use are overcultivation, overgrazing, deforestation and poor irrigation practices. These problems are of global dimensions, affecting more than 900 million people in 100 countries. About a quarter of the Earth's land areas (3.6 billion hectares) is being affected by land degradation. Desertification is

Encroaching dunes from the Sahara Desert threaten the Oasis of Ntekem-Kemt at Chinguetty, Mauritania.

occurring to some extent on 30 per cent of irrigated lands, 47 per cent of rain-fed agricultural lands and 73 per cent of rangelands. An estimated 1.5–2.5 million hectares of irrigated land, 3.5–4.0 million hectares of rainfed agricultural land, and about 35 million hectares of rangelands lose part or all of their productivity every year due to land degradation processes.

Although the central issue of desertification has been most frequently linked with the prolonged drought in sub-Saharan Africa since the late 1960s, it is a truly global issue. As such, both desertification and land degradation often are regarded as principally a climatic development. In truth, the causes are far more complex, involving interactions between physical, biological, political, social, cultural and economic factors. Population growth and the need for food production in ecologically fragile arid and semi-arid lands are putting too much pressure on many ecosystems. Too many development projects, for example, fail to appreciate the overall needs of the local population or to consider the sustainability of the natural resource base.

The challenge of expanding deserts has also affected more populous parts of the world. In China, where drought accounts for 50 per cent of all the natural disasters in agriculture, there has been a decreasing trend in precipitation over north China and the Yellow River Valley. The Yellow River (Hwang Ho) dried up in some places in 1972 for the first time, but since 1985, this phenomenon has been an annual occurrence, and in 1997, it was totally dried up in places on 226 days. While much of this change may be a consequence of overuse of water in the middle and lower reaches of the river, the decline in rainfall in northern China during the 1990s has had serious consequences for the capital, Beijing. In 2000 the city experienced a series of exceptionally heavy sandstorms, which were the worst in living memory. Large parts of north China are now under threat from desertification.

Attempts to keep the desert at bay take place at all levels. Here brushwood is used to stabilize dunes near Gour in Niger.

Salinization

Another form of land degradation, which has plagued agriculture since the dawn of history, is the build-up of salt in the soil (salinization). Initially, this was a consequence of inefficient irrigation processes that led to the accumulation of salt near the surface. More recently, deforestation in semi-arid parts of the world has been an additional major contributor to salinization. When trees are felled the water table rises and the topsoil becomes salty. Since European settlement of Australia started in the late 18th century, an estimated 15 billion trees have been cut down to make way for farms or urban development. The trees held the rising water levels in check, their roots sucking up millions of gallons of water, which then evaporated in the hot sun. Now large areas of formerly arable land are slowly dying, since removal of the deep-rooted trees has caused the water table to rise, lifting salt into the topsoil.

Salinization of deforested areas in Australia is making large areas arid wastelands.

2.29 Human influences on climate: (e) Stratospheric ozone depletion

The destruction of the ozone layer has been a hot scientific topic since the early 1970s. At various times scientists have pointed the finger at the exhausts of high-flying aircraft, the increased use of fertilizers and the use of chlorofluorocarbons (CFCs) as being potentially damaging. We now know that CFCs have created the ozone hole over Antarctica and led to the more recent depletion in the Arctic.

Satellite observations can be used to visualize the area covered by the Antarctic ozone hole and to monitor its changes over time. This image, for 12 October 2000, shows the area covered by the ozone hole to be approximately 16.5 million km² and extending over the southern part of South America. Earlier, in September 2000, the ozone hole had covered a record area of nearly 26 million km².

Ozone concentration (Dobson units – DU)
Dark grey <100, red > 500 DU

100 150 200 250 300 350 400 450 500

When Paul Crutzen, Mario Molina and F. Sherwood Roland were awarded the Nobel prize for their work on the photochemistry of ozone in 1995, it was the ultimate recognition of the scientific importance of the ozone hole and what it told us about the impacts of human activities on the global environment. Although there were isolated measurements in the early 1980s, including published observations made by a Japanese team in 1982, it was not until Joe Farman, of the British Antarctic Survey, published his paper in 1985 showing the decline of ozone levels over Antarctica during the early 1980s that the true scale of ozone loss was appreciated.

When the evidence of ozone depletion over Antarctica was published, the obvious question asked was why this had not been seen in the satellite observations collected since 1979. It subsequently emerged that the computer being used to analyse the data had been programmed to reject any ozone levels that were considered to be outside the bounds of physical probability, so the vital satellite data were stored away as being 'erroneous'. In fact, the development of the hole had been recorded but ignored. This oversight highlights a problem with the avalanche of data now being recorded on the climate, especially from satellites. If we rely on computers alone to sift out results, there is always a risk that unexpected but important observations might be discarded on the assumption that they are the product of instrumental error or malfunction.

The discovery of the Antarctic ozone hole had a profound influence on both scientific and political thinking. Here, for the first time, was clear evidence of human impact on a vital component of the atmosphere on a global scale. The response was dramatic: large-scale international scientific programmes were mounted to provide answers to explain the cause of the dramatic decline and to prove that CFCs (used as aerosol propellants, in industrial cleaning fluids and in refrigeration equipment) were the cause of the problem. Even more important was the immediate international action to reach agreement on action to curb the emissions of CFCs.

Ice clouds and sunshine

Plummeting ozone levels in the stratosphere over Antarctica during September and October are the result of complex chemical processes involving the effect of sunlight on various molecular species that have accumulated during the darkness of winter. The return of the Sun at the end of the sunless winter period triggers photochemical reactions that lead to the destruction of ozone in the stratosphere. The October values of ozone in the atmosphere over parts of Antarctica have declined by up to 70 per cent when compared with the pre-ozone hole years, and the size of the ozone hole had grown to more than 25 million km² (twice the size of Antarctica) by 2000. At altitudes of between around 15–20 km, where the concentration of ozone is normally greatest over Antarctica, levels commonly reduced to near zero in October.

The return of sunlight to the stratosphere over Antarctica each austral spring almost completely destroys the ozone layer at altitudes between 12 and 22 km, as these ozone soundings made in 1999 show.

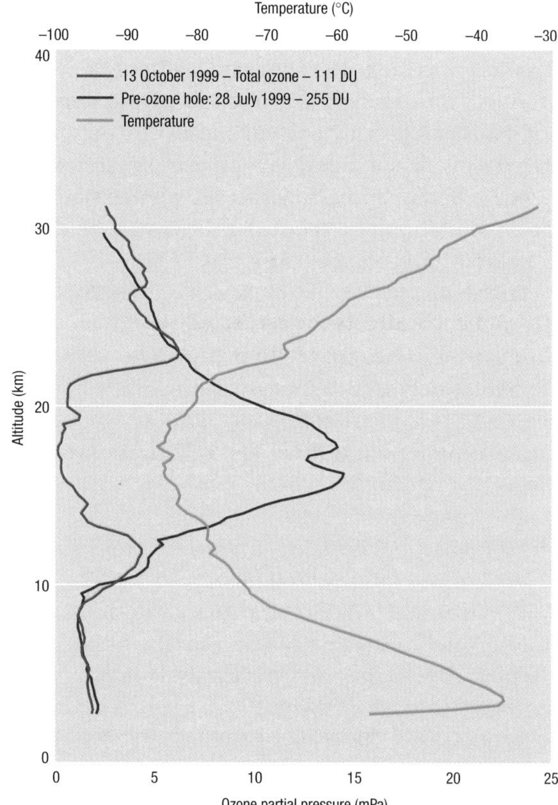

The explanation of these changes is based on how ozone is known to be created and destroyed in the atmosphere. It is formed by the absorption and action of ultraviolet solar radiation that breaks oxygen molecules into their two component oxygen atoms. These atoms then can combine with oxygen molecules to form ozone molecules that contain three oxygen atoms. Ozone is itself reactive and in the presence of certain other chemicals in the atmosphere it reacts with them to revert to oxygen. The balance between all the features of this chemical cycle produces a peak in the concentration of ozone in the stratosphere at altitudes of around 20–25 km. All this is critical for life on Earth, which has largely evolved in the absence of high levels of ultraviolet solar radiation.

CFCs can interfere with these normal processes because they break down in the upper atmosphere to form highly reactive chlorine compounds. This in itself is not enough to explain what is happening over Antarctica. The peculiar conditions that develop in the Antarctic winter vortex provide an additional ingredient. The intense cold, often below –90°C, at altitudes around 15–20 km, produces clouds of ice crystals which accelerate the depletion of ozone through complex chemical processes on their surfaces.

The dramatic difference between the chemistry of air effectively trapped in the polar vortex and that in regions outside it suggests that to a large extent the effects of CFCs on ozone will be confined to Antarctica. The chemical models, however, have had to be refined to take account of other factors. In particular, the nature of the larger-scale flow in the stratosphere needs to be incorporated to take account of how much the air over Antarctica is isolated in winter from the rest of the global atmosphere. The effects of changes in solar activity also have to be included, as there is evidence that the 11-year sunspot cycle affects the rate of ozone production in the upper atmosphere. What is more, volcanic activity may also increase the destruction; following the eruption of Mount Pinatubo in 1991, for example, there was an even greater loss of ozone.

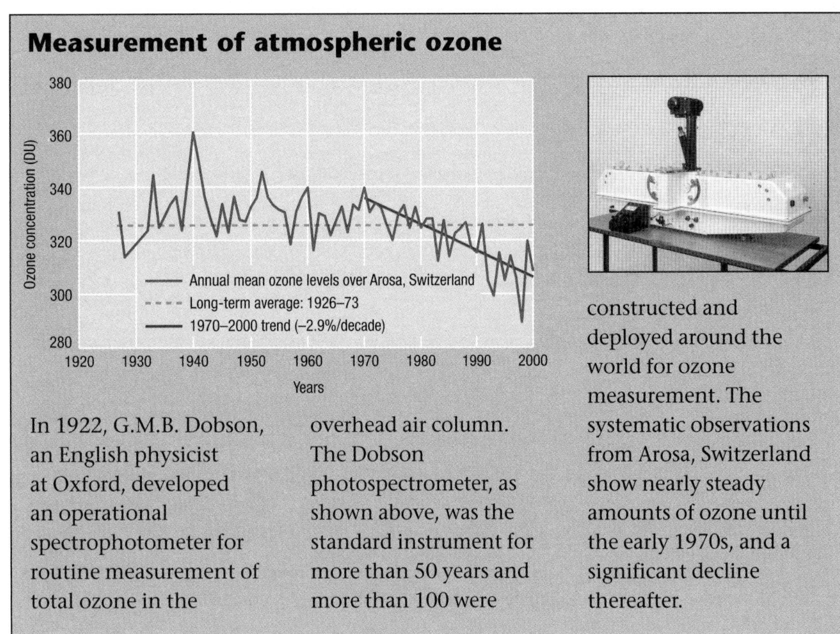

Measurement of atmospheric ozone

In 1922, G.M.B. Dobson, an English physicist at Oxford, developed an operational spectrophotometer for routine measurement of total ozone in the overhead air column. The Dobson photospectrometer, as shown above, was the standard instrument for more than 50 years and more than 100 were constructed and deployed around the world for ozone measurement. The systematic observations from Arosa, Switzerland show nearly steady amounts of ozone until the early 1970s, and a significant decline thereafter.

Other parts of the world

Over the Arctic the polar vortex is less intense and prolonged, so the temperature does not fall to Antarctic levels and the formation of stratospheric ice clouds is much less common. Thus, chemical conditions are less likely to lead to the rapid destruction of ozone. Also, the vortex is not so well defined and more ozone is quickly transported from low latitudes to replenish any that is destroyed when the Sun returns to the Arctic at the start of spring. Nevertheless, the gradual development of an annual decline during the 1990s is a significant trend. More generally, over northern middle latitudes the concentration of stratospheric ozone has decreased since 1979 by 5.4 per cent in winter and spring, and by about 2.8 per cent in summer and autumn. There has been no discernible trend in the tropics and subtropics.

International ozone agreements

The scale and suddenness of the ozone decline shocked the scientific world, and led to the 1985 Vienna Convention for the Protection of the Ozone Layer and the 1987 Montreal Protocol and subsequent amendments to eliminate certain CFCs from industrial production. As a result of this rapid action the global consumption of the most active gases fell by 40 per cent within five years and the levels of certain chlorine-containing chemicals in the atmosphere have started to decline. It will be decades before the CFCs already in the atmosphere fully decay. In the meantime, the substantial destruction of ozone in the stratosphere over Antarctica during September and October will continue.

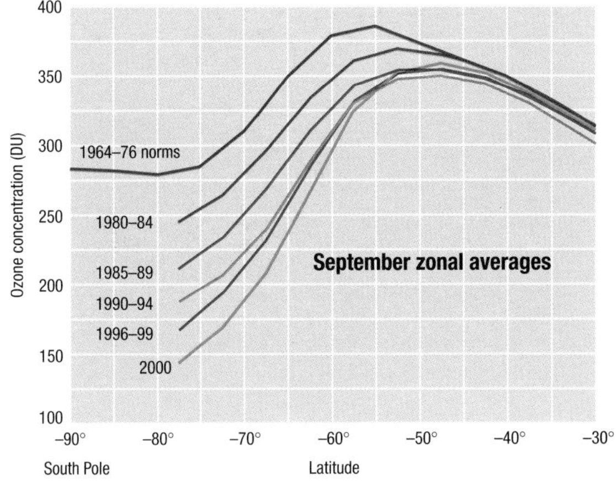

The intensification of the depletion of middle latitude and polar ozone in the Southern Hemisphere continued through the last decades of the 20th century.

2.30 Natural variations in the climate

One of the most difficult aspects of deciding whether current climatic events reveal evidence of the impact of human activities is that it is hard to get a measure of what constitutes the natural variability of the climate. This involves the combination of factors already considered in this section, but how they all add up is still a matter of great uncertainty.

We know that over past millennia the climate has undergone major changes without any significant human intervention. We also know that the global climate system is immensely complicated and that everything is in some way connected, and so the system is capable of fluctuating in unexpected ways. We need therefore to know how much the climate can vary of its own accord in order to interpret with confidence the extent to which recent changes are natural as opposed to being the result of human activities.

These proxy records of temperature provide broadly consistent indications of long-ago temperature variations that were global in scale, for example, during the periods indicated by the shaded vertical bands. In addition, there are some intriguing differences, which suggest that the spatial pattern of regional climates can also vary significantly over time.

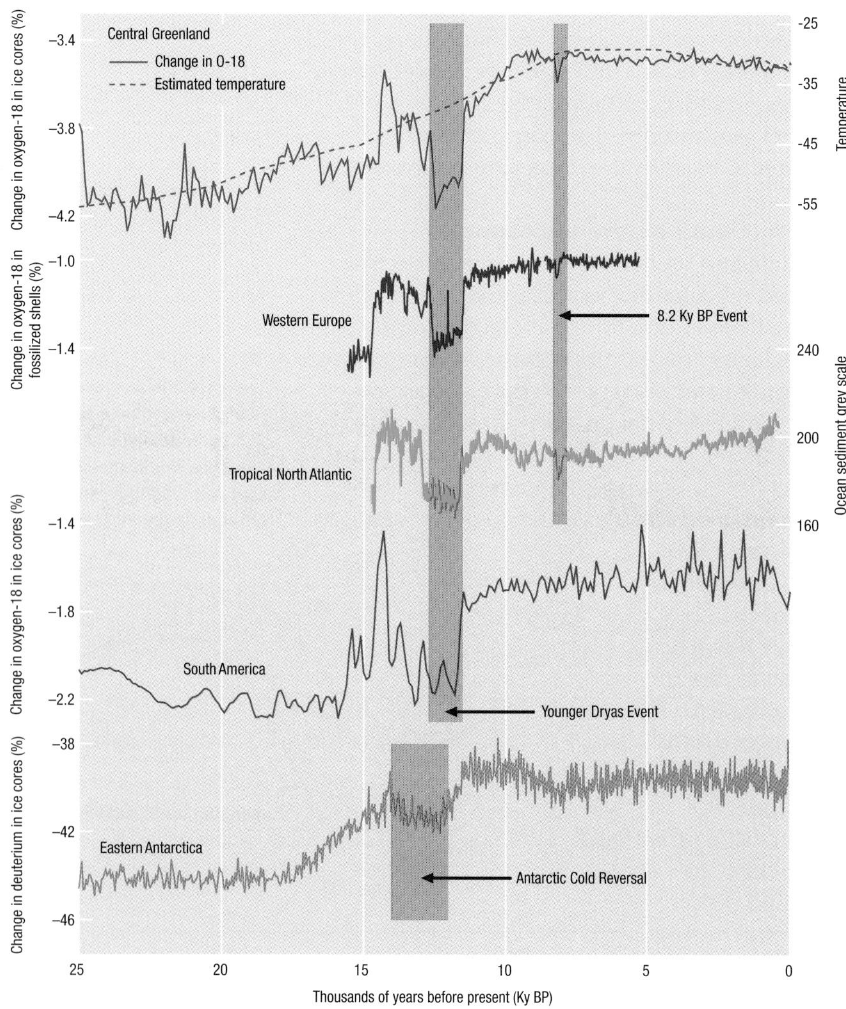

How much do we know about past changes?

Instrumental records do not go back far enough to provide us with reliable measurements of global climatic variability on timescales longer than a century. What we do know is that as we include longer time intervals the record shows increasing evidence of slow swings in climate between different regimes. To build up a better picture of fluctuations appreciably further back in time requires us to use proxy records.

Over long periods of time substances, whose physical and chemical properties change with the ambient climate at the time, can be deposited in a systematic way to provide a continuous record of changes in those properties over time, sometimes for hundreds or thousands of years. Generally the layering occurs on an annual basis, hence the observed changes in the records can be dated. Information on temperature, rainfall and other aspects of the climate that can be inferred from the systematic changes in properties is usually referred to as 'proxy data'. The proxy temperature records in the figure have been reconstructed from oxygen isotope ratios in ice core drilled out of the central Greenland ice cap (red), oxygen isotope ratios in calcite shells embedded in layered lake sediments in Western Europe (blue), ocean floor sediment cores from the tropical Atlantic Ocean (green), oxygen isotope ratios in ice cores from Peruvian glaciers (purple) and hydrogen isotope ratios in ice cores from eastern Antarctica (orange). While these records provide broadly consistent indications that temperature variations can occur on a global scale, there are nonetheless some intriguing differences, which suggest that the spatial pattern of regional climates can also vary significantly over time.

What the proxy records make abundantly clear is that there have been significant natural changes in the climate over timescales longer than a few thousand years. Equally striking, however, is the relative stability of the climate in the last 10 000 years (the Holocene period).

To the extent that the coverage of the global climate from these records can provide a measure of its true variability, it should at least indicate how all the natural causes of climate change have combined. These include the chaotic fluctuations of the atmosphere, the slower but equally erratic behaviour of the oceans, and changes in the land surfaces, and in the extent of ice and snow. Also included will be any variations that have arisen from volcanic activity, solar activity and possibly human activities.

What would we expect to see?

One way to estimate how all the various processes leading to climate variability that have

A good example of natural climate variability is the more frequent cold winters in Europe between 1550 and 1850, as portrayed in this 1608 winter scene by Hendrick Avercamp.

been described in this section will combine is by using computer models of the global climate (see Sections 4 and 5). As will be seen, they can only do so much to represent the full complexity of the global climate and hence may give only limited information about natural variability. Studies suggest that to date the variability in computer simulations is considerably smaller than in data obtained from the proxy records.

In addition to the internal variability of the global climate system itself, there is the added factor of external influences, such as volcanoes and solar activity. As has been made clear in earlier sections, there is a growing body of opinion that both these physical variations have a measurable impact on the climate. Thus we need to be able to include these in our deliberations. Some current analyses conclude that volcanoes and solar activity explain quite a considerable amount of the observed variability in the period from the 17th to the early 20th centuries, but that they cannot be invoked to explain the rapid warming in recent decades. Indeed, if anything, their contribution since 1980 would have been to slightly reduce the rate of warming.

Does what happened in the past matter?

There are two principal reasons why we need to learn from past data as much as possible about the natural variability of the climate. The first is so we put into a larger context the estimate of recent changes that are or could be the result of human activities. The more confident we are that what we are seeing cannot be explained away as natural fluctuations, the easier it will be to convince political leaders and voters that, for the sake of future generations, they should support perhaps unpopular and sometimes costly changes in lifestyles in order to reduce the impact of human activities on the climate. This goal requires us to extract the maximum amount of information from all the available sources of data on the climates of the past to build up as accurate a picture as possible about natural variability and hence derive better assessments of the recent impact of human activity.

The second reason is to get a better feel for whether the climate, as we currently know it, is capable of undergoing sudden and dramatic shifts. The more we know about past changes, the more likely we are to reach a position of knowing whether such changes are at all probable in current circumstances. This may then help us to address the question of whether the perturbations caused by human activities could make such 'flips' more or less likely. This particular question is of vital importance as we attempt to untangle the natural and human influences on climatic variability and change.

Growing concern about the rising social and economic costs of weather-related disasters, such as the terrible floods and landslides in Venezuela in December 1999, is the principal reason for needing to know more about how much human activities are affecting the climate.

The impact of extreme weather events should not be measured only in economic terms. Loss of life, permanent disruption of long-established lifestyles and setback to the development process carry a far greater cost to society than the matter of damage to property. Communities can be completely destroyed in climate-related disasters, an incalculable loss. Conversely, where communities recognize a hazard and band together to ward off a disaster, they can become stronger and develop more resilient ways of living.

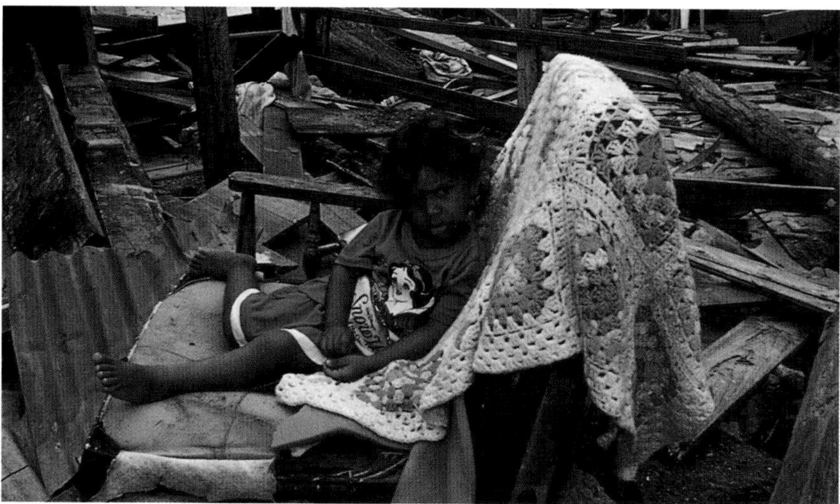

The impact of natural phenomena on communities can have devastating consequences, as in the case of the damage caused by Hurricane *Mitch* in Honduras in October 1998.

Is it possible to compare disasters that occur many years apart? What, if any, measure can we use to compare the hurricane that killed some 8000 people in Galveston, Texas, in 1900 with Hurricane *Mitch* which killed about 10 000 people in Honduras in 1998? Even more difficult to compare are different forms of disaster, as the long drawn-out consequences of a drought will not be the same as the immediate destruction wrought by floods. Furthermore, more accurate assessments are being made of the incurred damage done as a result of improvements in emergency services. The figures for human suffering caused by the failure of the monsoon rains in India in the second half of the 19th century, or disastrous floods in China at the beginning of the 20th century, are not easily comparable with more recent events. However, it is clear that mortality figures for these types of events have fallen dramatically through early warning and community assistance.

Estimating how the impacts of the varying climate have altered during the last 100 years is no easy task. It is especially difficult when we are looking at how societies have been disrupted by a wide variety of climate-related hazards. It is often impossible to isolate how much of the impact can be blamed on changes in the climate's variability or increases in extreme events because so many other features of our lives have changed over time. The fact of the matter is that the social and economic statistics are generally not sufficiently comprehensive to draw definitive conclusions about how the impacts of extreme events have changed over the last 100 years or so.

Mortality and human suffering

The most immediate and important human consequences of extreme weather events and climate variability are mortality and, in extreme cases, the disruption of an entire culture, such as occurred in the southwest of the USA in about AD 1270 when the Anasazi people were forced through drought and the cold temperatures of the Little Ice Age to abandon their homes and travel long distances to find better conditions for growing their crops. In the case of disasters where the immediate impact on the local population is loss of life, the media often see the number of people killed as a reasonable indication of the scale of the event. In one respect this is understandable. To most people, an event that kills thousands of people is much worse than one that kills hundreds. Nonetheless, there is a strong tendency to weigh deaths closer to home more heavily than those that occur elsewhere. Any major event must therefore be examined closely to see how it might fit into a possible pattern of longer-term changes. Take the example of a tornado that strikes the only populous area in an otherwise empty region. As a consequence the mortality is high. The question that must then be answered is whether, although a rare event, it was within the past climatological pattern or one of a rising number of similar storms.

The next stage in this type of analysis is to ask what fraction of the change in mortality is due to the climatic factors alone and what might be due to changes in social structures, including those designed to reduce loss of life, or perhaps demographic changes, which may have enhanced risk. As noted in previous Sections, the loss of life from weather-related disasters has fallen dramatically over the last 100 years in many countries, mostly as a result of better preparedness and better warning systems. More generally, there is evidence that as warning systems and emergency procedures improve, the benefits in human terms are enhanced enormously.

When it comes to prolonged periods of abnormal weather, mortality statistics are more complicated. In the extreme instance where drought destroys crops and leads to famine, the loss of life may be readily assessed. Where there are large-scale disaster relief activities, however, as has increasingly been the case over the last 100 years, the actual impact of the extreme conditions depends to a great extent on the timeliness and effectiveness of the aid, which is not without cost. Nonetheless, it is a measure of the success of the world's increasingly coordinated relief activities that the loss of life in many recent droughts has been significantly less than the disasters of the late 19th and early 20th centuries.

Another feature of mortality statistics that requires careful examination is the use of excess mortality to measure the impact of periods of abnormal weather that affect many aspects of our lives. Heatwaves or cold spells, for example, might be assessed for their severity in terms of the number of excess deaths that occurred compared with the average for the time of year. This is a valid measurement and of particular value when the excess mortality rates rise well above normal levels, but the figures do contain some hidden complexities. First, many of the people who died may well have been already in a vulnerable state and would have died within a few months in any case. The statistics must be corrected for below average mortality in the subsequent period. Second, put simply, we do not talk about the number of lives saved by a mild winter or a cool summer. This is understandable when talking about specific adverse events, but may be shortsighted when considering the overall impact of climatic shifts. A warmer climate could bring more mild winters as well as more heatwaves, and the benefits of one change must be considered alongside the disadvantages of the other.

Economic costs

When it comes to measuring the economic costs of specific types of extremes, there are even more corrections to be made. The way we have lived our lives during the last 100 years has changed dramatically. As personal wealth has increased, homes, offices and factories have become increasingly full of expensive electrical and electronic goods and other high-technology possessions. Thus, even a minor flood can destroy vast amounts of equipment. Even simple domestic changes such as the increased use of fitted carpets instead of throw rugs, or the rising numbers of refrigerators and air conditioners in use, serve to elevate our economic vulnerability, as such things cannot easily be moved out of harm's way. Even though in the developed world much of this property can be insured (a reflection of community resilience), premiums tend to rise to cover increasing losses. The high-profile publicity mounted by the insurance companies in the late 1990s about the rising cost of disaster losses and the possibility that this was related to global warming was a measure of their concern.

We will return to these issues as we consider the many aspects of impacts from a varying climate on human life. The important thing to remember is that whenever we make judgements about the economic and social impact of a varying climate, we must apply consistent criteria to each type of event to ensure that we are not introducing distortions into the analysis.

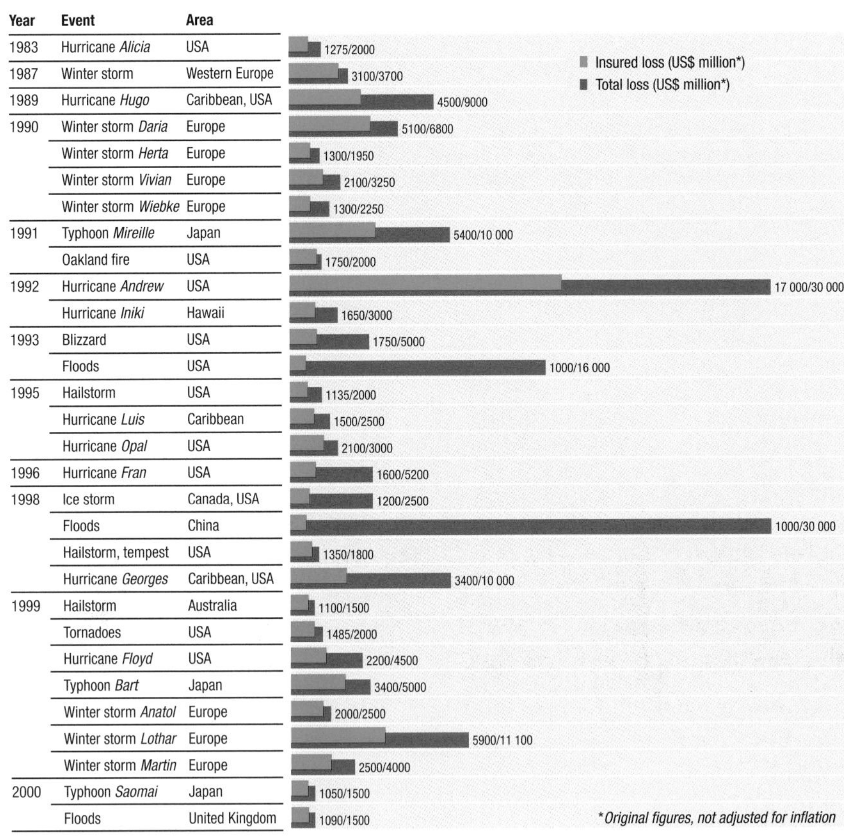

Year	Event	Area	Insured loss / Total loss (US$ million*)
1983	Hurricane *Alicia*	USA	1275/2000
1987	Winter storm	Western Europe	3100/3700
1989	Hurricane *Hugo*	Caribbean, USA	4500/9000
1990	Winter storm *Daria*	Europe	5100/6800
	Winter storm *Herta*	Europe	1300/1950
	Winter storm *Vivian*	Europe	2100/3250
	Winter storm *Wiebke*	Europe	1300/2250
1991	Typhoon *Mireille*	Japan	5400/10 000
	Oakland fire	USA	1750/2000
1992	Hurricane *Andrew*	USA	17 000/30 000
	Hurricane *Iniki*	Hawaii	1650/3000
1993	Blizzard	USA	1750/5000
	Floods	USA	1000/16 000
1995	Hailstorm	USA	1135/2000
	Hurricane *Luis*	Caribbean	1500/2500
	Hurricane *Opal*	USA	2100/3000
1996	Hurricane *Fran*	USA	1600/5200
1998	Ice storm	Canada, USA	1200/2500
	Floods	China	1000/30 000
	Hailstorm, tempest	USA	1350/1800
	Hurricane *Georges*	Caribbean, USA	3400/10 000
1999	Hailstorm	Australia	1100/1500
	Tornadoes	USA	1485/2000
	Hurricane *Floyd*	USA	2200/4500
	Typhoon *Bart*	Japan	3400/5000
	Winter storm *Anatol*	Europe	2000/2500
	Winter storm *Lothar*	Europe	5900/11 100
	Winter storm *Martin*	Europe	2500/4000
2000	Typhoon *Saomai*	Japan	1050/1500
	Floods	United Kingdom	1090/1500

Nearly all of the most costly disasters in recent years have resulted from extreme weather events. Of all the climate-related natural disasters since 1983, 30 incurred losses of more than US$ 1 billion. The extent to which these costs are absorbed by the insurance market or fall more widely on governments and individual citizens varies greatly, depending on the stage of development of affected countries.

The chaos caused by natural disasters is generally counted in terms of loss of lives and property damage, but losses that are less frequently quantified are the disruptions to work and lifestyles and the damage to personal goods. This young woman is struggling to clear the mess in her Santiago, Chile home following a flash flood.

3.2 The social and economic impacts of climate fluctuations

Extreme weather has always had a major impact on societies. Where these extremes have bunched together over a number of years, they may be part of a longer-term climatic variation. Whatever their cause, the consequences can be catastrophic. An important question is: have the scales of economic and social impacts from climatic fluctuations increased?

The power of flash floods and their capacity to destroy communities in a matter of moments is seen in this 1996 photograph of houses along the Saguenay River in Chicoutimi, Quebec, Canada.

The climate system supplies our basic needs of air to breathe, sunshine to grow vegetation and freshwater to drink. It also can bring challenges of all kinds to our lives. Most of these we can handle even if they are stressful, but every now and then our social and economic systems crack under the strain. More rarely, the scale of impact can be catastrophic and, whether it strikes a single household or an entire nation, the consequences can be devastating. How we come to terms with these disasters provides telling insights about the climate, our adaptation to it and our dependence on it.

Disasters great and small

Some events change communities in an instant, but others may be slow and insidious as they destroy ways of life forever. Flash floods can suddenly, and with little warning, sweep away towns and villages that have existed for centuries, killing many inhabitants, displacing those who survive and making it impossible for them to return. Tornadoes or hurricanes can wreak similar havoc.

Droughts are entirely different. They build up over time, generally covering a wide area, the extent of which can change with time as spasmodic and even heavy rain brings relief to some places while leaving others parched. The severity of their impact also depends on the stage of growth of crops. If lack of moisture kills off plants well after germination, but before any grain or produce has set, then the disaster is complete, as it is often too late to plant another crop. It is then that famine stalks the land if relief supplies cannot be provided. If, however, the rainfall is regular albeit reduced in amount the crop may survive to maturity, and even though yields may be low, starvation may be avoided. Belated rainfall following a long dry spell can revive prospects and even avert total failure.

Persistent droughts in the West African Sahel region have caused great suffering to nomadic peoples and their livestock.

Our perception of disasters

While we can carry out objective analyses to define the extent of a disaster, the reality, if not the scope of its impact, is all too readily apparent to those who have had the misfortune to be victims. In the case of a sudden and dramatic event, the reports of the loss of life and property provide clear evidence of the suffering of communities. The same applies to creeping (slow onset, incremental and cumulative) disasters like drought. A precise definition in terms of the scale of moisture deficit or how much below average rainfall has been for some period may prove to be an accurate measure of drought. However, analyses of press and other news stories can often give a better assessment of the relative importance of the impact of the event. What these reports show is that, when the rainfall for a month or more is very much less than normal, then local perceptions, based on effects on crops and pastures, etc. are just as important indicators of impact as the rainfall measurements. Nevertheless, some claim agricultural drought when, in retrospect, none has occurred. Much depends on when the shortfall occurs during the growing season and whether sporadic rainfall can resuscitate flagging crops or comes too late to do any good.

Disaster aid

The degree to which a natural hazard becomes a disaster depends on a variety of factors. First, there is the extent to which preventive action can mitigate the impact, whether as a consequence of normal planning or in response to predicted extremes. Then there is the matter of whether effective aid can be provided quickly after the onset or passage of the event. Many relief agencies can now spring quickly into action to reduce the short-term impact of disasters. The provision of emergency food supplies and medicines to prevent epidemics of contagious diseases spreading is all part of this process. As a consequence, the mortality in many disasters in the latter part of the 20th century, especially in the industrialized countries, was significantly lower than in the early part of the century.

The success of disaster aid does not end with providing relief for the immediate consequences of a specific extreme weather event. Equally important are the longer-term strategies to introduce changes to reduce damage and loss of life in future events. This is not just a matter of providing better flood defences, but also of repairing environmental damage, which might otherwise make the impact of future events even greater. For instance, in arid and semi-arid regions, the development of effective and sustainable uses of land and natural resources that do not endanger their future productivity is an essential part of reducing the impacts of future droughts. Better infrastructure, which is both friendly to the environment and capable of withstanding extreme events, is another part of the equation. To be successful these policies have to become part of national environmental and development planning.

When a community's infrastructure is badly damaged by a storm, hunger and disease can quickly devastate the survivors. In Nicaragua, food and medical supplies were brought in by the Mexican army to alleviate the suffering that followed Hurricane *Mitch* in 1998.

The statistics of disasters

Although human misery cannot be adequately represented by statistics, it is helpful to have some measure of the global scale of the impact of weather-related disasters. In the figures below, based on data from the International Federation of the Red Cross and Red Crescent Societies, the reported average annual number of deaths, of people affected and made homeless by various types of weather disaster are plotted for the five five-year periods from 1973 to 1997. The important features of these figures are:

- droughts killed more people than the others combined;
- droughts and floods affected about an equal number of people, and far more than high winds (including hurricanes, cyclones, typhoons, storms and tornadoes);
- floods were, however, by far the greatest cause of homelessness; and
- for the limited time covered, there were tremendous variations in the numbers of people affected by different forms of disaster in successive five-year periods. This makes it difficult to draw any definite conclusions about trends, apart from noting that the number of people affected by floods appears to be rising.

The figures do not adequately reflect the fact that developing nations are hit harder by extreme weather events. In particular, the impact of both drought and floods has had a disproportionate impact on some of the poorest countries in the world, notably in Africa.

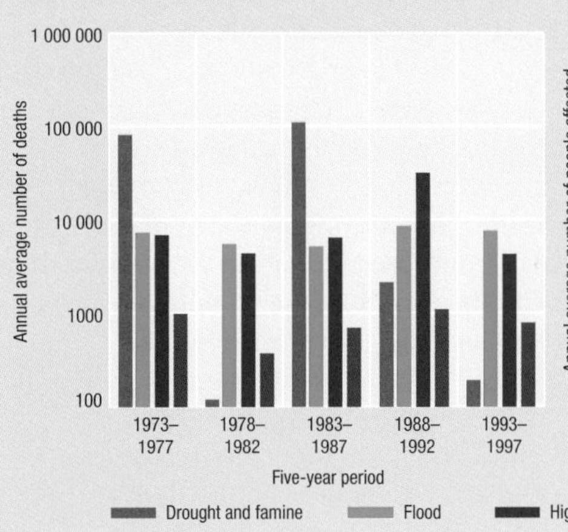

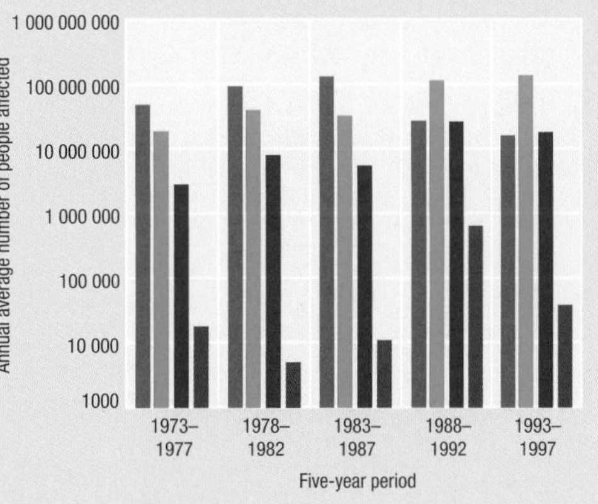

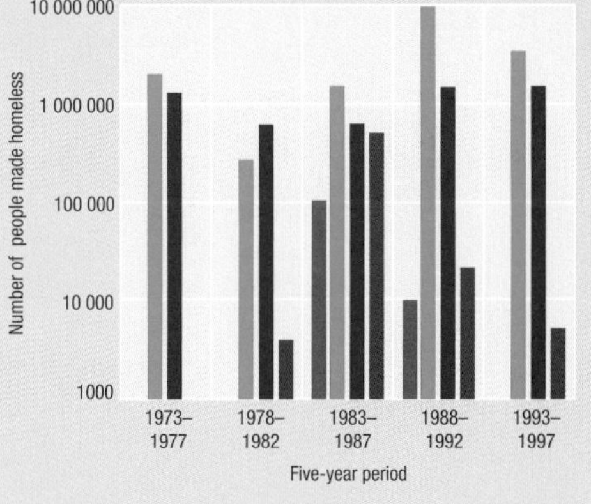

3.3 Learning from experience

Preoccupations with current extreme weather events make it all too easy to overlook events of the past. However, a balanced perspective of these earlier extremes is essential, both to put more recent disasters into context and to establish how effectively we have used the lessons of the past to manage more recent challenges.

Before looking at the early part of the 20th century, there is a need to consider the psychology of closing the books on the 20th century. In the 21st century we will likely view the experiences of the 1900s (even the late 1900s) just as people in the 1900s looked at the events and history of the 1800s. Many did not use those earlier experiences because they believed that technologies had improved, knowledge had advanced and, in some way, they had become smarter than their forebears. This hubris causes us to assume that recent experience is more important than that of long ago and can lead us to underestimate the challenges that we face. By so discounting the past we fail to exploit the lessons of history.

Same as it ever was?

It is all too easy to fall into the trap of thinking that recent events are more extreme than those of the past. Perhaps it is an obsession with statistics that makes the media keen to present everything new as something out of the ordinary. Whatever the reason, there is a risk that this attempt to establish the record-breaking nature of current weather events can give the impression that things are worse now than they were long ago. It is useful to remind ourselves of just some of the extremes that occurred in the first two or three decades of the 20th century.

For centuries the Chinese have repaired their dykes during floods with a combination of earth and bundles of kaoliang stalks. Here, frantic efforts were being made during floods in the 1930s to staunch a huge breach in the dykes.

Many major European cities sit astride rivers, but time and again we are surprised when famous downtown areas are flooded. In January 1910 Paris was hit by its worst floods in at least 200 years. More than 30 000 people were rendered homeless and troops were mobilized to prevent looting. In early January 1928, London experienced serious flooding when the combination of melting snow from a major storm over Christmas plus further heavy rain and a storm surge in the North Sea crested the flood defences.

The drought along the southern fringes of the Sahara during the 1970s and 1980s is one of the most poignant examples of recent climate fluctuations. Much earlier, around the turn of the century, a dry period had also gripped the region. Severe droughts in 1911, 1914 and 1915 reduced Lake Chad to half its size, and the flow of the Nile fell dramatically. In parts of northern Nigeria the population was reduced by nearly half between 1913 and 1914, and the number of cattle reduced to barely a third. Similar figures are recorded for Niger. In spite of a period of drought in the 1940s, generally more abundant rainfall between the 1920s and the early 1960s led to a large expansion in the country's population, more extensive herds and the development of agriculture on marginal land that had previously been used as rangelands for livestock grazing. When drought returned in the 1970s, the scale of the calamity was bound to be all the greater.

Flooding and drought in China are the cause of many of its greatest disasters. For instance, it is estimated that floods in 1931 affected a quarter of the population and killed over 400 000 people. These were China's most widespread and grievous floods of the century, and followed hard on the heels of the prolonged drought from 1928 to 1930 during which 1.4 million people starved to death. Extreme droughts during 1941 and 1942 killed more than 3 million people. Although the massive floods in the Yangtze River Valley and Northeast China in 1998 affected 100 million people, and direct economic losses were up to US$ 10 billion, the number of lives lost was less than 4000.

Are we condemned to repeat the past?

It was the historian and philosopher George Santayana who observed, "Those who cannot remember the past are condemned to repeat it". When we examine recent weather disasters there is a sense of going over the same ground as suggested in the examples above. This raises the awkward question: do we deliberately discount the past? The answer is not a simple one, and is mired in complex social responses to all disasters. In this Section, and the next, we will show that there are many ways in which past experience is built most thoroughly into our planning, and succeeds in reducing the impact of extremes.

At the same time, we must remember that disruption is bound to occur when extreme weather strikes. It is easy to fall into the trap of claiming current events are exceptional. This natural reaction is frequently reinforced by the media, which has a vested interest in dramatizing current events to sell newspapers or advertising space. The media then are often not neutral observers of the great and small weather events and their impacts on society.

Although the record of disasters seems never-ending, this does not alter the fact that we have learnt from experience. In much of the world there are now systems, including improved farming techniques, forecasting and early warning, flood defences, improved transport and national emergency plans, that draw on the lessons of the past. What has changed is that the lives we now lead are disrupted by the weather in so many different ways from those that affected our forebears at the beginning of the 20th century. This is particularly true where the assumption that technological advances have made us more independent of the weather is prevalent. Much of the continued sensitivity relates to new activities. The hustle and bustle of urban life is more easily disrupted by torrential rain, high winds and snow or ice storms. In addition, our expectations have risen, and we expect to be able to do what we want and when we want. In essence, we think we have neutralized many of the ways that weather can adversely affect our lives. This sometimes leads us to place too much reliance on our perception of the ability of technology to reduce the effects of natural hazards.

It is all too easy to forget the value that weather and climate information can play in improving the design of many of the systems on which we rely, until for example, a city traffic system is reduced to gridlock by an unexpected light snowfall. The same applies to agriculture, which can receive an extraordinary range of advice on likely conditions from ploughing and seeding to harvesting. Such services have led to major improvements in performance, but many systems remain subject to the vagaries of the weather, and in some instances may have become more vulnerable. For example, high-yield varieties of grain produce bigger harvests, but they are more sensitive to slight changes in rainfall. While we have learnt from the past, weather and climate services must continue to find ways to reduce weather-related surprises and help educate us on how to live our lives better in harmony with weather and climate.

Avalanches

The power of avalanches always strikes a chord of fear in the hearts of those who live in the mountains or visit them for recreation. Avalanches occur naturally as a consequence of accumulated snow pack, terrain slope and prevailing weather conditions, often triggered by the activities of people in snowfields. Approximately 50 people are killed each year in North America and about the same number in the European Alps. Efforts are made, particularly in recreational areas, to identify hazardous conditions, minimize the threat and provide warnings.

External forces can also trigger avalanches. In 1970, an earthquake-induced rock and snow avalanche on Mount Huascarán, Peru buried the towns of Yungay and Ranrahirca. The total death toll was 66 700. The avalanche started with a sliding mass of glacial ice and rock that swept downslope about 14 km to Yungay at an average speed of over 150 km/h. At that time the avalanche was estimated to have consisted of about 2 200 000 m^3 of water, mud and rocks.

During World War I, avalanches were cynically used as a weapon of war. The South Tyrol of Austria had accumulated by far its heaviest snowfall on record during the winter of 1916–17. The Italian and Austro-Hungarian forces, locked in combat in the region, each used artillery and other means to trigger avalanches to crash down on their foes. Between 50 000 and 100 000 troops were killed by this stratagem.

Exceptional weather can occur in the most unlikely of places from time to time, as shown by this picture of the 35 cm of snow that fell on Laghouat on the edge of the Sahara in Algeria on 10 February 1935.

3.4 Climate and biodiversity

The diversity and distribution of all forms of life on the Earth reflect the wide range of climatic conditions in which they have evolved. This biodiversity is now threatened by human activities.

Over the last 600 million years the diversity of all types of marine and continental life, including microbes, algae, fungi, protists, plants and animals, has increased. Despite interruptions by mass extinctions at various times in geological history, the number of species has risen into the current era and there are currently many millions of species co-existing on this planet. This biodiversity, however, is now threatened by human activities, including a wide variety of environmental alterations of which climate change is but one.

Climate controls

All natural habitats are controlled by the climate, and most forms of life have evolved to survive in the characteristic climate of their normal habitat. As climatic conditions have changed over time, species have adapted, migrated or become extinct. Animals are generally better able to adapt than plants as they can more easily move to new areas where the climate is more amenable. For plants that are squeezed into isolated pockets (often termed refugia) by shifts in the climate, the consequence can ultimately be extinction.

The same pressures exist in the oceans. Episodes of abnormally high sea surface temperatures, often associated with El Niño events, have caused widespread damage to coral reefs throughout the tropics. In the North Atlantic, changing sea surface temperatures have had a major impact on fluctuating cod stocks on the northern and southern boundaries of their range. In the cold waters off Labrador and Newfoundland the stocks declined dramatically as a result of the lower temperatures in the late 1980s and 1990s, associated with the positive phase of the North Atlantic Oscillation. Conversely, rising temperatures in the North Sea and overfishing have led to a sharp fall in stocks there since 1988.

Cod stocks have crashed in both the North Sea and the western North Atlantic in recent decades as a result of both climatic factors and unsustainable levels of fishing.

Global patterns of biodiversity

Biodiversity, the variety of life, is not distributed uniformly across the Earth. Some areas teem with biological variation (e.g. some moist tropical forests and coral reefs), others are virtually devoid of life (e.g. some deserts and polar regions), and most fall in between. Global patterns of biodiversity depend on a variety of factors, many of which are climate-related. Broadly speaking, biodiversity is greatest where it is warmest, in the tropics, and least where it is coldest, at high latitudes; greatest at low altitudes and least in high mountains; and greatest in regions of high rainfall and least in the driest areas. Some regions of the world have been identified as biodiversity 'hotspots': areas with very high numbers of endemic species that are also experiencing critical loss of habitat. The 25 terrestrial regions that fit these criteria represent only 1.4 per cent of the Earth's land surface, but comprise the remaining habitat for as many as 44 per cent of the world's plant species and 35 per cent of the four vertebrate groups, birds, mammals,

Refugia

The threat of climate change to biodiversity is greatest where plants or animals are isolated in places without hospitable adjacent areas to which they can migrate. This phenomenon is evident around the world where unique species are found, for example in mountainous areas. On Mount Olympus in Greece, there are 20 species of alpine plants found nowhere else in the world, relics of the climatic fluctuations of the last Ice Age. Here, and elsewhere in mountainous areas, a significant warming could mean that the climate to which these species are adapted might disappear from that location. There are other well-known examples of this vulnerability. The Adélie penguins living on the northern part of the Antarctic Peninsula will have nowhere to go if the climate warms appreciably. The tree frogs of the cloud forests of Costa Rica are also at great risk. The World Wide Fund for Nature reports that the critically endangered golden toad (*Bufo periglenes*) and the harlequin frog (*Atelopus varius*) have probably already disappeared. They have not been seen since 1986–87 when Costa Rica suffered exceptionally low rainfall and high temperatures during an El Niño event.

The concept of refugia may, however, offer the prospect of effective interventions in the biodiversity of hotspots (see map, right). By concentrating efforts on areas of high biodiversity, where the impact of climate change can best be accommodated, it may be possible to offset the impact of human activities elsewhere. In addition, these areas may offer the best opportunities for exploring the true extent of biodiversity and identifying flora and fauna that may provide economically valuable products such as new medicines. Any such initiatives, however, do need to take full account of plausible future climate scenarios. They also require better knowledge of what climate factors are currently most important in controlling animal and plant distribution.

reptiles and amphibians. Predominant are tropical forests, appearing in 15 hotspots, Mediterranean-type zones in five hotspots, and island groups in nine hotspots; 16 of the world's most vulnerable biodiversity zones are in the tropics.

It is important to identify these hotspots so that conservation planners can concentrate their efforts on areas where there is the greatest need and the greatest potential pay-off. In this way, it may be possible to produce a systematic plan to mitigate any large-scale extinctions ahead, some of which may result from climate change caused by human activities. Without a targeted effort, damage to hotspots will be highly likely, given that many are under threat from development activities.

Climate change may affect the living conditions of the Adélie penguins on Shirley Island in Antarctica.

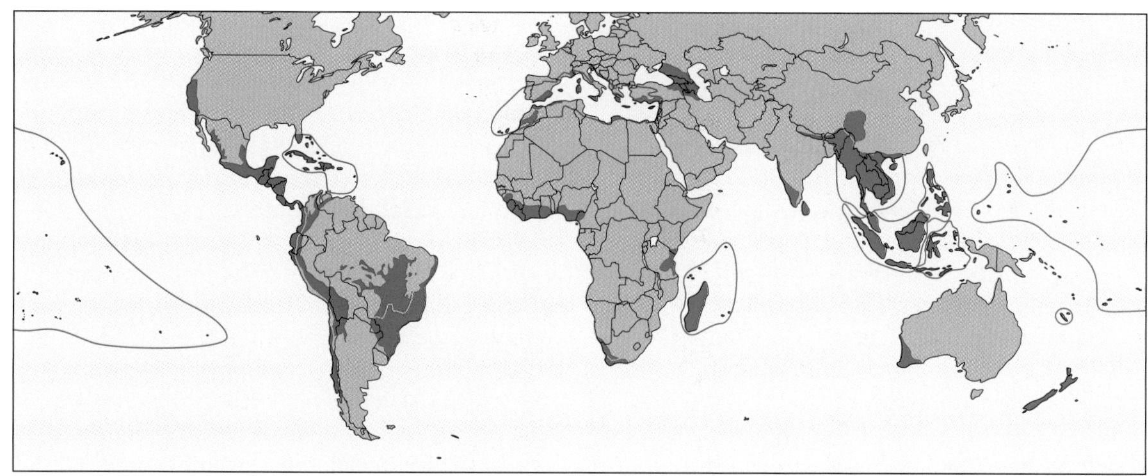

Biodiversity hotspots are principally in regions with tropical or Mediterranean climates. It is these areas that are under greatest threat of environmental destruction and loss of biodiversity as a consequence of human activities. The creation of refugia within these endangered areas is one proposal to safeguard the planet's biodiversity.

Phenology

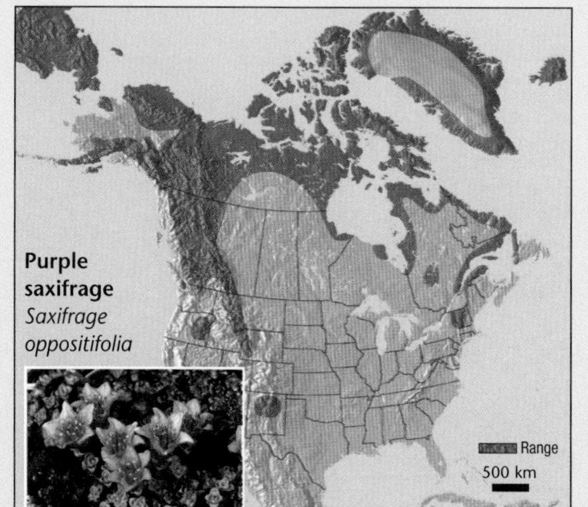

The life cycles of many types of wildlife depend on seasonally changing signals, such as temperature or day length. Changes in the climate can disrupt these life cycles. Phenology is the study of seasonal plant and animal activity (e.g. the dates when the leaves of different trees open, when they flower and fruit, or when various species of birds lay their eggs). Phenological observations contribute to our understanding of the controlling influence of climate and hence of the likely impacts of climate change on ecosystems. Records of the behaviour of plants, such as the flowering of cherry trees in Japan and the emergence of the leaves on many trees in England, have been maintained for many centuries. During the 19th century many phenological networks were set up in Europe to complement standard climatological observations and now provide valuable information about climate change during the period.

Analyses of phenological data obtained more recently from horticultural gardens across Europe indicate a lengthening of the growing season by up to 11 days over the last 30 years; about six days in spring and five days in autumn. The extended growing season has been linked to rising temperatures. Similar changes have also occurred in parts of northern Canada where the range of some plants has extended appreciably in recent years. The same types of changes have been observed in both the earlier laying of eggs by songbirds and the earlier return of migrating birds to both Europe and North America.

The purple saxifrage is found in Arctic and Northern Hemisphere alpine regions and is one of the earliest flowering plants in such habitats. It begins to flower very soon after the snow melts (April in the mountains, June in the Arctic). The temperature increase following snowmelt triggers flowering.The purple saxifrage is one of several Arctic and alpine plants being studied for the impact of a warming climate.

3.5 Weather, climate and warfare

The success and failure of many military campaigns have often rested on the ability of one side or the other to exploit weather and climatic conditions more effectively.

These troops struggle to move a cannon through the mud at Pilkem Ridge, Ypres in August 1917.

In the history of humankind, the outcome of war has often hinged on the weather and climate of the time. A classic example is Napoleon's Russian campaign and retreat from Moscow in 1812, immortalized in Peter Tchaikovsky's 1812 Overture. Napoleon set out in June of 1812 with his *Grande Armée* of over half a million soldiers, intending to conquer Russia in a matter of weeks. The Russian strategy was to lure the invading troops deeper into Russia, and the fierce Russian defence, which included burning the city of Moscow, led to Napoleon's defeat. The weather events of the time certainly worked against him as well. Tens of thousands of Napoleon's army had fallen from heat exhaustion in the summer surge eastward and, during the retreat to France, the bitter cold of December contributed significantly to the destruction of the remnants of the shattered army.

In more recent times, from the armies of Europe bogged down in the mud of the Western Front in World War I, to the Russian steppes in World War II, and to the jungles of Viet Nam in the 1960s and 1970s, weather and climate conditions have continued to play a major part in military matters. Some military planners have even considered that changing the weather might be a feasible strategy to winning otherwise intractable conflicts.

World War II

After several decades of generally mild winters, the combination of three consecutive severe winters (1939–40, 1940–41 and 1941–42) in the heart of Europe had no equal in the previous two centuries. The influence of these winters on the progress of World War II was first felt in January 1940, in what was then the coldest winter in northern Europe since 1830. For much of December, January and February, the fields of northern France and Belgium were covered by deep, powdery snow, and the roads by sheets of ice. General Alan Brooke, who commanded the British forces in France in 1940, later concluded that the bitter winter prevented Germany from launching what might have been an overwhelming attack against an ill-equipped and ill-prepared Anglo-French army.

The remnants of Napoleon's great army make frantic efforts to cross the River Berezina in late November 1812, as the savage winter weather intensifies.

The intense winter cold in Finland in early 1940 was also a major factor in limiting the advance by the Russian army in the Finnish War. The successful defence by the much smaller Finnish army, which exploited the extreme conditions, was widely seen as evidence that the Soviet army was ill-equipped to fight in winter. It also affected the USSR's own preparations for future conflicts, as was to be seen in the German invasion of Russia the following year.

By early October 1941, the German offensive depended on the success of *Unternehmen Taifun* (Operation Typhoon) – the attack on Moscow. Germany had retained enough strength to defeat the Russian armies, providing the weather remained mild enough not to complicate the plan. The German tactics relied on expertise in seasonal forecasts based on a set of rules built up from climatic records existing at the time. The prediction of a mild winter was also influenced by the statistical improbability of having three very cold winters in a row. Nonetheless, in the European part of Russia, the winter of 1941–42 broke all records. Around Moscow the five-month period from November to March was one of the coldest in the past 250 years and this was disastrous for the German offensive.

The cold virtually paralysed all military operations. As temperatures fell below –20°C weapons and machinery failed, firing pins shattered, hydraulics and lubricants froze, rifles, machine guns and artillery jammed, and tanks and supply trains ground to a halt. Frostbite was a major problem for the German troops, and about 100 000 were lost through frost-related deaths between October 1941 and April 1942. During the same period some 155 000 died or went missing as a result of actions in battle.

The Allied invasion of Europe

One of the best examples of where weather forecasts had a large bearing on a military decision and hence on its outcome was the timing of the Allied invasion of Europe in 1944 (D-Day). An American group of meteorologists prepared all forecasts for US forces and this work was combined with the predictions of the UK Meteorological Office and the Admiralty to produce consensus advice for the Allies. The advice was coordinated by Group Captain James Stagg, meteorological adviser to General Eisenhower.

There were only two three-day periods in June 1944 (5–7 June, and two weeks later) when the tides were sufficiently low to offer a good chance of negotiating the German beach defences. The Allies needed a forecast five days ahead that the weather would be sufficiently calm and clear to enable the naval and aerial operations to proceed.

The head of the US group of meteorologists was Irving Krick, who had developed an analogue technique for extended forecasts. This matched current conditions to similar meteorological situations from the past 40 years, and then predicted future behaviour on the basis of these sequels several days ahead. The British forecasters' work was based more on physical principles of the atmosphere but did not lead to confident predictions beyond 48 hours. Krick's forecasts were more detailed and were presented with greater confidence, but appeared to do less well than the more cautious British efforts.

D-Day was planned for 5 June and the forecasting groups remained in disagreement as the date approached. By late on 3 June it was becoming clear that the weather was deteriorating, as an exceptionally deep low for that time of year had moved toward northern Scotland. At 4.15 am on 4 June, Stagg predicted confidently that the weather would be too bad on 5 June and Eisenhower decided to postpone the invasion for 24 hours. By early on 5 June all the forecasters agreed that the storm would abate by the next morning, and conditions would enable the invasion to go ahead.

Although Krick's analogue technique for medium-range weather forecasts has not stood the test of time, his drive and enthusiasm subsequently drove serious efforts to provide more scientifically based strategic forecasts up to five days ahead.

3.6 Major temperate zone droughts

Drought has always been part of the human environment. Some parts of the world are persistently arid, but even in mid-latitudes where westerly winds provide more reliable rainfall, dry spells have, over the centuries, caused major problems for farmers and led to food shortages.

The record-breaking drought and heatwave in England and France in 1976 greatly reduced the flow in such mighty rivers as the Rhone.

Hot dry summers can have a positive effect on some sectors of agriculture. In the wine-producing regions of Western Europe, for instance, the best vintages are usually associated with the warmest summers. The heat and drought of 1921 resulted in a magnificent vintage, especially in Germany. For the most part, however, prolonged hot dry conditions are detrimental to agriculture.

European droughts

Over Europe the summers of the 1930s and 1940s were exceptionally dry. The most extreme year by far was, however, 1921, during which severe drought extended from Europe's Atlântic seaboard to the Caspian Sea. At Cologne, Germany, measurements of the flow of the Rhine show that both the mean annual figure and the minimum flow rates in that year were the lowest in a record stretching back to 1817. In the European part of the former USSR it followed the severe drought of 1920, which afflicted all its grain-producing regions. This sustained dry period combined with the hardship of a civil war to cause a dreadful famine.

Other notable European droughts include 1947, which was the driest summer across Europe as a whole during the 20th century, and the dry summers of 1949, 1950 and 1989. In addition, the dry summer of 1976 seriously affected the British Isles and northern France. In Britain, the 1976 drought raised serious doubts about the reliability and adequacy of the country's water supplies.

Russian grain harvests

Famine is not a stranger to what is largely now the Russian Federation. Russian film director Sergei Mikhailovich Eisenstein's 1929 film 'The Old and the New' has a memorable scene in which peasants, led by priests, form a great procession to pray for an end to the drought that afflicts their land. The film contrasts this scene with the success of the technology of the modern state: the triumph of modern technology over ancient superstition. The film was aimed at reassuring people that famines, such as those of 1920 and 1921, were a thing of the past. As with so many famines, weather conditions were only part of the story. The widespread armed conflict that followed the revolution in 1917–18 and the collapse of many aspects of social order also contributed to the terrible conditions. It was the drought, in combination with these political and social factors, that significantly reduced agricultural production.

The drought across the grainlands of the Russian Republics in 1972 presented another crisis for agriculture. After World War II, productivity rose as a result of a series of relatively wet summers. In the 1960s and 1970s rainfall fluctuated more dramatically. For the region as a whole, yields varied by nearly a factor of two between good and bad years. Drought led to poor harvests in 1963 and 1965, when for the first time, the former USSR had to import grain. Better harvests in the late 1960s lifted prospects, but the failure in the traditional grainlands of the Ukraine and the European part of the

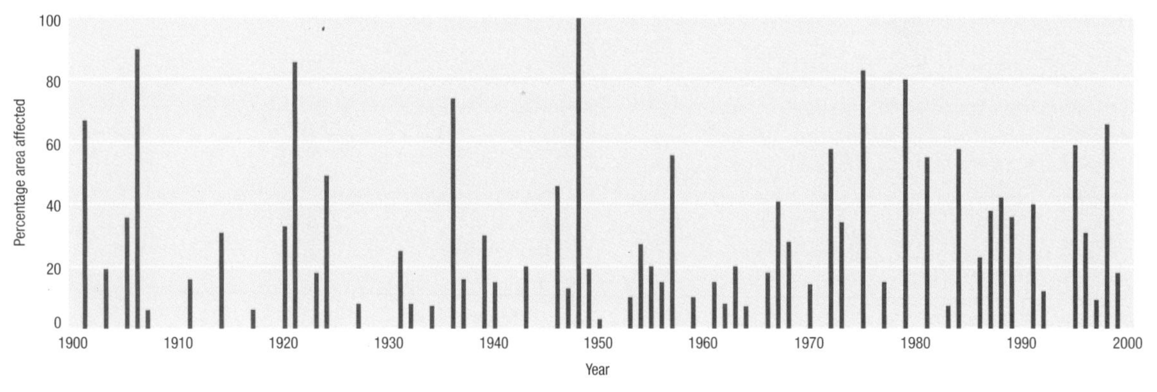

The drought index, a measurement of combined rainfall and hot days for Asian grain-producing regions of the Russian Federation in May and June, shows that hotter and drier conditions have become more frequent and pervasive in recent decades.

former USSR in the hot, dry summer of 1972 dealt the country a serious blow. The drought and its impacts were not as severe as 1921, but the summer was the hottest in the region for the century. The drop in grain production meant that Russia again had to import large quantities of grain, which had a significant impact on the world market.

North American droughts

The climatic conditions associated with the drought that afflicted the grain belt of North America in the 1930s were described earlier in 2.10. The human impact of these dry years can be well measured in terms of the drop in agricultural productivity, the high level of migration from the region and the abandonment of farms. The years had a cataclysmic impact on farming over much of the Midwest USA and the Canadian Prairies. The frequent dust storms – the terrifying 'Black Rollers' – made life nearly unbearable and blotted out the Sun in the cities of the east coast, 3000 km away. In Washington, DC, the dust settling on the desks of congressmen brought the plight of the farmers directly to the attention of the government.

In 1934 and 1936 the wheat yields across the Great Plains dropped by about 30 per cent. There was massive outward migration from the hardest hit parts of Kansas and Oklahoma, with more than half the population migrating to cities and to the west coast. This huge social disruption was the catalyst for government action, including the provision of disaster aid that enabled many people to survive the drought, and a major review of land use practices in the region was undertaken. It also set the precedent for similar actions in the future whenever severe weather was to strike the USA.

As a consequence, agricultural practices were introduced into the Great Plains that were more appropriate to such a semi-arid region. Previously, farmers had rendered the land surface vulnerable to strong winds, scorching heat and hence the loss of huge quantities of topsoil, by not leaving grass or trees in place to hold the soil together. Subsequently they were encouraged to plant trees as shelter-belts in order to reduce wind erosion, to grow crops better suited to dry land conditions and to introduce conservation methods (e.g. contour ploughing, water conservation in irrigation ponds and strip ploughing to allow part of the land to lie fallow). Marginal land was purchased by the government, retired from cultivation and seeded with grass.

More recent droughts have been less dramatic in their consequences. The droughts and heatwaves in the midwest USA in 1980 and especially in 1988 attracted a great deal of comment as possible indicators of global warming. They did not, however, constitute climatic stress of the same order as the Dust Bowl years of 1934 and 1936. Nevertheless, the cost of the 1980 drought was put at some US$ 20 billion and the estimated cost of the 1988 drought was about US$ 40 billion, the most costly natural disaster in US history.

The Great Grain Robbery

As a consequence of the 1972 drought, the former USSR had to purchase grain from the world markets. Some 30 million tons of grain had to be imported, even though in 1973 a record-breaking harvest was in the making. Buyers managed to achieve these large purchases without initially causing a sudden and large jump in prices because few outside the country were aware of the seriousness of the drought. Moreover, no one knew that the buyers had been secretly negotiating separately with private grain merchants. This coup is commonly referred to as the 'Great Grain Robbery'. When the scale of the purchases became apparent, prices rose threefold during 1973 to levels that would not be seen again until 1996. The result was a worsening of the global food crisis and a scramble for available grain reserves in which, inevitably, the poorest countries suffered most.

Since then Russia has been a major importer of grain, but never again would the Chicago Futures Market be caught unawares as in 1972, taking the even more severe drought and purchases of 1975 in its stride. In addition, more information from comprehensive monitoring from space means that the markets can now anticipate poor harvests, such as the dramatic shortfall in the Russian harvest of 1998.

3.7 Northern Hemisphere cold winters

Severe winters repeatedly affected the economic and historical development of many mid-latitude nations in the Northern Hemisphere during the 20th century.

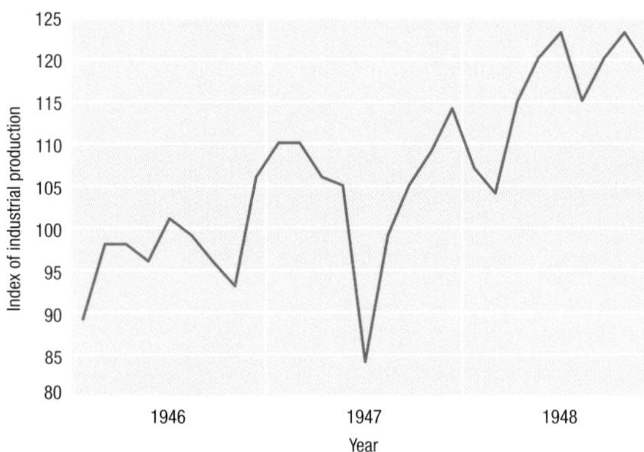

The intense cold and associated fuel crisis of February 1947 caused a major setback to British industrial production.

Europe, emerging from the trauma of war and having experienced three consecutive severe winters in 1939–40, 1940–41 and 1941–42, had to survive yet another bitterly cold winter in 1946–47. The cold weather was immensely damaging to agriculture. In Britain, some four million sheep died. The subsequent melting snow and record rainfall in March 1947 led to massive floods and further crippled agriculture.

1947 Fuel crisis in Europe

In the post-war period, industry in much of Europe had been either run down through lack of investment or destroyed by warfare. As a consequence of the cold winter and floods of 1946–47, supplies of coal were insufficient to generate enough electricity to meet demand and therefore industry was shut down in many areas for several weeks to save energy. That resulted in a temporary but severe 25 per cent drop in production. The underlying damage was profound and accelerated economic problems. In Britain, the loss of £200 million in exports contributed to a record balance of payments deficit in 1947 (£630 million – an enormous figure in those days) that ushered in the currency devaluation of 1948 (sterling was devalued from US$ 4 to US$ 2.80). As a consequence of its own perilous economic state in 1947, Britain was no longer able to provide financial aid to Greece and Turkey, considered, at that time, likely areas for the expansion of communism.

In March 1947, US President Harry S. Truman enacted the 'Truman Doctrine', whereby the USA would support free peoples in their resistance to attempted subjugation, and persuaded his government to take over provision of aid to Greece and Turkey. This effectively marked the beginning of the USA bipartisan Cold War foreign policy. As well, the USA recognized that Britain had lost the capacity to maintain an independent role as a world power. With much of Europe suffering badly from the extreme winter, the USA intervened with the Marshall Plan to prevent a complete economic collapse. The fuel crisis was an added stimulus for the formation of the European Coal and Steel Community in 1950, designed to tie the economy of Germany to the other economies throughout the continent. This led eventually to the establishment of the European Economic Community.

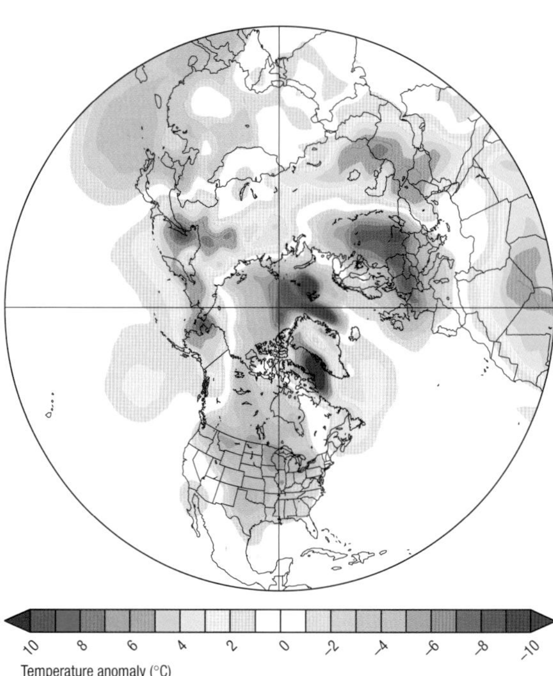

The pattern of anomalous temperature around the Northern Hemisphere in January 1963 produced regions of exceptional cold over northern Europe, China, Japan and North America, and equally exceptional warmth over Central Asia, Alaska and Greenland.

The winter of 1962–63

The sustained cold spell of January and February 1963 broke many records in northern Europe. In central and southern England it was the coldest since 1740, while across other parts of Europe it

Along the Netherlands coast, the winter of 1962–63 was exceptionally bitter, as it was throughout most of northern Europe.

was the coldest since 1830. There was also cold weather across North America and in Japan. This event is often cited as a classic example of atmospheric blocking, the phenomenon known to be responsible for some of the most extreme winter weather experienced in parts of the Northern Hemisphere (see 2.17).

Despite the record-breaking cold of early 1963, the economic and social impacts were relatively small compared with the chaos that occurred in 1947. This was because the economies of Europe, North America and Japan were growing strongly. This response shows how the behaviour of economics is driven by interaction between the weather, the state of the economy, and the underlying vulnerability and resilience to climate extremes.

The late 1970s in the USA

The eastern USA was hit by three exceptionally severe winters in a row at the end of the 1970s. Each of these caused substantial disruption, but it was the winter of 1976–77 that had the greatest impact. It came after five mild or very mild winters and was preceded by a very cold autumn. There were also shortages in natural gas, and by early 1977 the USA was beginning to experience economic problems.

Many records were broken in January 1977 as it was probably the coldest month experienced in the eastern half of the USA in the previous 200 years. Virtually all the USA east of the Mississippi had a monthly temperature anomaly of at least 5°C below normal, and in the upper Ohio Valley it dropped to 10°C below normal. The intense cold affected the most populous areas of the country and it led to record fuel demand, which added to the problems of supplying natural gas. The upper Mississippi and its tributaries froze solid. Barges supplying heating oil and salt for the roads were trapped in the ice.

The government enacted emergency legislation to enable rationing of natural gas supplies for three months. However, the value of this temporary measure was not fully tested as the weather relented in early February, and people soon forgot about the vulnerability of the energy system. Economic losses were estimated at US$ 25 billion.

The following winter (1977–78) overall was only marginally less cold than the previous one. A series of major snowstorms paralysed the larger cities of the east coast. Eighteen severe snowstorms brought the greatest accumulation of snow across Illinois, Ohio and western Pennsylvania since these areas were settled at the end of the 18th century.

In meteorological terms, the following winter of 1978–79 was even more exceptional than the previous two. In December the cold hit the west of the country, where in many places it was the coldest on record. The cold moved eastwards in January and for the country as a whole it was the coldest on record. The cold persisted over the east coast where many places experienced either their second or third coldest February on record. In spite of it being statistically by far the coldest winter across the country as a whole in the previous 100 years, it did not affect the most populous regions as much as the 1976–77 winter.

North and south

Cold winters in the Southern Hemisphere (apart from Antarctica) are generally less disruptive and less severe. This is because the middle latitude land masses are smaller and they have the moderating influence of the Southern Ocean as a buffer between them and Antarctica. Most major Southern Hemisphere cities and agricultural areas tend to have a more maritime influence on their climate. Their average wintertime temperatures are more moderate than their counterparts in the Northern Hemisphere and they are rarely subject to frigid polar air masses.

Disruption over northern Germany

The snowstorm that paralysed the northern German state of Schleswig-Holstein between 28 and 31 December 1978, when it snowed for more than 85 hours, was a bitter reminder of the disruption such storms can cause. Snowfall amounts were equivalent to around a metre in depth and the strong easterly winds whipped up drifts 6–7 m high. Transport was disrupted over a wide area. A second snowstorm between 14 and 17 February 1979 produced slightly less snow but had stronger winds and again caused widespread disruption.

3.8 Storm surges

Storm surges are the curse of many coastal areas. Where shallow seas and severe storms combine, the sea can rise suddenly and sweep away whole communities.

The storm surge that breached the dykes in the Netherlands on 1 February 1953 flooded large areas of the country, causing immense damage and killing over 1800 people.

Along many coastal shores and on large lakes the variations in surface air pressure and winds can cause significant short-term fluctuations in water level, particularly during storms. These storm surges (called, somewhat erroneously, tidal waves) cause inundation to low-lying near-shore areas and, in addition to an immense amount of damage, can cause considerable loss of life. The threat of surges is greatest when strong on-shore winds, low pressure and high tide coincide. Emergency services need reliable forecasts of flooding to enable evacuation of vulnerable areas. Planners also need assessments of the likely highest level in 50 or 100 years to provide a basis for making decisions about coastal land-use and the design of flood defences.

North Sea surges

Over the centuries many severe storms have caused great damage around the North Sea coasts, and this tale of destruction continued through the 20th century. Perhaps the most notable was the intense depression that produced a storm surge on 31 January 1953, flooding the east coast of Britain and breaching the Dutch dykes the following day. Over 300 people died in Britain, and over 1800 in the Netherlands.

This killer storm showed just how vulnerable the coasts around the southern North Sea are. While London was spared the worst of this particular surge, the fear of a devastating combination of high tides and an exceptional storm surge led to the building of the Thames Barrier. This structure was eventually opened in 1983 and was almost immediately pressed into service when, precisely 30 years after the disaster of 1953, a similar storm developed. In the Netherlands, the loss of life and the massive damage to farmland and property in 1953 were followed by decades of strengthening of the dykes and the building of major flood control systems in the Rhine and Scheldt estuaries.

Calculating the size of a storm surge

The extent to which a lake or sea level rises during the passage of a depression depends principally on the air pressure reduction, the speed of the depression, wind speed, wind direction, proximity to land and the depth of the water. In the right circumstances, usually involving fast-moving storms over waters of less than about 100 m depth, a resonance can occur which produces an exceptional rise in sea level. Regions prone to high

When a deep depression moves towards southern Scandinavia with high pressure to the west of Ireland, the strong north to northwesterly winds and the funnelling effect of the shape of the North Sea combine to produce a major surge. Analyses of tidal records show the height of the surge building up as it swept southwards down the east coast of Britain in the wake of a deep depression crossing the North Sea in 1953.

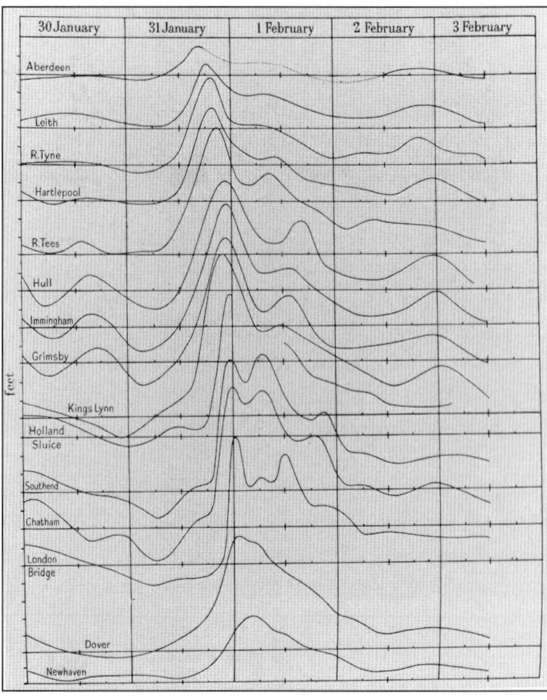

The massive flood control system on the River Scheldt in the Netherlands is the final component of a coastal defence system that provides protection from storm surge events and coastal flooding of the type that occurred in February 1953.

storm surges include the Bay of Bengal, the Gulf of Mexico, the North Sea, the Yellow Sea and the Gulf of Carpentaria.

Assessments of the probability of extreme storm surge events have been prepared for many locations. For example, computer models of the North Sea show how a surge builds up in the southeast corner with the area around the mouth of the Elbe River being particularly vulnerable. Here, a one-in-50-year surge may exceed 5 m.

Forecasting storm surges

Many damaging storm surges can now be predicted more accurately using increasingly complex statistical and computer models. In Britain the forecasting system has evolved from long-established rules for how much the sea level will be raised in any part of the North Sea for a given combination of atmospheric pressure, and wind speed and direction. These have been combined with standard numerical weather forecasts to estimate the size of the surge and its timing.

The large tidal range in the North Sea means that the surge only really matters if it coincides with high tide. A surge typically lasts between 6 and 12 hours, therefore predicting its timing is an essential part of useful warnings. Computer-based forecasts can identify potentially dangerous conditions more than 36 hours ahead, but the important operational work is concentrated in the last 24 hours. Warnings of a flood event have to be issued at least 12 hours in advance to enable emergency services to react effectively. To protect London the Thames barrier is closed whenever water levels are predicted to rise above a certain threshold.

Storm surge models are an essential part of hurricane warning in the USA. The National Hurricane Center in Miami, Florida provides predictions of surge height as a component of its hurricane warning system. In addition, research institutes are testing more detailed computer models for local surges that include inland rainfall in the prediction of inundation. In the USA flooding is the principal cause of deaths when hurricanes make landfall. Emergency plans to ensure effective evacuation of low-lying areas rely heavily on forecasts of how rapidly water levels will rise.

Bangladesh: planning for storm surges

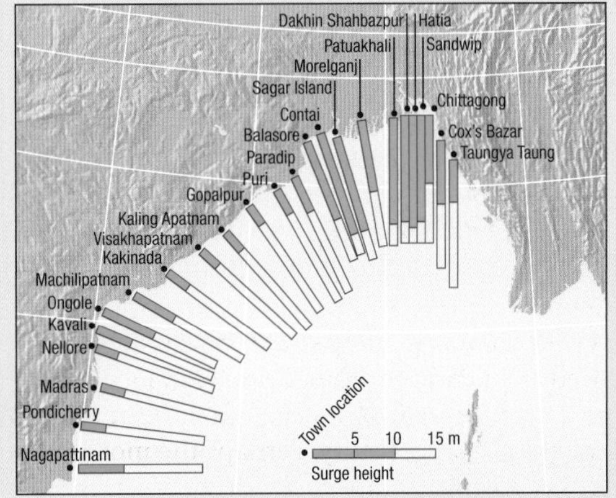

More than anywhere else in the world, Bangladesh is at the mercy of storm surges from tropical storms. Throughout its history the low-lying areas of the country have repeatedly been hit by cyclones and storm surges sweeping in from the Bay of Bengal. Here the combination of shallow seas and high winds can produce a maximum surge in excess of 12 m around the Ganges Delta as depicted in the map. The disaster of 1970, when 300 000 people perished in a storm surge that swept across the deltas of the Ganges and the Meghna Rivers, prompted national authorities into action. An innovative preparedness programme was developed using local volunteers to disseminate cyclone warnings and guide people to constructed mounds or 'cyclone shelters'.

During the 1980s the low level of casualties was seen as evidence that this programme was working. The loss of life in the cyclone of April 1991, when 130 000 people died, provided a stark reminder of the continuing chronic vulnerability of Bangladesh to coastal surges. Although there is no clear trend in the incidence of cyclones in the Bay of Bengal, the combination of population pressure to exploit the rich soil deposited in the delta regions and rising sea levels underscore the need for effective warnings and civil defence plans.

3.9 Tropical storms

Communities around the world have developed strategies to handle the ever-present threat of tropical storms. The responses vary from place to place and reflect not only the local climate but also social conditions.

Tropical cyclones (also known as hurricanes or typhoons, depending on the region) occur most frequently in five key regions of the world: in the northwestern Pacific, the tropical North Atlantic and the Caribbean, the northeastern Pacific, the southern Indian Ocean and the southwestern Pacific. By far, it is the northwestern Pacific that sees the greatest activity each year.

Japan

Over the years, tropical storms have caused extensive damage in Japan. The most extreme case during the 20th century occurred when more than 5000 people were killed in the Nagoya area by a typhoon in 1959. There has been, however, a dramatic decrease in the number of lives lost in meteorological disasters (mostly tropical cyclone disasters) in the latter half of the century. This has largely been as a result of the intense application of advanced meteorological technology, in the form of radar and other early warning systems, together with the implementation of disaster prevention measures. Even so, typhoons can still pack a formidable punch. In September 1999, Typhoon *Bart* lashed Japan with winds of 150 km/h, 240 km/h wind gusts and heavy rains. The storm surge crashed over coastal homes, knocked cranes into buildings, snapped trees and toppled power poles, leaving 26 dead and more than 300 others injured. It also generated a major tornado. In response to early warning and emergency procedures, 8000 people fled from their homes to seek refuge in shelters.

The threat of typhoons in Japan has led the Japan Meteorological Agency to develop a computer-based storm surge model, which has been in operation since July 1998. When a typhoon is threatening, the Japanese Meteorological Agency regularly updates its prediction for the three-hourly sea levels and highest tides 24 hours ahead; these are made available to the public and authorities through local meteorological observatories.

Australia

Much of the coast of Northern Australia is sparsely populated, and many cyclones in the region do little harm. Cyclone *Tracy* showed, however, what a devastating impact a direct hit on a city can have. The cyclone was relatively small, with the region of gale-force winds only about 100 km across and lowest central pressure 950 hPa. However, it destroyed or damaged almost every building in Darwin on 25 December 1974; 49 people lost their lives on land and 16 at sea. Total loss was some A$ 800 million. It transformed

Cyclone *Tracy* brought almost total devastation to Darwin, Northern Australia, in 1974.

Effective exploitation of technology?

Technology can be a two-edged sword. It can be very beneficial if used competently by those who know its capabilities and limitations. However, if used indiscriminately, it can render parts of society more vulnerable to extreme weather. Taking the obvious benefits of technological advances for granted without recognizing the limitations can lead to complacency. We may leave it to the last minute to take shelter from a severe thunderstorm, or to move inland as a hurricane approaches. When we over-rely on technology we may suffer greater losses when it unexpectedly fails us.

Recognition of these possible shifts in human behaviour may also lead to a defensive response by forecasters. As more people need to respond to warnings, forecasters are under increasing pressure to be more accurate. They recognize, however, the limitations of their scientific knowledge and thus may compensate by extending the domain of the warnings to err on the safe side. For instance, there is some evidence for the USA that the area forecast likely to be affected by hurricane landfalls has not declined in line with improvements in predicting the tracks of storms. This conservative response to a critical decision may reduce the potential benefits of an accurate warning, if the disruption caused turns out to be greater than need be. Responding to forecasts affects people's lives before the actual impacts do, and as a result, in the 'calm before the storm', some people question details of the forecast and do not respond to the warnings. When Hurricane *Georges*, a Category 4 hurricane at its peak, threatened the lower Florida Keys in 1998 as a still powerful Category 2 storm, 40 per cent of the people remained at home.

The cost of tropical cyclones

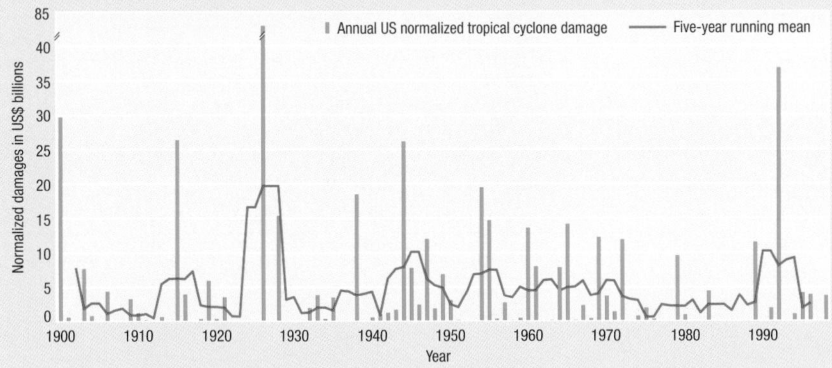

There has been a sharply rising trend in the damage loss caused by tropical cyclones. This has been most noticeable in the USA. As a consequence, the insurance industry has changed its attitude to catastrophic weather-related losses. Before 1986 no insured loss had exceeded US$ 1 billion. Hurricane *Hugo* in 1989 incurred losses five to six times this figure, while for Hurricane *Andrew* in 1992 the insured loss exceeded US$ 15 billion. The upsurge in insurance payouts led some to the unjustified assumption that global warming was producing an increase in the number of intense hurricanes.

Recent analysis of hurricane damage in the USA indicates that neither the frequency nor the intensity of Atlantic hurricanes increased over the 20th century. When the data are normalized, they take into account changes in economic and social circumstances over the period of the analysis, and it can be seen that relative damage values tended to decline after the 1940s with the exception of Hurricane *Andrew* in 1992. Of course, it must be remembered that these results are sensitive to the accuracy of the reported damages for each original event. Several hurricanes prior to Hurricane *Andrew* caused relatively high damages, including the Galveston hurricane in 1900, but it was the unnamed storm in Miami in 1926 that topped the list with an estimated total damage bill of more than twice that of Hurricane *Andrew*.

thinking in this region about building codes and emergency response plans for tropical cyclones, as well as completely altering the community's attitude to warnings of such storms.

USA

The impact of Hurricane *Floyd* on the Carolinas in the southeast USA in September 1999 highlighted both the progress that has been made in responding to the threats of tropical storms and the challenges that remain. Hurricane *Floyd* was the biggest storm to threaten the USA since *Camille* in 1969. Falling just short of a Category 5 hurricane, its track across the Bahamas towards the Carolinas was well predicted days in advance. This enabled a large-scale evacuation of the coastal region to be executed in good time and, in spite of massive traffic jams, the loss of life caused by the high winds and storm surge was relatively low. Inland, however, flash flooding swept people away in their cars as they sought to escape from the vulnerable shoreline areas and caused over 60 deaths. This does not mean that it would have been safer to stay by the coast, but it does mean that the management of disasters must recognize how warnings may alter people's behaviour and create a potential for secondary problems.

Central America

Hurricane *Mitch* was the deadliest Caribbean Sea hurricane since 1780. It caused immense damage to Honduras and Nicaragua in October 1998. It formed in the southwest Caribbean on 21 October 1998, and five days later the pressure at its eye had fallen to 905 hPa, the fourth lowest in the Atlantic region during the 20th century, and the most intense ever recorded in October. Maximum wind speeds were 290 km/h. It made landfall on the north coast of Honduras on 29 October, and its wind speeds declined rapidly. It was as a weakening tropical depression, however, that it did the greatest damage. Between 100 and 150 cm of rain fell over a wide area within 48 hours. The torrential rains brought havoc, killing upwards of 10 000 people in landslides and floods. The total damage to Central America was estimated at around US$ 5 billion.

Although Hurricane *Mitch* was deadly, it could have been worse. A similar, slightly less extreme storm in September 1974, Hurricane *Fifi*, killed 8000 people in Honduras. As a result, in towns in northern Honduras, new construction in flood-prone areas was banned. Hillsides were reforested, bridges strengthened and storm drains maintained. In addition, improved evacuation schemes in this region enabled many people to move to safety before Hurricane *Mitch* struck. Compared with Hurricane *Fifi*, loss of life was greatly reduced in the areas where evacuations had taken place.

In this satellite image Hurricane *Floyd* can be clearly seen approaching Florida, USA, in August 1999. Accurate predictions of its path enabled people to evacuate the vulnerable shoreline areas.

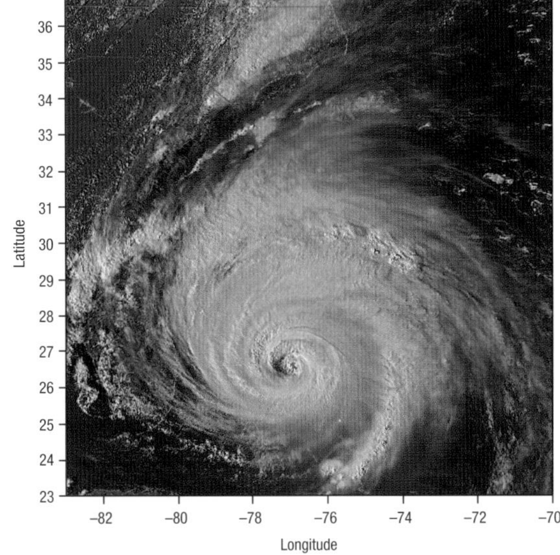

3.10 Good and bad monsoon years

For centuries, the failure of the monsoon in India has led to famine and death. The droughts of 1877, 1899 and 1918 instigated efforts both to understand and forecast the variation of the monsoon from year to year and to alleviate the social consequences of crop failure.

Variations in monsoon rainfall from year to year across the Indian subcontinent are part of the normal pattern of global climate variability. For example, between 1917 and 1918 the monsoon switched from causing floods to causing severe drought conditions. Around the world there were other dramatic swings in the seasonal weather between these two years, linked to the sudden shift from La Niña to El Niño conditions.

What constitutes a 'good' or 'bad' year?

The great size of the Indian subcontinent makes it difficult to make simple statements about total monsoon rainfall. In many years the amounts falling in adjacent regions vary appreciably. An index of the summer monsoon (covering June to September) has been constructed from measurements made across the whole of India, which is divided up into 35 climatically homogeneous subdivisions. These measurements form the All-India Southwest Monsoon Rainfall Series.

It is usual to consider the dry years as being the bad years. Indeed, 1918, 1877 and 1899 were the three worst drought years on record. During each of these years over 60 per cent of the total area of India was drought-affected (defined as occurring when an area receives less than 75 per cent of its normal seasonal rainfall). These conditions led to widespread famine in India. More recently, the driest years have been 1965, 1972 and 1987, when well over 40 per cent of the country was drought-stricken. It is a measure of the improved food production and supply systems that the extent of famine during these later droughts was greatly reduced.

Indian famine relief

Since the 1870s, successive Indian governments have responded to drought-induced need by offering employment. This approach has been hotly debated over the years; is it cheaper to feed people minimum maintenance diets, or organize public works of questionable social benefit? Both national and local governments have concluded that while food aid might be cheaper, work was better. Experience has shown, however, that if employment is to address needs it must meet two conditions. First, relief must begin early, before the population has exhausted its reserves and started to suffer from acute lack of food. Second, food supplies should flow through well-controlled distribution systems to ensure they reach those in need, and at a fair price.

Although years with abundant rains are usually regarded as good, they can cause widespread flooding, loss of life and property, and damage to crops. The floods of 1961 affected nearly 60 per cent of the country (a flood year being defined as one where the summer rainfall is more than 125 per cent of normal). Interestingly, other notable wet years either immediately followed severe drought years (e.g. 1878 and 1988), or came immediately before a drought year, as was the case in 1917.

Year-to-year fluctuations

The strength of the summer monsoon circulation is linked to the movement of the intertropical convergence zone each year, so it is hardly surprising that its fluctuations are intimately linked to other variations in tropical weather patterns. Phases of the El Niño/Southern Oscillation (ENSO), for example, have frequently been linked to the amount and distribution of Indian summer monsoon rainfall, with some of the most severe droughts coinciding with El Niño

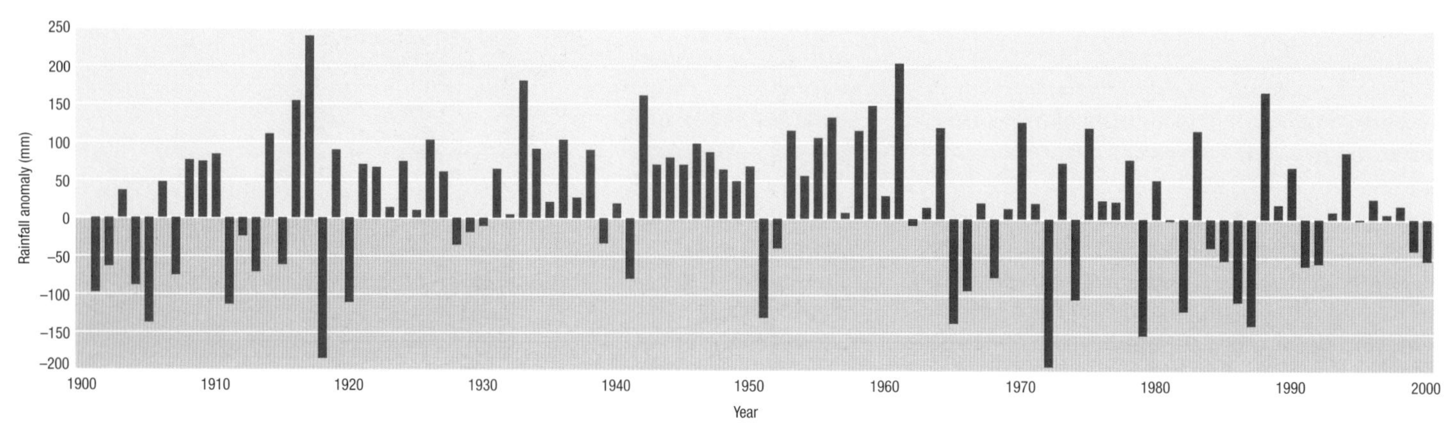

The All-India Southwest Monsoon Rainfall (June–September) Series where data are anomalies from the 1961–90 Normals, shows that the monsoon precipitation was generally above normal from the 1920s to the 1960s, and more variable since. The worst drought years in the 20th century were 1918 and 1972.

years. The droughts of 1972, 1982 and 1987 illustrate the point. Similarly, the flood years of 1917, 1978 and 1988 are associated with La Niña years. A tendency for a see-saw pattern between successive years of drought and flood has also been attributed to occasions when there has been a sudden shift from El Niño to La Niña episodes.

A more detailed examination of the statistics shows that the link between Indian rainfall and ENSO fluctuations is more complicated than might be assumed from selecting a few extreme years. While many of the extreme years have been linked to ENSO events, others have not. The record wet year of 1961 coincided with neither an El Niño nor a La Niña event. There are three principal reasons for this more complicated behaviour. First and foremost, the Indian subcontinent lies on the fringe of the patterns of strong correlation within ENSO events, the western part of which is centred on Indonesia. Second, El Niño or La Niña timing could be critical, since El Niño events tend to peak around December, i.e. midway between successive Indian summer monsoons. The variation in timing of each ENSO event may influence India differently. Third, the more proximate sea surface conditions in the Indian Ocean must also be taken into account and not just the sea surface temperatures prevailing in the more distant equatorial Pacific Ocean, which are the principal driving mechanisms for ENSO.

Trends in the Indian monsoon

The All-India Southwest Monsoon Rainfall Series shows no appreciable trend over the last 100 years, but there have been considerable changes in the variability from decade to decade. The monsoon was less reliable before 1920 and the variability seems to have been less between the 1920s and 1960s. The latter may be linked to the behaviour of the ENSO, which was also less variable in this period. Since the mid-1970s both the monsoon and ENSO have been more variable, although the links with El Niño events in the 1990s have not been as clear-cut as earlier in the century.

Chinese patterns

The great interannual variability of the East Asian monsoon in China brings excessive rains and causes floods and destruction in some years, and only sparse precipitation and severe drought in others. The onset and retreat dates, the speed of advance and retreat, and the intensity of summer monsoon in any given year are directly associated with the location of the main rain belt.

The most favourable monsoon for China's agriculture brings more precipitation in the north, which normally receives less rainfall than the south. This pattern occurred in 1984 and produced an abundant harvest. Conversely, in 1978, the monsoon failed in the Yangtze River Valley with disastrous consequences. Rainfall was 40–60 per cent below normal during July through August, the key part of the growing season. Over a quarter of farming land in China was hit by drought and there was a serious shortfall in food production. It was the worst drought since 1949, when 57 000 people died.

The success of agriculture in India is dependent on the monsoon, with the output of grain being closely correlated with rainfall across the country.

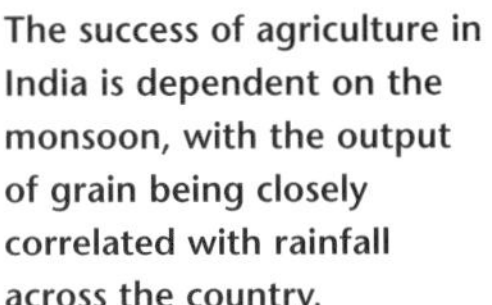

The pattern of drought across India when the monsoon fails can vary appreciably with different parts of the country suffering in different bad years.

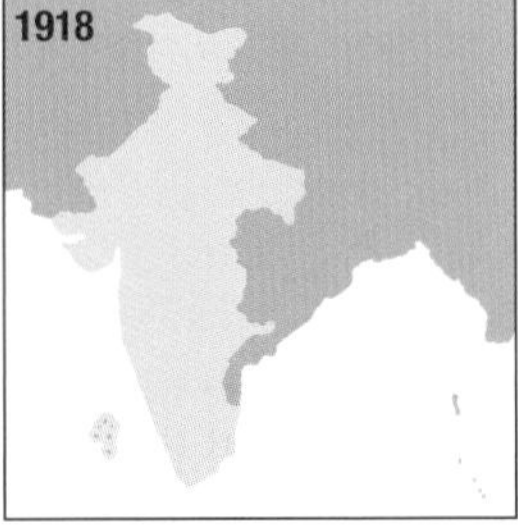

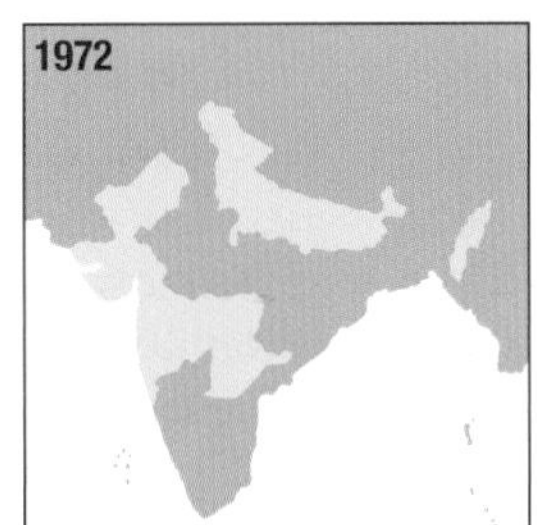

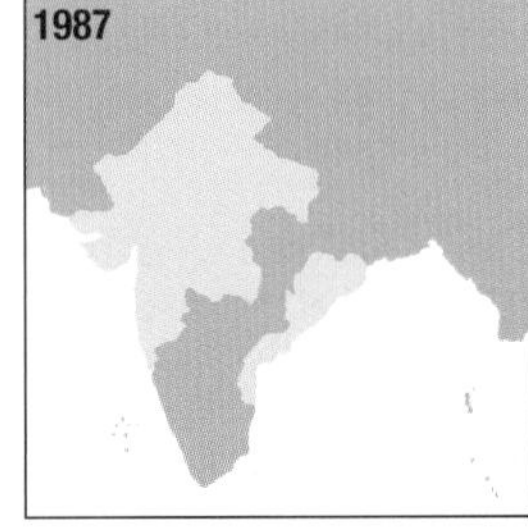

Areas affected by drought where the rainfall was less than 75 per cent of the long period average

The Gharial crocodile times its egg laying to coincide with the onset of the monsoon, as this enhances successful breeding. Indeed, legend claims that success is so linked to the monsoon that it is the sound of thunder that cracks the eggs.

3.11 Sahel – the human consequences

Perhaps the most significant example of systematic and persistent regional climate fluctuation in recent decades has been a shift to drier conditions in the sub-Saharan region of Africa, known as the Sahel. Since the late 1960s this arid to semi-arid region has suffered frequent multi-year droughts which were unprecedented in the years before.

In the mid-1980s, famine took the lives of over 800 000 people in the Horn of Africa. Years of drought took a terrible toll on the livestock in the region as well.

The clearest example of a persistent and large-scale regional decline in precipitation during the last 100 years is in the Sahel – the southern fringe of the Sahara Desert. Much of this drying occurred after the 1960s, with the drought reaching its first peak in 1972. On average, about 50 per cent of the region has been in severe drought since 1970, which is about twice the area affected in the first half of the century. The drought abated to a certain extent in the late 1970s but returned with greater vigour during the 1980s before easing off towards the end of the century.

Consequences of drought

The Sahel has often suffered years when food was in short supply. While lack of rainfall has been the principal cause of famine, locusts have devastated crops in wetter years. There were severe droughts in the early 1910s and again in the early 1940s. What made the events in the early 1970s so important was not just the scale of the drought, but the impact that it had on many peoples' perceptions about climate. Drought in sub-Saharan Africa came to represent first the demands that growing populations place on limited natural resources and later the threat of global warming. The images from the Sahel in the early 1970s, when over 200 000 people died, and from the Ethiopian famine in 1984, which had even greater mortality (estimated at near 1 million), brought home the human consequences of year-to-year climate variability. The recurrent droughts in Africa have provided a formidable challenge for climatologists; to explain the many fluctuations in the climate throughout the region and the part played by the El Niño/Southern Oscillation, thousands of kilometres away in the tropical Pacific.

The benefits of forecasting

Empirical studies link the drought in the Sahel with various climatic parameters and have led to the development of seasonal forecast techniques for the region. These show a useful level of accuracy and have implications for managing agriculture, foreign aid and humanitarian assistance in this region, but raise important questions about how best to optimize the value of forecasts.

Statistical methods, using established empirical links between tropical sea surface temperature and Sahel rainfall, work best but, so far, they cannot yet predict precisely where the rains will fall, nor when. The research does show,

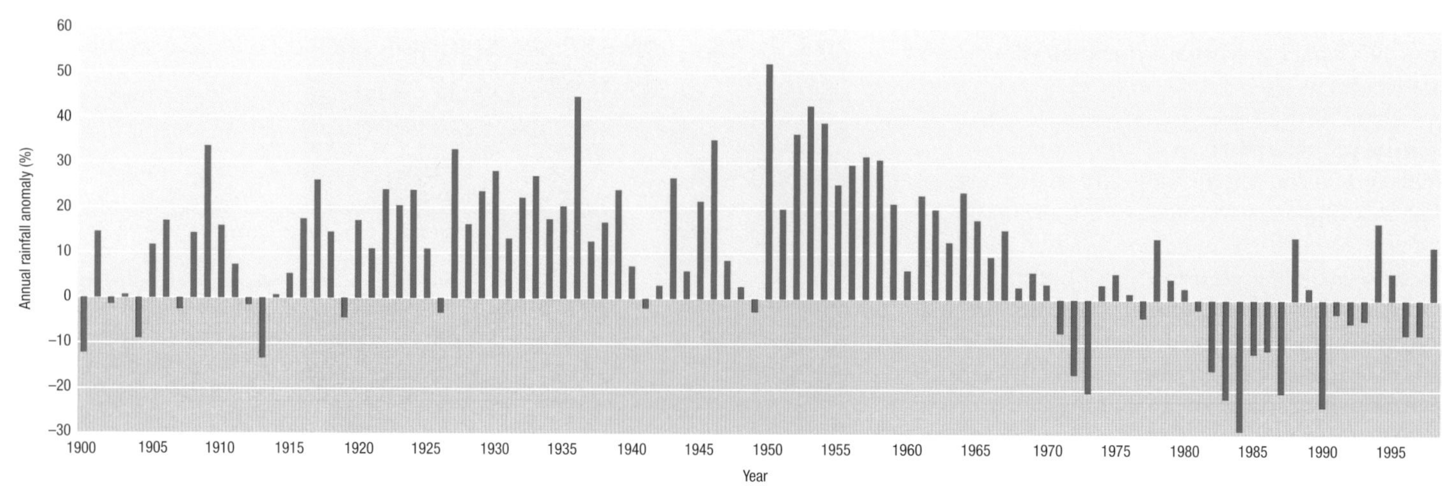

The decline in rainfall in the Sahel region since around 1970 contrasts starkly with the relatively moist conditions in the first half of the 20th century.

however, that there is a prospect of being able to produce more useful long-lead forecasts of overall seasonal rain patterns. Another approach has been to explore how computer models handle the rainfall across the Sahel as a function of changing sea surface temperature patterns across the tropical oceans. This approach is shown to have the potential for useful long-lead forecasts of seasonal rainfall based on data after the month of June. The practical shortcoming to this current constraint is that forecasts made in July are too late to influence agricultural planting decisions in the Sahel.

The economic benefits of seasonal forecasts depend on who gets the forecasts, when they get them and how prepared they are to use the information. The users range from international aid organizations, through various government agencies in the different countries and, indirectly, the farmers, pastoralists and freshwater fishermen who are most affected by the rainfall fluctuations. Rain-fed agriculture is concerned about the difference between water input (rainfall) and water loss (evapotranspiration), as it is this water balance that is available to the crop throughout the growing season. Decisions for ploughing, sowing, weeding, fertilizer application and so on are also dependent on water balance considerations.

Annual rainfall forecasts are of limited utility to farmers. What they require is information on expected changes in the normal wet season structure, including the timing and regularity of rainfall. These affect dates of onset, length and ending of the growing season, plus false starts and dry spells within the season. Furthermore, to meet the needs of small farmers, rainfall forecasts need to be produced in a form that can be readily broadcast by radio, reflect local calendars, fit in with other factors influencing agriculture, and stress the strengths and weaknesses of the prediction scheme. In addition, the forecasts must be couched in terms that reflect local communities' perceptions (for example) of what constitutes a drought. This is no easy task.

Current forecasts fall well short of these demands and in the foreseeable future it may be that only large commercial growers can benefit from them. It may be that small-scale farmers, who lack the flexibility to respond to such general forecasts, could become more vulnerable to how the markets respond to the information and be further disadvantaged.

The potential to exploit the forecasts may also be greater for donor agencies and governments. The information must still be timely, as donor agencies must allocate their aid resources early in each financial year. Aid and developing national responses clearly matter most in extreme years, therefore reliable early predictions of very dry or very wet years are what are most needed. The benefits of forecasting an exceptionally dry season show up not only in planting decisions but also in livestock management. The ability to draw up a national plan of herd destocking in a staggered manner to avoid a collapse of livestock prices, because of a glut on the market, could make all the difference to the lives of pastoralists.

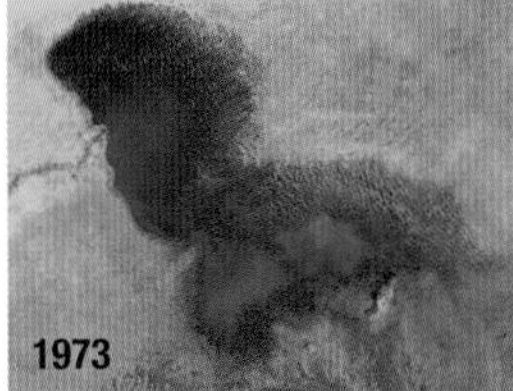

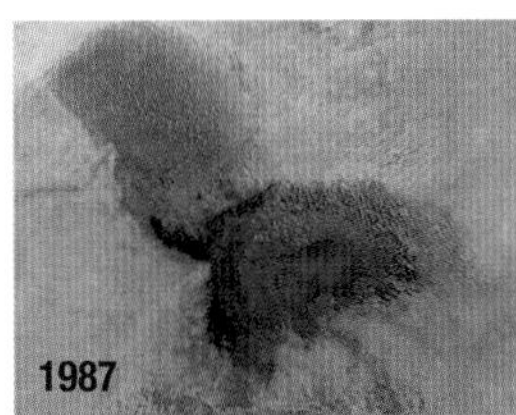

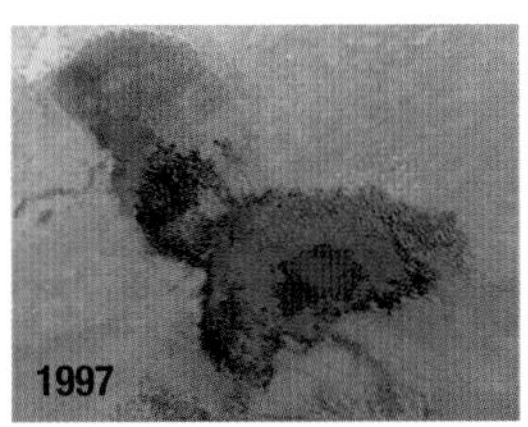

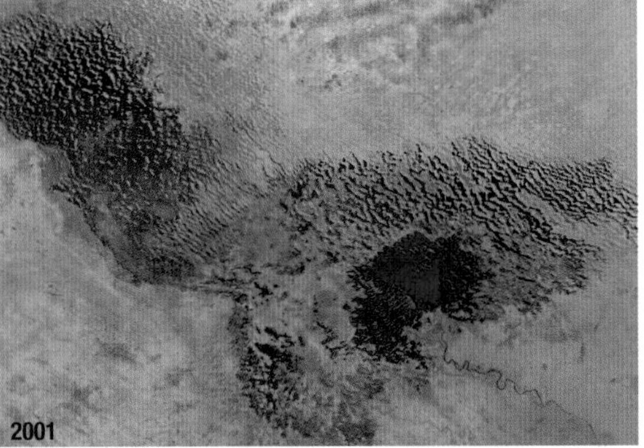

Images from space show the dramatic decline in the extent of Lake Chad, once one of the largest bodies of freshwater on the African continent. Situated between four countries (Chad, Niger, Nigeria and Cameroon), Lake Chad has been the source of water for a number of large-scale irrigation projects, during a time when the region has experienced a considerable decline in rainfall. The rippled areas on the images are sand dunes formed by the wind. Vegetation on the lake bed is shown in red (top images) and green (2001).

A young girl stands in the refugee camp in the Kassala district of Sudan. Regular dust storms make visibility very difficult. Fending off famine in times of drought is a major challenge for aid agencies.

3.12 Sahel – the nature of desertification

The drought in the Sahel during the 1970s led to international action. The apparently inexorable advance of the Sahara Desert stimulated discussion on desertification, and the extent to which it was the product of human activities, as opposed to natural climatological factors. It also resulted in international efforts by governments to curb the advance of the world's deserts.

Rain that falls on the arid and semi-arid regions of the world is characteristically sporadic and difficult to predict. Moreover, except in the most severe cases when rains completely fail for an extended period, the impacts of the sporadic rainfall are highly variable. Consequently, the concepts of aridity, drought and desertification are not easy to define succinctly.

Causes of desertification

Deserts do not advance or retreat along a continuous front line at a steady rate each year. An advance first occurs in patches where specific soil and vegetation types are in a fragile equilibrium that is easily disturbed. Elsewhere, what looks like serious damage may prove to be reversible, as the flora and fauna of arid regions are well adapted to the harsh environment. What seems like permanent change may prove to be nothing more than normal year-to-year variability or even fluctuations on decadal timescales.

Initial speculation about the advance of deserts in the 1970s centred on the Sahel and, because the scale of the changes seemed so great, the possibility of a permanent shift in the intertropical convergence zone (ITCZ) over West Africa. Terrifying figures became accepted environmental statistics. It was estimated that more than 20 million hectares of once-productive soil (an area about the size of Belgium) was being reduced to unproductive desert each year. Many aid agencies were alarmed by estimates of the desert marching southwards at up to 50 km a year. The global concern culminated in the United Nations Conference on Desertification, held in Nairobi in 1977. Later, in the mid-1990s, a Framework Convention to Combat Desertification was negotiated.

In the Sahel, particular emphasis was placed on human activities. The Sahel population was increasing in areas that, in the 1950s and 1960s, had been wet and less drought-prone. For example, in Niger the population rose from 250 000 in 1960 to 3 million in the mid-1990s. As the wetter areas of the rangelands were taken over by farmers, the herders moved towards the desert's edge where rainfall is even more variable. When the region became more arid from the late 1960s to the 1990s, the land-use change, particularly overgrazing, was singled out as a key cause behind the desertification. It was argued that once the vegetation was stripped away the albedo (reflectivity) of the land surface would increase. This, in turn, would produce a positive feedback to reinforce the process of desertification. The basis for the proposal is that denuded soils reflect much more sunlight than savannah vegetation, so the amount of heat absorbed by the desert would be correspondingly less. This would then reduce convection and rain showers when the ITCZ moved northwards into the Sahel during the rainy season from June to September. It was also suggested that the removal of vegetation in the region, both by browsing livestock and by the cutting of brush for firewood, would reduce the amount of biogenic aerosols which act as cloud condensation nuclei and help raindrops form. Overall, the reduction in vegetation would mean that once the desert-like conditions were created the rainfall would remain low until more dominant rain-producing atmospheric processes returned to the region.

Data collected during the 1980s suggest, however, that the role of larger-scale fluctuations in

Deep wells: an inappropriate technology?

During the drought of the early 1970s the cattle population of the Sahelian region fell by about 40 per cent from a figure of around 25 million. The initial reaction was that this decline was from a previous level of overstocking for the region. In practice, the problems of drought may have been compounded by the introduction of deep wells in the 1960s. These permanent sources of water disrupted traditional pastoral migration that followed the seasonal pattern of rainfall and vegetation growth. Permanent water supplies led to herds lingering around wells and putting pressure on available vegetation. Cattle then had to travel farther to find new fodder or fodder had to be transported to the herds. Many cattle died from shortage of fodder, rather than lack of water and, where fodder was transported in, severe soil degradation occurred in the vicinity of wells.

Satellite imagery is proving to be a powerful tool to monitor vegetation growth. The African drought of 1984–85 shows clearly in this image from August 1984. Reddish-brown areas indicate unhealthy vegetation relative to a normal year. Crops from the Sahel (along the southern border of the Sahara desert) to East Africa were withered during the drought that hit Ethiopia, Sudan and Somalia especially hard. These data can be used in conjunction with routine land-based climate information for early warning of potential disaster, to avert desertification and to maximize the benefits of land resources.

the climate had been underestimated. Many of the observed shifts in vegetation along the desert fringes were largely due to decadal fluctuations in rainfall. Furthermore, satellite observations showed that the extent of vegetation during the rainy season was closely linked to changes in rainfall, with extensive re-establishment in wetter years. In the period 1980 to 1995, satellite observations of the western Sahel showed that the desert expanded and contracted up to 300 km in response to changes in rainfall from year to year. The seeds of many arid land plants are able to lie dormant for several years, then when the rain returns they germinate and what looks like desert landscape springs back to life.

As for explaining why the drought in the Sahel has gone on so long, the answer will probably be found in a better understanding of decadal and longer timescale fluctuations of the climate. It is likely to include the role of the deeper oceanic circulations coupling through air–sea interactions with large-scale atmospheric oscillations. However, the positive feedbacks associated with desertification cannot be discarded as a contributing factor.

Preventing the expansion of the deserts

The 1977 United Nations Conference on Desertification launched a Plan of Action, which funded projects amounting to some US$ 6 billion over the subsequent 15 years to prevent desertification. However, questions about the concept of desertification and its causes remain. In particular, there is no adequate distinction between degradation due to human activities (e.g. overgrazing by pastoralists' herds, collection of firewood and inappropriate farming) and the effects of drought. If fluctuations in the global climate are an underlying factor then what matters is developing strategies that can best respond to these variations. Taking action to reduce the impact of human activities should be part of any strategy, but we also must be realistic about what we can do when natural changes pose such a major challenge to any planning. At the very least, how climate variability and longer-term change interact with human activities must be considered as part of any planning activities.

To the extent that changes in the severity of natural hazards are part of the process of global warming, they should be addressed within international agreements designed to reduce this climatic threat. In addition, the desertification experienced over the last three decades has taught us that, when considering the complex web of life in even the harshest environments, we must not overestimate the impact of a single factor.

A nursery in Gour, Niger, growing young trees to be used to help keep the desert at bay.

3.13 Floods around the world

In some regions of the world, flooding is part of the way of life. In areas of Bangladesh and China it is a feature in summer, but elsewhere may be more episodic. Floods often cause monumental damage and loss of life. On the plus side, however, floods can be a vital source of freshwater and are an essential element of some ecosystems. Flooding can also replenish nutrients in the soils, reduce the build-up of agricultural chemicals and flush out pollutants from waterways.

Floods in Bangladesh swamp a large proportion of the countryside, and also inundate the major cities.

Rapidly growing populations in hazard-prone localities, such as on flood plains or unstable hillsides, are at great risk; hence the need for developing emergency preparedness and response measures. In Harin, China soldiers are dropping sandbags to prevent further erosion of the river bank during the floods of August 1998.

Anyone living in the vicinity of a river knows that heavy rain upstream can lead to the water rising over the banks and covering the landscape downstream. Prolonged rainfall, sometimes abetted by rapidly melting snow, can inundate vast areas of land, destroying crops and killing large numbers of people and livestock. In hilly country a flash flood with fast-flowing water, often carrying mud, debris and rocks, can sweep away everything in its path. In the most arid places a single heavy thunderstorm can produce floods that may devastate a locality.

Bangladesh

Nowhere is the chronic threat of flooding more evident than in Bangladesh. Here annual inundation is a fact of life. Some 80 per cent of the country, with no point higher than 30 m above sea level, is in the flood plain of the Brahmaputra, Ganges and Meghna Rivers. In heavy monsoon years such as 1987, 1988 and 1993, virtually all of this land was flooded but even in years with moderately above average rainfall, widespread flooding is experienced.

There is little evidence of a marked trend in rainfall in Bangladesh over the 20th century. There has been, however, a marked increase in flooding, which appears to be a result of upstream deforestation and hence more rapid runoff in the foothills of the Himalayas. Thus the rate of clearance of the forests of Nepal, Bhutan and Assam, India for cultivation, timber and firewood in recent decades may be a major factor in the increased incidence of flooding and the more rapid runoff that has also greatly increased topsoil erosion. The absence of reliable long-term statistics makes it difficult to identify which particular aspect of the land- and water-use changes upstream is most adverse and how best to address it.

Following the floods of 1987 and 1988 the Government of Bangladesh considered a plan to build a huge set of embankments to tame the Ganges and Brahmaputra Rivers to protect cities and increase crop production. Costing US$ 10 to 15 billion, this scheme was opposed by local political groups, western environmentalists and hydrologists. Opponents argued that not only would the scheme make matters worse for the vulnerable communities in the delta regions, but it would isolate the rivers from their flood plains and hence greatly increase the damage if the embankments were ever breached. The scheme would also cut off river flow to the extensive wetlands, damaging fisheries that are the principal source of animal protein in Bangladesh.

The associated issue of soil erosion is an interesting example of the complications in river management. Leaving aside the damage caused by erosion to the local environment, it also has a number of impacts downriver. First, in silting up waterways, the sediment plays a significant part in exacerbating floods. In the Ganges delta the enormous quantities of material swept downriver (1.5 to 2.5 billion tonnes per year) creates new land. The net balance is a complicated mix between deposition, settlement and coastal erosion. It is not possible, therefore, to accurately estimate either future changes in land area or how they will be affected by sea-level rises.

China

Flooding is one of the most severe forms of disaster that China experiences. Some written records about floods in China date back 4000 years. Floods mostly occur in the middle and lower reaches of the major rivers that are both the cradles of its culture and a source of its sorrows. An ancient Chinese saying is that "to manage the country, we must first control the waters". During the 20th century, China suffered major flood disasters in 1931, 1935, 1954, 1991 and 1998. In the Yangtze River Valley there has been a rising trend in rainfall in recent decades. Comparing the scale of flooding between events is not easy because of the size of China; the extent of flooding can vary from region to region depending on the precipitation amount, and intensity, spatial coverage and duration of the storm.

The floods in 1998 in the Yangtze River Valley were the worst since 1954, when 30 000 people died and the rail network between Beijing and Guangzhou was paralysed for 100 days. In 1998 more than 4000 people died and over a quarter of China's 1.2 billion people were affected by the flood waters. Newspapers around the world carried pictures of the heroic efforts of soldiers and civilians mobilized to mend the dykes with sandbags to keep the rising waters at bay.

Extensive deforestation in China has been a critical factor in making the impact of heavy rainfall more devastating downstream.

Many of the problems in China can also be attributed to deforestation and to the consequent silting up of the rivers. The government has now developed programmes of reforestation in the mountains and is carefully controlling logging. Downstream, a combination of massive dredging and building of dykes is essential to prevent further silting up of the rivers.

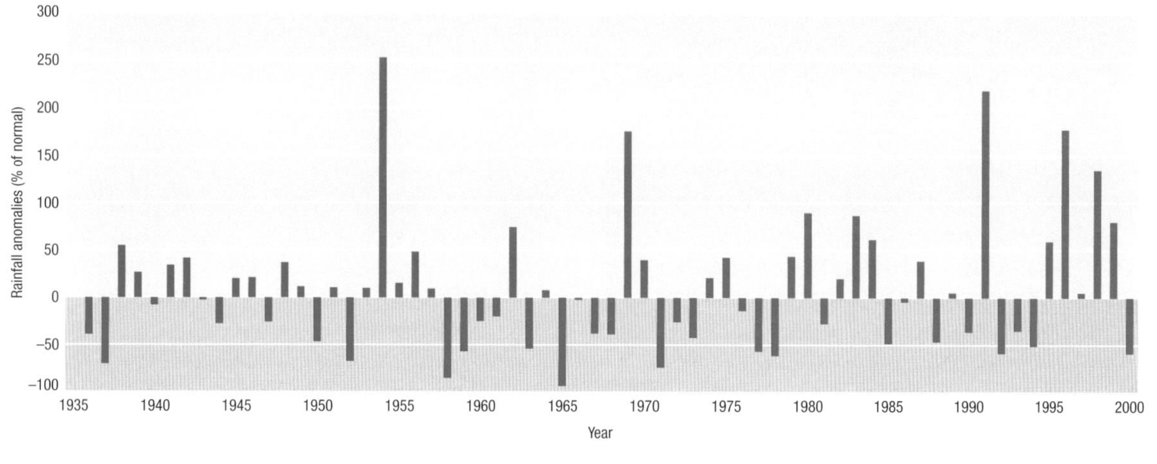

The fluctuations in the monsoon rains (mei-yu) along the Yangtze River show exceptionally wet years: 1954, 1991 and 1998. In some years, the heavy precipitation combined with other factors led to major floods.

Flash floods

Flash flooding is the largest threat to life from severe thunderstorms. Heavy rainfall in small river basins can cause rivers to rise rapidly and result in great destruction. In 1935, following five days of continuous storm rains, a flash flood on the lower Yangtze River in China killed 142 000 people.

In the USA, the three biggest thunderstorm-related death tolls, apart from plane crashes, in the last 40 years have all been associated with flash floods. In 1996, over 80 people were killed when a flash flood destroyed a campground in the Pyrenées Mountains of Spain. The only way to reduce these tragic losses is to provide better warnings of the dangers. However, it will always be difficult to cover isolated regions. Where single thunderstorms in sparsely populated desert areas cause sudden floods, as occurs from time to time in Yemen and south-western Saudi Arabia, the prospect of providing adequate warning is slim. Moreover, the floods can rise with terrifying rapidity as shown in these pictures, taken 15 minutes apart, of the Zervraggia River, Switzerland in 1987.

As more and more people choose either to live or take recreation in the mountains, the dangers from flash flooding will increase. As with other measures of trends in losses due to severe storms, the figures for flash flooding are difficult to interpret. Many events occur in isolated spots, and if no property is damaged or lives lost, no record is made.

3.14 European floods

In Europe flooding can occur at almost any time of the year. During the 1990s there was an unusual number of particularly serious floods that led to much comment about climate change. Whether these extremes were the result of climate change or changing flood-control systems remains a current debate.

Flooding in Europe can result from a variety of weather conditions. Britain experienced its worst floods of the 20th century in March 1947, a combination of the melting of deep snow that had accumulated during an exceptionally cold winter and the wettest March of the previous 300 years. Huge areas of low-lying land were inundated when crops should have been planted. The consequences for agriculture were disastrous.

Other disasters have struck more rapidly. Devastating floods hit Florence, Italy, in November 1966 when a third of the average annual rainfall fell in 24 hours. The Arno River rose catastrophically and parts of the city were flooded to a depth of 5 m causing losses estimated at US$ 1.3 billion in current terms. Flooding of the Uffizi Gallery destroyed countless art treasures, in spite of the courageous rescue efforts of the local populace.

The Rhine River floods of 1993 and 1995

The huge floods along the Rhine in the 1990s raised issues of climate change and flood-control strategies. In both December 1993 and January 1995, much of the region drained by the Rhine and its tributaries experienced exceptionally heavy rain. During a wet spell in 1993, parts of the region had three times the average December rainfall. German towns, including Cologne and Koblenz, suffered extreme flooding with water levels coming within 6 cm of the level reached in the great flood of 1926, the highest in the last two centuries. Belgium, eastern France and southeast Netherlands were equally badly hit. The costs in Germany alone were estimated at US$ 580 million.

Just 13 months later a similar situation occurred. The Belgian Ardennes and parts of northern France had their wettest winter of the 20th century. In Cologne the flood level equalled the record of 1926. Downstream the peak flows were less extreme, but were sustained for longer. Conditions in the Netherlands were very serious. For a while it was uncertain as to whether the floods would undermine the dykes holding back the North Sea and lead to a much greater disaster. For safety, 200 000 people were evacuated from their homes. Fortunately the dykes held, but two such severe floods so close together raised many questions about whether this was part of the warming trend in the climate. However, they also demonstrated how people adapted to the threat. Although the flood levels were higher in Germany in January 1995 than December 1993, the economic costs were halved because people heeded the warnings issued at the time and protected their property more effectively.

The interpretation of these events has to be made in terms of the combination of meteorological trends and greater exploitation of major rivers. In northwest Europe there has been a clear upward trend in rainfall in the winter half of the year, while there has been an almost equal and opposite downward trend in summer rainfall. Other social developments have amplified the economic impact of heavy rainfall. While it took a

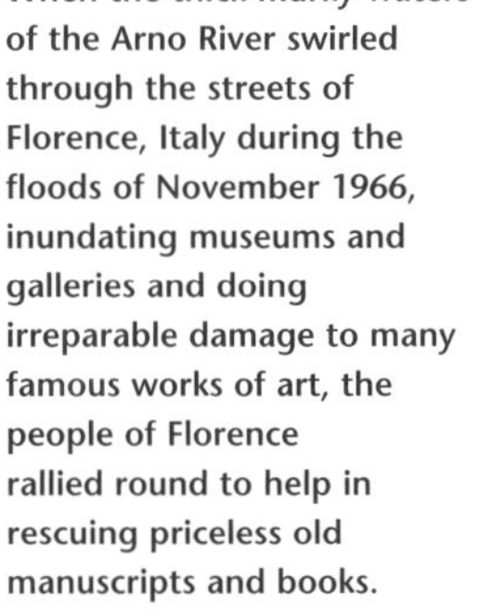

When the thick murky waters of the Arno River swirled through the streets of Florence, Italy during the floods of November 1966, inundating museums and galleries and doing irreparable damage to many famous works of art, the people of Florence rallied round to help in rescuing priceless old manuscripts and books.

A bitter winter followed by an exceptionally wet March led to severe flooding in parts of the United Kingdom in 1947.

lot of rain to produce both floods, the changes made in the drainage of the Rhine watershed and of river channels to improve the handling of barge traffic played as big a part in the disasters. The speeding up of the runoff has approximately halved the time that heavy rainfall now takes to flow through the system. These figures are supported by measurements of the flow rates at Cologne since 1817. These show that the peak flow in 1993 exceeded that of 1926, but the 1995 figure did not. The highest flow rates were actually recorded in 1824 and 1876.

The European floods of July 1997

Some of the worst European floods in the 20th century hit the basin of the Oder River in southwestern Poland and parts of neighbouring countries in July 1997. Two separate bouts of rain were the cause. Between 4 and 9 July torrential rain hit the region, with places in the Czech mountains receiving over 500 mm of rain. Up to 150 mm of this fell within two hours during thunderstorms on 9 July.

The flood from this first drenching had not cleared the lower reaches of the Oder River before more heavy rain fell between 18 and 21 July. One estimate put this flood at a one-in-a-thousand-year event. The loss of life exceeded 100, and the economic cost ran to some US$ 5 billion. In the Czech Republic and Poland some 3000 towns and villages were flooded, and 3000 km of railway and 5000 km of roads, with 100 bridges were damaged. About 190 000 people had to be evacuated and more than 50 000 buildings were affected.

The unprecedented rainfall in the Czech Republic and Poland in July 1997 produced torrential floods that caused immense damage, as shown in Nysa in southwest Poland.

During the floods of December 1993, the Rhine River rose to record levels in the city of Cologne, Germany.

More flooding in a warmer world?

Global warming has the effect of increasing evaporation of moisture from oceans, soils and plants. This additional moisture-holding capacity in the atmosphere provides an enhanced reservoir of water that can be tapped by all weather systems, be they tropical storms, thundershowers, snowstorms or frontal systems; i.e. an enhanced water cycle.

During the 20th century, it is estimated that rainfall increased up to 1 per cent per decade over parts of the middle to high latitudes of the Northern Hemisphere continents and up to 0.3 per cent per decade over tropical land areas (10°N to 10°S). By contrast, rainfall has likely decreased by up to 0.3 per cent per decade over the Northern Hemisphere subtropical (10°N to 30°N) land areas. No comparable systematic changes have been detected over the Southern Hemisphere, and data are insufficient to determine trends in precipitation over the oceans. Over the latter half of the 20th century in middle to high latitudes of the Northern Hemisphere, there was an increase of 2 to 4 per cent in the frequency of heavy precipitation events.

What will the future bring? Climate models project continuation of this enhanced water cycle in a warmer world, which will mean, on the one hand, that increased summer drying and the associated risk of drought are likely over mid-latitude continental interiors. On the other hand, more frequent intense precipitation events are likely over many areas, particularly over the Northern Hemisphere middle to high-latitude land areas. Where this occurs, it could well lead to more flooding.

3.15 Mississippi floods

The Mississippi River drains about 40 per cent of the land mass of the continental USA. Ever since the great flood of 1844 people living along the river have grappled with the challenges of flood control. How they have done this and the compromises they are now making provide an object lesson in coming to terms with climate variability along a mighty river.

In 1879 the Mississippi River Commission was formed to coordinate flood defences along the river. It has been fighting a losing battle ever since then. As Mark Twain observed in his 1882 autobiographical narrative, *Life on the Mississippi*, "one... cannot tame that lawless stream, cannot curb it or confine it, cannot say to it, Go Here, or Go There, and make it obey; cannot save a shore which is sentenced; cannot bar its path with an obstruction it will not tear down, dance over and laugh at". Over the next 100 years the earth levees built alongside the river banks have typically grown from modest affairs 2 to 3 m high and 15 to 20 m wide, to massive structures, often to 10 m in height and up to 100 m wide.

One flood after another

In spite of these massive construction projects, areas beside the river have frequently been flooded. The worst floods were in 1927, when 214 people were killed. Over the following half-century, the US Army Corps of Engineers built 300 dams and reservoirs to prevent such catastrophes in the future. While, on the whole, they succeeded in protecting the main towns, there were, nevertheless, major floods in places along the river system in 1937, 1945, 1950, 1973 and 1983. A consequence of channelling the flood waters, rather than letting them spread out over the floodplain, is that the natural ability of the floodplains to absorb a flood is denied. As tributaries feed the flow it increases in height as it

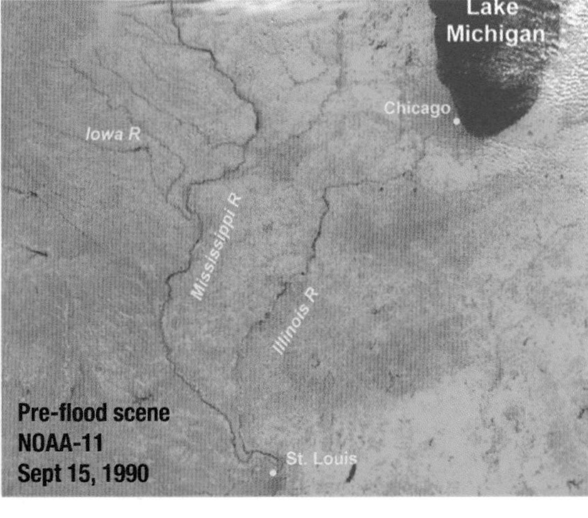

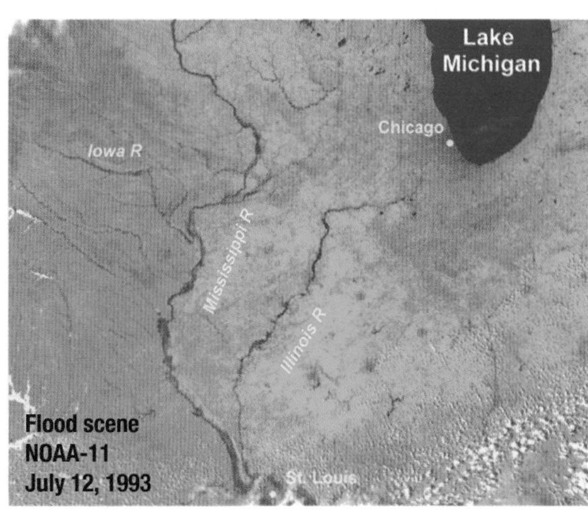

Two images from space show the swollen rivers and flooding along the Mississippi in the midwest floods of 1993 (right), quite a contrast from the dry year in 1990 (left).

Optimum flood control

The long-term management of floods is undergoing subtle changes. As the economic impact of flooding has grown in many countries, serious questions are being asked about the strategy of flood control. In particular, the assumption that the best response to successive meteorological extremes is to build bigger and better defences is being called into question. The alternative is to accept the inevitability of occasional floods. Controlled inundation of the flood plain could lead to better watershed management, including slower runoff and reduced downstream peak flows and flood levels. These issues are still the subject of intense debate and involve difficult political decisions about individual freedom to live in vulnerable situations, the obligations of the state to provide people with adequate protection, and the ability of authorities to forcibly remove them from danger.

In the case of the 1993 Mississippi floods, the scale of federal disaster aid led to prompt action. The US government decided to finance a scheme to 'retire' the most vulnerable riverside properties. Local towns were funded to purchase and demolish the most frequently flooded properties and turn the areas into parks or recreational land. In the case of Valmeyer, Illinois, it was decided to abandon the original site of the town and rebuild on higher ground away from the river. In the state of Missouri alone US$ 100 million was used to purchase 2000 residential properties. It was reckoned that this measure alone would save US$ 200 million over 20 years even with no exceptional flooding. At the same time, it became a requirement that people with federally backed mortgages were adequately insured. Furthermore, the ability to take out insurance to provide cover within five days, when flooding was imminent, was blocked. Positive benefits from these policies were realized when near-record floods struck some of these areas only two years later, in 1995.

moves downstream. As the levee heights are raised the risk to the surrounding land grows. If the water tops a levee, or breaches it, the subsequent damage can be immense.

Managing the river's flow is also more complex, because sometimes it can run low during drought and impede shipping, or even dry up as in the summer of 1987. Managing the river level has to include plans for low as well as high flow levels.

The floods of 1993

The challenge of controlling the Mississippi came to a head with the floods of 1993. These were caused by sustained abnormal precipitation. A wet autumn combined with a snowy winter left Iowa with its greatest snow pack since the spring of 1979. Then the rain started in April. For the Mississippi watershed the 1993 April to July precipitation was the greatest since records began in 1895. Many places in Iowa and Kansas had more rain in those four months than in a normal year, while from North Dakota to Illinois the totals for June and July broke all-time records, often by wide margins. The chance of these extremes recurring was estimated to be less than once in every 200 years.

Inevitably, the flood levels broke all records. Severe flooding began in May on the Redwood River in Minnesota and in June on the Black River in Wisconsin. Record levels followed on the Kaman, Mississippi and Missouri Rivers in July. At St Louis the water crested 1.9 m above the previous record set in 1973 and exceeded the earlier record figure for over three weeks. Just north of the city the flood waters reached a width of 32 km where the Missouri joins the Mississippi. Over 7 million hectares were flooded across nine states, and at least an equal area was saturated, which further added to crop losses. At least 50 000 homes were damaged or destroyed and 85 000 residents had to be evacuated. In Des Moines, Iowa, normal water supplies were unsafe for drinking for 12 days. Some flooding was caused by levees collapsing under the sustained pressure of water, whereas in other places it simply flowed over the top. Overall about 60 per cent of the 1400 levees on the Mississippi and Missouri Rivers were overrun or breached by the water.

The immediate economic consequences were estimated at about US$ 15 billion and the death toll was 48. Crop losses exceeded US$ 5 billion. The national soybean yield was 17 per cent below the record level of 1992 while the corn (maize) yield dropped by 33 per cent. Many farm animals perished. The damage to housing, property and business made up the remainder of the estimated losses.

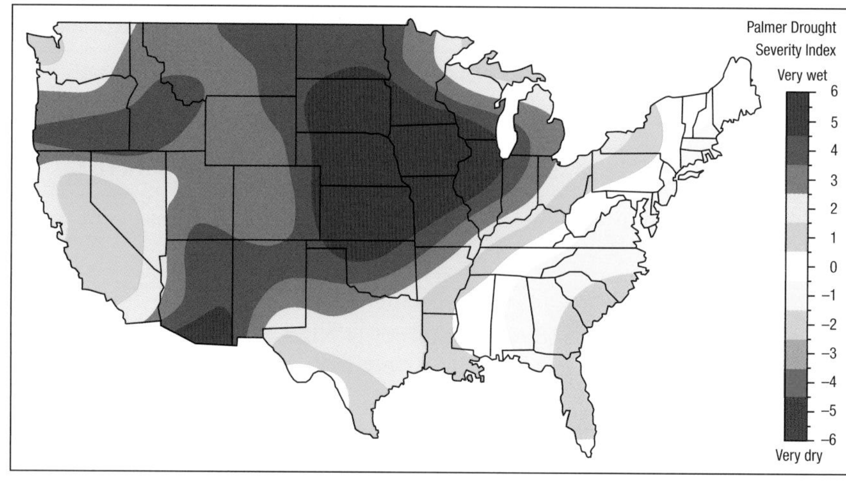

A Palmer Drought Severity Index map showing the extreme wetness over much of the USA during the summer of 1993.

A riverboat sails down the flooded Mississippi River in Illinois in July 1993.

3.16 ENSO – El Niño

At times the surface temperatures of the central to eastern equatorial Pacific Ocean are warmer than normal. These warm episodes, usually referred to as El Niño events, are the best-known feature of the El Niño/Southern Oscillation (ENSO). Their occurrences have global consequences.

An El Niño event produces a consistent pattern of above- and below-normal temperatures and rainfall around the world that changes with the seasons. In the early stages of development (June–August) the effects are strongest in the tropics and on the Southern Hemisphere winter weather systems. Near the peak of development (December–February) impacts remain within the tropics but at higher latitudes, shift to affect Northern Hemisphere winter weather systems.

Although not mentioned by name until the early 1890s, features of El Niño were common knowledge in Peru for centuries. The name arises because, near the end of each calendar year, ocean surface temperatures warm along the coasts of Ecuador and northern Peru. This normal seasonal warming around Christmas time became known as 'El Niño' (Spanish for 'the little boy' – alluding to the Christ Child). Every few years a much stronger seasonal warming appears which extends across much of the equatorial Pacific. Over time the term 'El Niño' has been used for these very warm episodes. The importance of the phenomenon was noted more generally by scientists following the 1972–73 event, but it was not until the record-breaking event in 1982–83 that El Niño started to become part of global popular culture (see also 2.15).

Warm episode relationships June–August

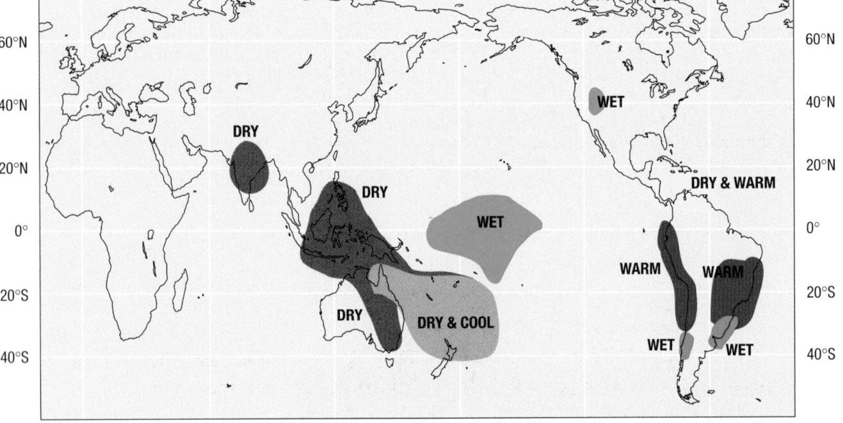

Warm episode relationships December–February

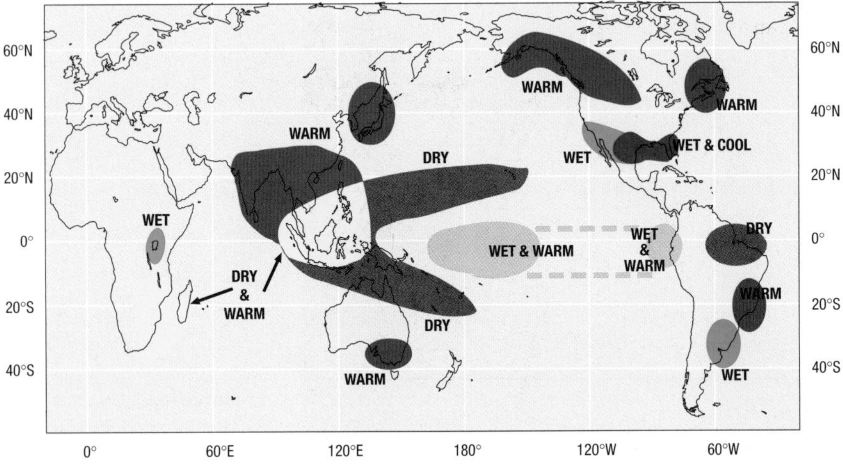

Principal consequences

During El Niño episodes, the normal patterns of tropical precipitation and atmospheric pressure are disrupted. Tropical convection and rainfall extend eastward across the Pacific Ocean, reaching the Americas toward the end of the year. As the El Niño event develops from June through September rainfall over parts of southeastern Australia is often well below normal. Drier than normal conditions may also be observed over southeastern Africa during the austral winter. The summer monsoon rains of India and China tend to be more erratic and often do not penetrate into northwest India and northern China.

Over South America, during El Niño events, unusual storm activity brings above-normal rainfall to the central coast of Chile, particularly in the austral winter and spring periods. Later, during the austral summer, frequent and often heavy rains deluge the subtropics east of the Andes (particularly in the catchment areas of the Paraná and Paraguay River systems). During the austral summer the usually dry coastal regions of southern Ecuador and northern Peru often experience torrential rainfall. In contrast, much of the Amazon Basin and the northeastern region of Brazil become drought-stricken, and rainfall is reduced over Indonesia, Malaysia, the Philippines and northern Australia. Traditional slash and burn practices in northern Amazonia and Indonesia have been the catalyst for large forest fires and serious health problems from smoke and haze.

Another consequence of El Niño events is their influence on tropical cyclone development around the world. The decline of hurricane activity in the tropical Atlantic and the Caribbean is most obvious. Over the western Pacific Ocean, on both sides of the equator, the number of tropical storms does not vary appreciably with ENSO but there is a tendency for them to form and recurve to higher latitudes further to the east than usual during El Niño events. As a result some regions, such as the Philippines and islands of the South-West Pacific, that rely on passing tropical storms for summer rainfall suffer deficiencies during El Niño events. Also, tropical cyclone activity usually rises in the eastern Pacific and in the South-West Pacific east of the dateline, while around Australia activity declines.

The increased heating of the tropical atmosphere over the central and eastern Pacific also affects atmospheric circulation patterns at higher latitudes. The jet streams shift their location

> **The Pineapple Express**
>
> One consequence of an El Niño event is that some mid-latitude depressions, instead of approaching the US Pacific coast from the northwest, will track eastwards from the vicinity of the Hawaiian Islands. This produces the greatest precipitation anomaly in southern California that can also extend up into British Columbia, Canada. Passing over warmer water, these storms pick up moisture and produce heavier rainfall. The warm episodes of 1982–83 and 1997–98 each produced a stream of these storms that were dubbed the 'Pineapple Express' because of their origins near Hawaii.

and mid-latitude depressions are steered along different courses and tend to be more vigorous than normal in the eastern North Pacific. Abnormally warm, moist air is pumped into western Canada, Alaska and the extreme northern USA. Thunderstorms also tend to be more frequent in the northern Gulf of Mexico and along the southeastern coast of the USA resulting in wetter than normal conditions there.

Impacts of the 1997–98 event

The El Niño which started in March 1997 was an equal to the major 1982–83 event. The pattern of extreme weather events (e.g. floods, droughts, fires and tropical cyclones) was generally characteristic but shows how each event can have somewhat different impacts. In India the 1997 monsoon produced near-average rainfall. In contrast, much of eastern Africa had much heavier rainfall than was expected and over 15 000 East Africans died as a direct result of floods and disease. Over Southern Africa rainfall was higher than anticipated.

In Australia, rainfall was low, but not as low as expected, during the 1997–98 winter and spring. This was in marked contrast to disastrous drought that coincided with the warm episode in 1982–83 which led to the worst outbreak of bushfires in Australian history (in February 1983, the 'Ash Wednesday' fires killed 76 people in Victoria and South Australia). It is a measure of the growing awareness of these conditions and the value of public education that, during the 1997–98 event, firefighting authorities took additional preparedness measures and, despite many small outbreaks during the hot summer, there were no serious conflagrations.

During 1997, tropical cyclone behaviour was broadly in line with expectations for an El Niño event. In the tropical Atlantic it was a notably quiet season, with the August to October period being the quietest on record. In the Pacific west of 125°E it was very quiet with southern China having the lowest number of storms in nearly 50 years, and the Philippines had the first September in 30 years with no typhoon entering the region. In contrast, farther east it was much busier than normal and the eastern Pacific was relatively active. Fiji was affected by three 'off-season' tropical cyclones – an exceedingly rare occurrence.

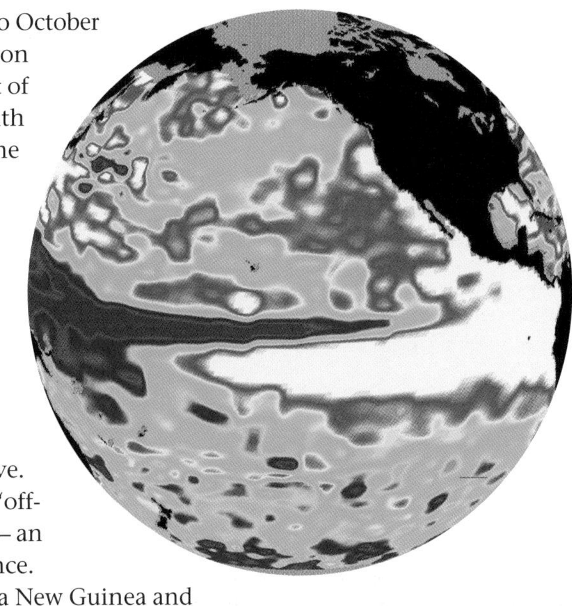

An image from the TOPEX/Poseidon satellite which measured the increased sea level of the huge tongue of warm water extending from South America across the Pacific during the 1997–98 El Niño event.

Over Indonesia, Papua New Guinea and northern Australia the wet season came to an abrupt and early end in March 1997. By August fires were burning out of control in Indonesia, with Borneo experiencing particularly heavy damage, and the associated haze spreading over a huge part of the region. The smoke and haze disrupted air transport and drastically reduced the tourist trade. The effects on human health were widespread, with dramatic increases in hospital admissions for respiratory infections. In Indonesia, the worst drought in 50 years destroyed the bulk of the country's palm oil crop, and rice yields plummeted, prompting imports totalling 5.1 million tonnes. The price of rice rose fourfold, precipitating an economic crisis and the rupiah was devalued by almost 80 per cent.

Associated with the intense anticyclonic conditions which developed over Australia during 1997, strong subsidence, clear skies and dry air extended virtually to the equator and brought devastating frosts to the New Guinea highlands in July. Combined with the worst drought of the century this exceptional cold caused enormous damage to crops and led to 1.2 million people suffering severe food shortage that required a major international aid effort. In addition, hydroelectricity supplies were greatly reduced, and mining operations halted by lack of water, leading to a major loss of employment and export earnings.

Drought in Indonesia during the 1997–98 El Niño event was a catalyst for huge forest fires that produced dangerous levels of smoke and air pollution over cities in the region.

3.17 ENSO – La Niña

At times the sea surface temperatures of the central and eastern equatorial Pacific Ocean are cooler than normal. These episodes are now referred to as La Niña events. Although less well known than El Niño events, they are an essential part of the El Niño/Southern Oscillation (ENSO), and also have global consequences.

Cold episode relationships June–August

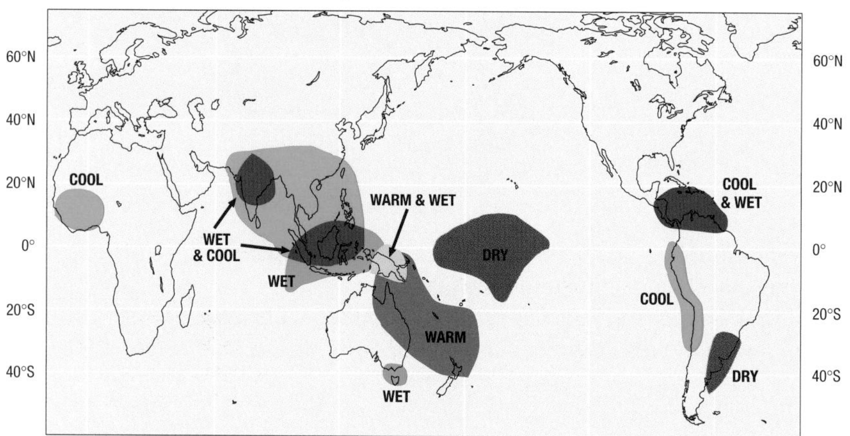

Cold episode relationships December–February

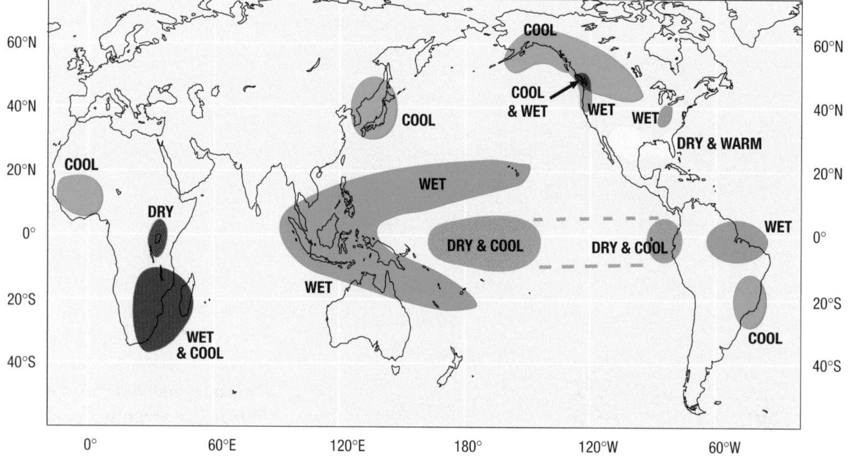

A La Niña event also produces a consistent pattern of above- and below-normal temperatures and rainfall around the world that changes with the seasons. In the early stages of development (June–August) the effects are strongest in the tropics and on the Southern Hemisphere winter weather systems. Near the peak of development (December–February) the impacts remain within the tropics but at higher latitudes, shift to affect Northern Hemisphere winter weather systems.

As with the El Niño events, the La Niña events disrupt the normal patterns of tropical convection and occur as the opposite phase of the Southern Oscillation circulation. The core feature of these events is that cooler than normal ocean temperatures develop in the central and eastern equatorial Pacific Ocean. Atmospheric pressure over the cooler water rises but to the west over Indonesia and northern Australia it falls below normal. Overall, the pressure difference across the Pacific increases and strengthens the easterly Trade winds. The pressure difference between Tahiti and Darwin is a measure of the Southern Oscillation (the Southern Oscillation Index (SOI)).

ENSO and global warming

There has been a close link identified between changes in global surface air temperature and the variations in sea surface temperature in the equatorial Pacific. El Niño conditions are generally associated with warmer than average surface air temperatures, with the peak in global temperatures following a few months after the episode has peaked. La Niña events are followed by colder than average surface air temperatures. This is consistent with what happened in the late 1990s. The record-breaking warm event in 1997 contributed to 1998 having the highest observed global average surface air temperature in the instrumental record, even though La Niña developed during the later part of the year. However, 1999 and 2000 global air temperatures were cooler, although remaining well above the 1961–90 average, and still clearly among the 10 warmest years on record since 1860. However, any influence that global warming of the atmosphere might have on El Niño or La Niña frequency, intensity and duration is as yet unclear.

Principal consequences

When the Pacific shifts into a La Niña mode, the effect is to suppress cloudiness and rainfall over the central and eastern equatorial Pacific, especially in the Northern Hemisphere winter and spring seasons. At the same time, rainfall is enhanced over Indonesia, Malaysia and northern Australia during the southern summer, and over the Philippines during the northern summer. Wetter than average conditions are also observed over southeastern Africa and northern Brazil during the southern summer season, while southern Brazil and central Argentina tend to be dryer than normal in their winter season. During the northern summer, the Caribbean and northern South America are usually cooler and wetter than normal.

La Niña events have a pronounced influence during the austral winter and spring over most of Australia, producing well-above-average rainfall, especially in the east of the continent. In the northern summer, the monsoon over India tends to be stronger than normal, especially in the northwest.

In many locations, the impact of La Niña events on tropical storm activity around the world tends to be the reverse of El Niño events, with the number of storms increasing in the tropical Atlantic and the region of genesis in the Northwest Pacific shifting westwards. Farther to the north, mid-latitude low-pressure systems tend to be weaker than normal in the region of the Gulf of Alaska. This favours colder than normal air over Alaska and western Canada, which often penetrates into the northwestern USA. The southeastern USA, by contrast, becomes warmer and drier than normal.

A consistent pattern of extremes?

The lengthy La Niña event in 1998, 1999 and 2000 has provided some interesting insights into the global implications of these cold episodes. However, comparisons with earlier events have to recognize that the latest La Niña followed a somewhat different pattern than normal. The El Niño decayed rapidly in 1998. During May and June sea surface temperatures in the central Pacific cooled rapidly to around 1°C below normal, but contrary to the standard model, the waters off the coasts of Ecuador and Peru remained warmer than normal until the end of the year.

During 1998 and 1999 the hurricane seasons were, as expected, very active in the tropical Atlantic. In the western Pacific, following an extremely busy season in 1997, both 1998 and 1999 were much quieter. However, as the storms originated farther west, as expected, they had relatively greater impact on the Philippines and eastern Asia. La Niña also had a strong influence on tropical storms in northeastern Australia, with a pattern of increased winter and spring rainfall occurring over much of Australia between 1998 and 2000. The expected colder than normal conditions over Alaska, western Canada and down into the northern Great Plains and the western USA did not occur. Most of Canada was exceptionally warm during the winters of 1998–99 and 1999–2000.

An image from the TOPEX/Poseidon satellite which measured the depression of sea level by the huge body of cool water covering much of the equatorial central Pacific Ocean during the 1998–99 La Niña event.

The footprint on Australia of winter and spring rainfall anomalies for the 12 major La Niña events between 1910 and 1988 is a pattern of wetter than normal conditions over most of the country.

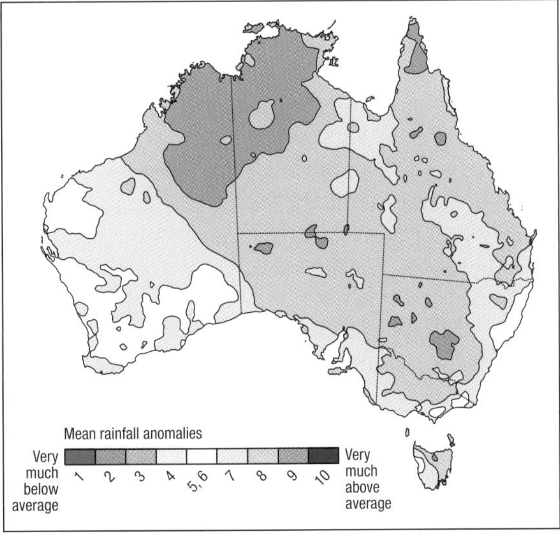

Severe flooding in early 2000 in Mozambique forced people to take refuge on rooftops. A succession of tropical cyclones, *Leon-Eline*, *Gloria* and *Hudah* devastated Madagascar and other parts of Southern Africa as well.

Other oceans

The fact that different El Niño and La Niña events are associated with somewhat different combinations of weather extremes around the world is a reflection of the many factors that contribute to such extremes. Observations over recent years have identified long-term fluctuations in the behaviour of the Atlantic and Indian Oceans that can also exert a major influence on weather patterns in Africa, India and South America. In some instances these effects can cancel out the influence of ENSO fluctuations. In particular, a pattern of fluctuations in the Indian Ocean seems to play an important part in the variety of responses to ENSO phases that occur in Southern Africa and the monsoon in India. For instance, in 1961 a warm anomaly developed in the western Indian Ocean while in the east the temperature was well below normal, but the ENSO was effectively neutral. During this period, East Africa had torrential rain and India experienced the highest monsoon rainfall in the past 150 years.

In 1997, in association with the El Niño, temperatures in the western Indian Ocean were well above average while in the east they fell more than 2°C below normal. This triggered the wettest year on record in East Africa, while Southern Africa had considerably more rainfall than was expected given the strength of the El Niño event in the Pacific.

3.18 ENSO – impact on agriculture

It was the impact of the 1982–83 El Niño event on food production around the world that first brought the global implications of the El Niño/Southern Oscillation (ENSO) to wider public notice. Once this connection was recognized, many earlier examples of this global phenomenon were identified.

Although the consequences of the influence of ENSO on global weather patterns and agricultural production only began to be truly appreciated in the 1980s, there were many earlier straws in the wind. The work of Sir Gilbert Walker and other meteorologists in the 1920s and 1930s had paved the way for future progress. Seminal papers by Jacob Bjerknes in the 1960s identified many of the important links in the oceanic–atmospheric patterns. Then the widespread problems of fisheries decline and grain production in 1972, including a poor monsoon in India, were linked with an El Niño event of that year. Thereafter, the hunt was on, but it was only with the important event of 1982–83 that the global nature of ENSO was realized.

An integral part of Australian history

Fluctuations in climate, linked partly to ENSO, were a vital contributor to the long prosperity of the 40 years up to 1890 in Australia and of the leaner decades that followed. Optimism, born of wealth from primary production during the run of good years, had even nurtured the idea that Australia, in resources, population and economic power, could become a second USA. The unexpected droughts in the late 1890s and very early years of the 20th century eroded that idea, marking a step in the nation's slow discovery of the real fickleness of climate and its effect on the agricultural resources of the nation.

The 1895–1903 droughts, amongst the worst on record for Australia, are known as the Federation Droughts. The number of sheep almost halved, falling from 106 million in 1891 to 54 million in 1902, with a significant impact on wool revenue. Cattle numbers declined by 40 per cent from 12 to 7 million, so that most working families ceased to eat beef.

Flood waters near Lima, Peru swamped this farmer's fields, trapping his cattle during the El Niño event of 1998. Damages in Peru, not counting the human toll, were nearly US$ 200 million.

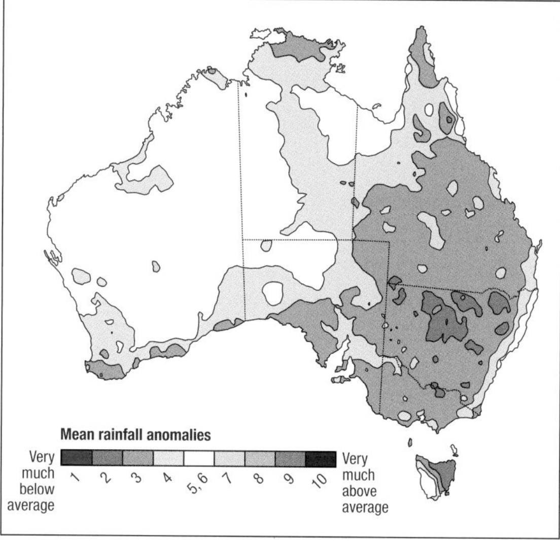

The footprint on Australia of winter and spring rainfall anomalies for the 12 major El Niño events between 1905 and 1997 is a pattern of drier than normal conditions over eastern and southern parts of the country.

Similar but not so severe stories have emerged from more recent El Niño events. During 1972–73 much of eastern Australia was afflicted by drought. The 1982–83 event, with drought reckoned to be the most intense in terms of area affected, resulted in total losses estimated in excess of US$ 3 billion. Then, during the prolonged event of 1991–95, the average production by rural industries fell about 10 per cent, resulting in about US$ 5 billion cost to the Australian economy and US$ 590 million of drought relief being provided by the Australian government between September 1992 and December 1995.

The African connection?

Research into the influence of El Niño events around the world identified Southern Africa as a region of significant impact. Several of the major events appeared to correlate with droughts in the region, especially in South Africa, Mozambique and Zimbabwe, and in the mid-1990s researchers documented the connection. This region was seen as one where forecasts of El Niño events could enable decision makers to encourage farmers, water resource managers and others to prepare for the possible onset of a drought.

During the 1997–98 El Niño event, however, the rainfall in much of Southern Africa was near average, as was food production. As a consequence, the reaction to the forecasts, particularly decisions not to plant crops in the expectation of drought, had a significant adverse impact on grain production. Missing from the forecasts was an adequate appreciation of factors other than eastern Pacific Ocean sea surface temperatures on seasonal rainfall. More recent studies point to the role played by the nearby abnormally warm western Indian Ocean in the overall outcome for 1997–98, and this experience has provided a salutary lesson on the complexities of seasonal forecasting.

Preventive action

The effects of El Niño events, when taken with the phase of the tropical Atlantic Ocean north to south sea surface temperature dipole, are well predicted across South America, particularly over northeast Brazil and along the Pacific coast. When an event develops, the case for taking early action is strongest in these areas. In Peru, in 1997, storage capacity along several affected rivers was increased to provide additional scope to capture flood waters, thereby reducing damage and saving lives.

Governmental and intergovernmental organizations in Peru have also tried to influence farmers' decisions on planting either cotton or rice in northern coastal areas according to rainfall predictions. This is because as rice requires plentiful rainfall it does best in El Niño years, whereas cotton is more drought tolerant. As a consequence of this strategy, during the 1986–87 El Niño event the overall output of the agricultural sector remained virtually steady as compared with the dramatic fall during 1982–83.

Similar success was achieved in northeast Brazil in 1992, which was also an El Niño year. Rainfall in the region was approximately 70 per cent of normal, about the same as in 1987, when no mitigating action was taken and grain production was less than 20 per cent of normal. In 1992, El Niño was predicted and drought-resistant crops were planted, with the result that grain production achieved a much higher figure of nearly 80 per cent of normal.

In parts of Australia the same difficult decisions are being addressed in terms of whether to plant wheat or chickpeas. When dry conditions threaten, chickpeas represent a more appropriate crop than wheat. These decisions hinge on the phase of the Southern Oscillation and prevailing soil moisture levels. Research, involving the integration of climate variability with crop modelling, is looking for practical solutions to these choices in order to assist farmers develop sustainable and profitable cropping methods.

The cumulative precipitation in Papua New Guinea between March 1997 and February 1998 was less than 30 per cent of average. This shortfall, together with crippling frosts in the highlands, destroyed much of the food supplies and led to 1.2 million suffering food shortage.

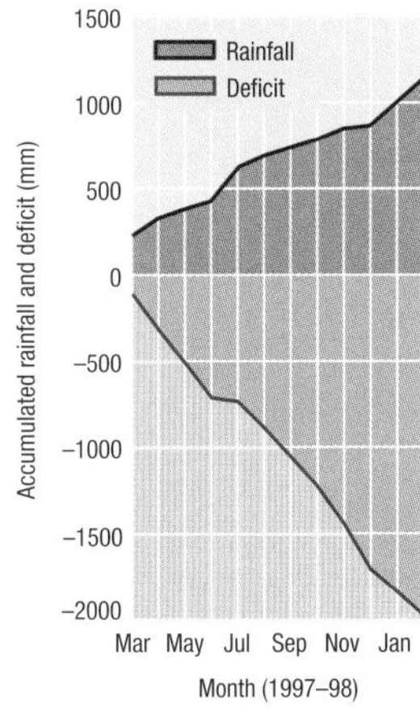

Is timing everything?

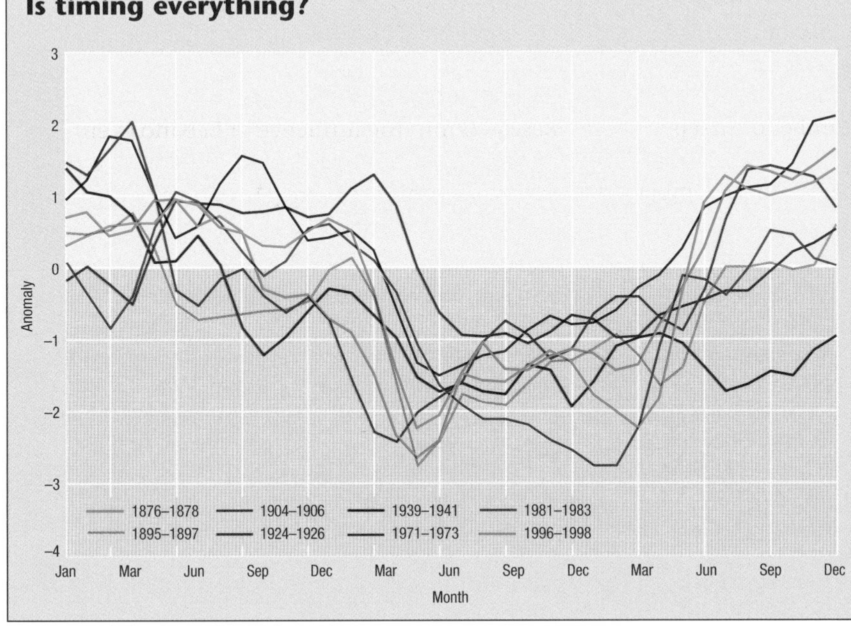

Although broad features of ENSO fluctuations are similar, the different consequences from event to event raise the question about what might be causing the differences. One interesting feature is the precise timing of the onset of El Niño events. The annual warming of the eastern Pacific arrives around the turn of the year but, for the major warming events, the time of onset and the rate of warming vary appreciably from event to event. Since each event pumps so much energy into different parts of the tropical atmosphere, the timing of this huge impulse with respect to the march of the annual cycle can be crucial when it comes to influencing seasonal weather patterns. A similar set of issues surrounds the timing of the onset of La Niña events. Attempts have been made by researchers to integrate a number of ENSO characteristics in order to enable more direct comparisons between different events.

3.19 Other ENSO impacts

The impacts of extreme events resulting from the El Niño/Southern Oscillation (ENSO) have, thus far, emphasized the negative aspects. By understanding the nature of the climatic anomalies in different parts of the world, however, it is possible to plan a more measured response to these extremes.

Tuna represent a major fisheries resource in the oceans. The sustainable level of catch depends on understanding the impact of climatic variability on their numbers and migratory patterns.

The impacts of droughts and flooding during strong ENSO events on all aspects of agriculture have become well recognized. Less well known, however, are the ENSO effects on other natural systems.

Fishing

The local consequence of El Niño events off the coast of Peru is to reduce the upwelling of the nutrient-rich waters that normally provides the basis for a high level of pelagic fish stocks. This has happened many times in the past. The best-known example is the collapse of the catch of Peruvian anchovy during the El Niño event of 1972–73. In 1970 the catch was 13 million metric tons (19 per cent of the world total) and Peru was the number one fishing nation in the world (by tonnage). This figure fell to a little over 100 000 tons in the late 1970s and the fish population did not recover until the mid-1990s. The Peruvian fishing industry now routinely uses El Niño forecasts to reduce fishing pressure in order to protect stocks.

In the Gulf of California and off the west coast of Mexico, abnormally warm water also has a major effect on the fishing industry. As in earlier events, there were dramatic shifts in fish populations between 1997 and 1998. Catches of lobster, sardines and squid fell sharply. In contrast, catches of shrimp and tuna increased. Indeed, the large rise in the value of the shrimp harvest more than balanced out the losses in other fisheries. It is doubtful, however, that the shrimp bonanza adequately compensated for the widespread social and economic disruption of the other fisheries.

Planning sustainable fisheries

Commercial fisheries also exploit knowledge of how tuna are influenced by the depth of the thermocline. As purse seine nets can only fish to a depth of about 200 m, whereas long lines can reach to twice this depth, net fishermen will move their operations to where the thermocline is nearest to the surface. For instance, the changes caused by ENSO in the western equatorial Pacific can cause the areas of highest abundance of skipjack tuna (*Katsuwonus pelamis*) to migrate over a range of as much as 6000 km. Nearly 70 per cent of the world's

Bleaching coral

Since the early 1980s reefs throughout the tropics have been subjected to unusually high sea surface temperatures resulting in a series of coral bleaching events. These unusually high warming events lead to the loss of the corals' pigmented endosymbiotic algae which play an essential role in maintaining healthy corals. Since 1983 reefs worldwide have incurred moderate to high levels of bleaching and subsequent mortality. In 1997 and 1998 the combination of a major El Niño event and warming in much of the western Indian Ocean caused mortalities between 70 and 90 per cent in the Seychelles, Mauritius and the Maldives. Losses were also high in the western Pacific from Viet Nam to the Philippines and Indonesia. In early 2000, several sites in the Southern Ocean – particularly Fiji, Rarotonga and Easter Island – bleached for the first time, a result of warming coincident with strong La Niña events which led to unusually high sea surface temperatures in these regions during the austral summer. The image shows corals at Easter Island intact in 1999 and severely bleached in 2000. Subsequent surveys revealed that these reefs had incurred very high levels of mortality by 2001.

annual tuna harvest, currently over 3 million tons, comes from the Pacific. The catch of skipjack tuna in this region approaches 1 million tons per year. Studies of how the distribution of catches of skipjack is influenced by ENSO's extremes suggest the regions of highest abundance can be predicted several months in advance. Predictions of regional and global changes in sea temperature therefore have to be included in scientific analyses of future fish stocks in order to determine sustainable harvesting levels.

Samoan fishermen know that during El Niño events the extensive warm water makes tuna scarcer. They also know that some fish do not migrate over great distances, as they need to stay close to shore. These fish can often be found at greater depth than usual, in the vicinity of the deeper than normal thermocline. Although the fish are easier to catch when the thermocline is near the surface, using longer lines can restore food security.

The same experience applies in the Indian Ocean with yellowfin tuna. Here the thermocline moves similarly to but not to the same extent as ENSO causes it to move in the Pacific Ocean. Usually fishermen do not travel east of 70°E and they land their fish in the Seychelles or Madagascar. During the 1997–98 El Niño event, the tilt of the thermocline meant it was deeper in the western Indian Ocean and few fish were caught there. Instead the fishermen travelled up to 3000 km eastwards to the Indonesian coast and landed their catches in Thailand.

Health

The dramatic swings in rainfall throughout the tropics during different phases of ENSO have had a significant effect on the incidence of some tropical diseases, notably malaria and dengue fever. In Colombia the number of cases of malaria has been rising in recent decades and the peaks in occurrence have coincided with El Niño episodes. In Kenya and Somalia in early 1998, following the heavy rains, an outbreak of deadly, mosquito-borne, viral Rift Valley fever affected 90 000 people. Worse still, there were around 100 000 cases of cholera and over 5000 people died.

These health problems extend to animals. In South Africa there is evidence that outbreaks of African horse sickness are linked to ENSO. This most lethal of horse diseases, with a mortality of up to 95 per cent in susceptible animals, occurs once every 10 to 15 years in the western region of South Africa. These outbreaks seem to require the combination of first drought closely followed by heavy rains, which tends to occur during El Niño episodes in this part of South Africa. Since the beginning of the 19th century, 13 out of the 14 epidemics of the disease have coincided with El Niño episodes.

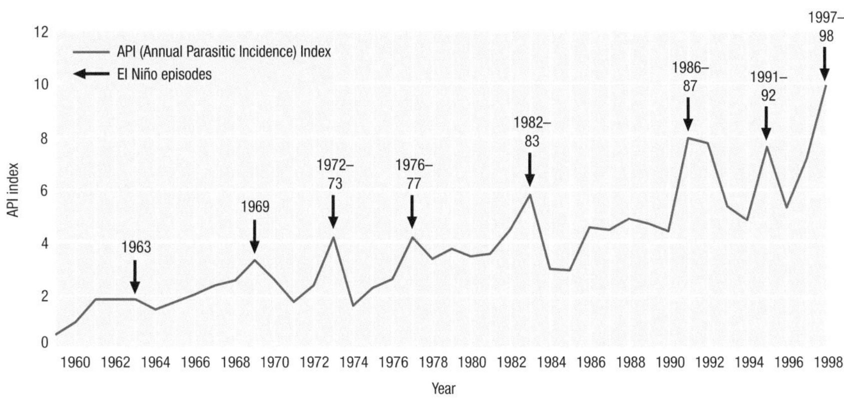

The incidence of malaria in Colombia increases in El Niño years when the more humid and generally wetter conditions enable mosquitoes to breed in greater numbers. The Annual Parasitic Incidence (API) Index is the ratio between the number of malaria cases reported and population at risk per 1000 inhabitants.

The North American balance sheet

Despite the vast amount of adverse publicity in the USA that greeted the arrival of the 1997 El Niño, the overall economic impact was surprisingly favourable, for two reasons. First, El Niño years are almost always quieter than normal hurricane seasons in the Atlantic. Given the scale of damage caused by hurricane events in the USA, this represents a substantial benefit (although a single severe hurricane can still cause tremendous damage). Second, as widely forecast in 1997, El Niño years usually lead to milder winters in western Canada and northwestern USA. This was the case in 1997–98 and the resultant reduction in winter mortality and in heating bills represented a significant benefit.

On the downside, the Pacific northwest coast was battered by storms throughout the autumn and winter. Much of California had record or near-record rainfall between December and March with floods and mudslides, while the southeast USA, from Louisiana to Florida, also had near-record amounts of rainfall during this period. Even so, California had the benefit that the Sierra Nevada ski resorts had exceptional snow conditions and a long and lucrative season. Offshore, abnormally warm waters had a major impact on marine life and fishing industries. A comprehensive review of the true costs and benefits of El Niño and La Niña events has yet to be done.

3.20 Heatwaves

Heatwaves are a part of life in many areas of the world during the summer half of the year. In many cases they are no more than a nuisance, but when they reach extreme proportions they are major killers. With the threat of global warming, heatwaves may become an even greater health threat in the 21st century, especially in urban areas.

Children plunge into a lake near Calcutta for much-needed relief from the 1998 heatwave in which thousands died.

A heatwave is not easy to define in precise terms but, like an elephant, you know one when you see it. What would constitute a heatwave in London or Stockholm might be just a bit warm for St Louis or Sydney and quite normal in New Delhi or Riyadh. Although heatwaves almost inevitably involve maximum temperatures well above 30°C, the important aspect of any definition is to emphasize how much above normal the temperature is and for how long. It is after several days of exceptional warmth that the heat has the greatest impact on people. Furthermore, the impact is heightened if the humidity is also high, which increases discomfort and keeps temperatures high at night. Conversely, conditions are somewhat alleviated if the air is very dry since the stifling conditions tend to ease through more radiative cooling overnight.

Part of the difficulty with defining the danger of heatwaves is that humans are capable of adapting remarkably well to hot conditions. Even people who have lived all their lives in temperate regions can acclimatize to tropical conditions over a matter of weeks. It is a sudden increase in temperature, therefore, that matters most. In the USA, the most vulnerable cities for heat-related mortality are in the northeastern quadrant of the country, whereas southern cities are far less sensitive. This suggests that the residents of the more southern cities are better acclimatized to hot weather.

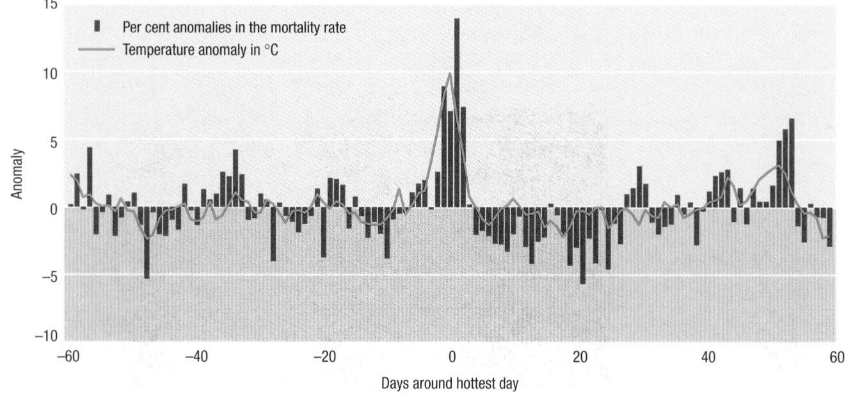

Sixteen of southwest Germany's strongest heatwaves (1968–97) have been analysed to show that mortality is highest within a few days of peak temperatures and then drops to below average levels for a period after the event is over. This suggests the hastening of the time of death of susceptible (i.e. very ill or elderly) people.

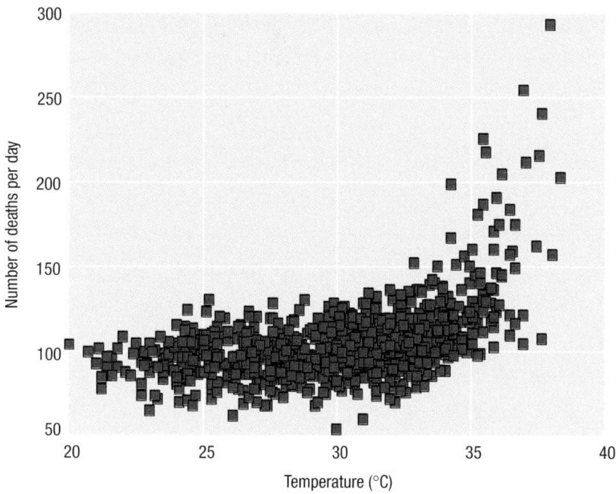

Mortality statistics for Shanghai show that the daily death rate rises sharply when the temperature rises above 34°C.

As a general rule, when daytime highs rise above 35°C in those vulnerable cities, the excess mortality rises sharply. In India, heatwaves are most likely to occur in the dry season before the monsoon (March to July), with over half occurring in June. A severe heatwave in India is defined either as a period when the prevailing maximum temperature remains 7°C or more above the long-term normal for the location, with an average maximum temperature at the time less than 40°C, or when the average maximum temperature is greater than 40°C, if the prevailing temperature is 5°C above the normal. This latter recognizes that, at 45°C, the heat stresses everybody. As in the USA, the impacts of heatwaves are not as serious in the hottest parts of the country (e.g. Rajasthan) as they are in the less torrid areas (e.g. Orissa, West Bengal and Bihar).

The difficulty in defining precisely what constitutes an extreme heatwave was illustrated by the event in Chicago in July 1995, which killed over 500 people. Although the peak combinations of temperature and humidity were not that exceptional, the fact that high heat and humidity lasted day and night for over 48 hours contributed to the high mortality.

Trends

In India there was an increase in heatwave mortality in the 1990s compared with the 1980s. Official figures show that on average some 250 people died each year, with dehydration,

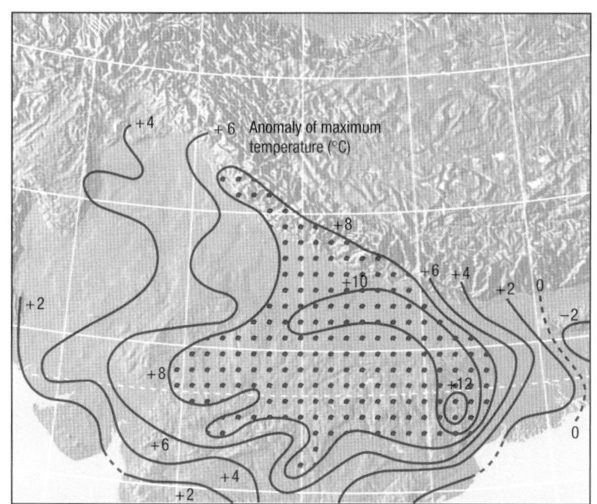

During the heatwave of June 1926 large areas of India experienced days when the maximum temperature was at least 8°C above normal (stippled area) for the time of year.

high fever and sunstroke being the major causes of death. The death toll was estimated to be 1300 in 1998, an extreme year.

The mortality from heatwaves in US cities may also have risen in recent decades, especially in the northeast of the country. This is surprising for two reasons. First, the incidence of heatwaves has if anything fallen, the worst decade being the 1930s. Second, the widespread introduction of cooling devices might be expected to have reduced the impact of the hottest days. The reason for the unexpected rise may be that the heat hits the poor, the infirm, infants and the elderly very hard. In the inner cities, where the impacts of urban heating are the worst, there are many who are unable to afford cooling devices or escape to cooler climes.

In Australia, where extreme high temperatures have caused many deaths, a different picture has emerged. Death rates have declined dramatically throughout the century. The most intense heatwave was in 1939 when 420 deaths were registered as being attributable to extreme heat. During the last two decades, however, the number of deaths has averaged less than 10 per year.

Only mad dogs and Englishmen go out in the midday sun

An obvious way to reduce the damaging effects of heatwaves is to minimize the amount of outdoor activity during the hottest part of the day. This is not simply a matter, as Noel Coward wrote of "staying out of the midday sun". Climate services can help businesses, local authorities and employers to plan their activities to minimize the health consequences of extreme heat. This is a matter of both issuing sensible advice on avoiding the hottest part of the day and also exploiting improving weather forecasts to provide warnings in time for avoidance strategies to be implemented effectively.

Prolonged heatwaves build an environment for wildfires. In January 1939, as a result of extreme drought and a sequence of days with record-breaking high temperatures, large areas of Victoria, Australia experienced devastating bush fires, known as the Black Friday fires, in which 71 people died.

The increasing prevalence of air conditioning

The use of air conditioning can transform living conditions in the hotter parts of the world. In the USA use of air conditioning began in the 1910s in the textile and paper industries to control temperature and humidity, and to cut down dust in the air. The first air-conditioned cinema was opened in Los Angeles in 1922 and was an instant success. Soon many public venues were offering air conditioning, and by the 1930s it was installed as a matter of course in many apartment buildings in the hotter parts of the country. Around the same time, the price of domestic window units had fallen to US$ 150. It is estimated that the effect of air conditioning in New York between 1968 and 1988 was to reduce the number of heat-related deaths by 20 per cent.

The widespread availability of air conditioning was a major factor in the move to the southern states where the proportion of the total population has risen from less than a third in 1960 to well over half by the end of the 20th century. This is not simply a matter of making living in the south more comfortable, but making it more economically competitive. Once the summers are bearable, the milder winters become a major benefit. This was a significant factor influential in the decision to build a Nissan car factory in Tennessee, USA. Traditionally the US car industry had always been sited farther north, where bitter winters could disrupt production. With the changed perception of conditions in the south, an important factor was the benefits of more reliable winter production.

As air conditioning became more widely available, it altered the living patterns in other parts of the world, essentially creating temperate-zone climates in the tropics. It is already an essential component of many tourist developments in tropical regions. The downside is that the additional demand for electric power may well place a heavy burden on the local environment, and makes it harder to reduce global emissions of greenhouse gases.

3.21 Urban air quality

Ever since the deadly 'Great Smog' of 1952 in London, and the photochemical smogs of the 1950s and 1960s in Los Angeles, governments have been seeking legislative and technical solutions to improve urban air quality. Progress has had to take account of shifting social conditions and, in many cases, the changing nature of the air pollution.

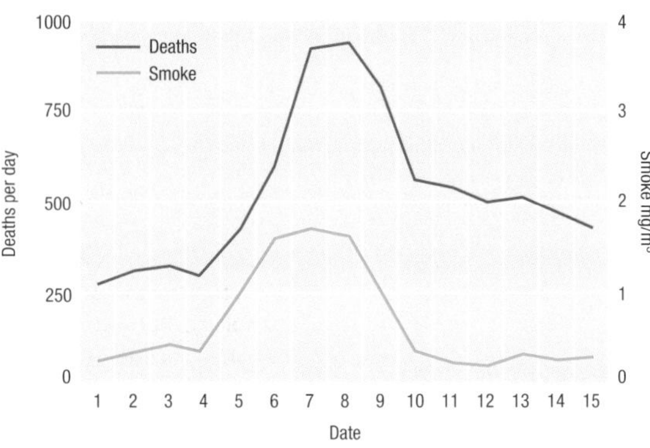

During the London smog of December 1952 the death rate rose dramatically as dense smoke and gaseous pollutants became trapped over the city.

Throughout the first half of the century air pollution in many towns and cities in industrial countries grew to unacceptable levels. The worst conditions occurred during calm anticyclonic conditions when dense fog, known as 'pea-soupers' in London, formed near the ground. Trapped pollutants, notably sulphur dioxide and smoke particles from burning coal, then accumulated in large quantities over a limited area. This dirty mixture tended to form sulphuric acid droplets that did great harm to people, particularly those with respiratory ailments, and was corrosive to materials and buildings. This mixture of smoke and fog became known as 'smog'.

In later decades motor vehicles became a dominant factor in urban air pollution. These brought a more exotic cocktail of pollutants to the cities, including oxides of nitrogen, unburnt hydrocarbons, carbon monoxide and very fine particulates from diesel engines. This deadly mixture, in the presence of ultraviolet radiation, reacts to form a 'photochemical smog'. The most important consequence of this process is the formation of ground-level ozone, which is especially corrosive and a particular health hazard.

Looking down from a distance, the brown cloud of photochemical smog trapped over the city of Denver, Colorado by stable atmospheric conditions and topography, is plain to see.

A worldwide problem

In spite of high levels of pollution in many cities and some significant pollution episodes, notably in the Meuse Valley in Belgium in December 1930, and at Donora, Pennsylvania, in October 1948, governments were slow to act. The 'Great Smog' in December 1952 changed all that. It blanketed London in dense fog for four days and was responsible for at least 4000 excess deaths. This disaster led the UK government to pass the Clean Air Act in 1956. The transformation was dramatic. In 1962 a similar smog killed about 750 people. By the 1970s winter sunshine levels had nearly doubled, and the incidence of fog in the city, which had been much higher, dropped below that of the surrounding countryside. This switch was the result of both declining pollution levels and a side benefit of the urban heat island effect.

In Los Angeles the villain of the pollution drama was the motorized vehicle. A combination of climate and geography means that for much of the year pollutants are trapped in the Los Angeles basin by a temperature inversion (a situation where the cooler air near the ground is effectively trapped and

The success of control strategies to cut down vehicle emissions in Los Angeles, California has led to a dramatic reduction since the 1970s in the number of days with dangerous levels of smog.

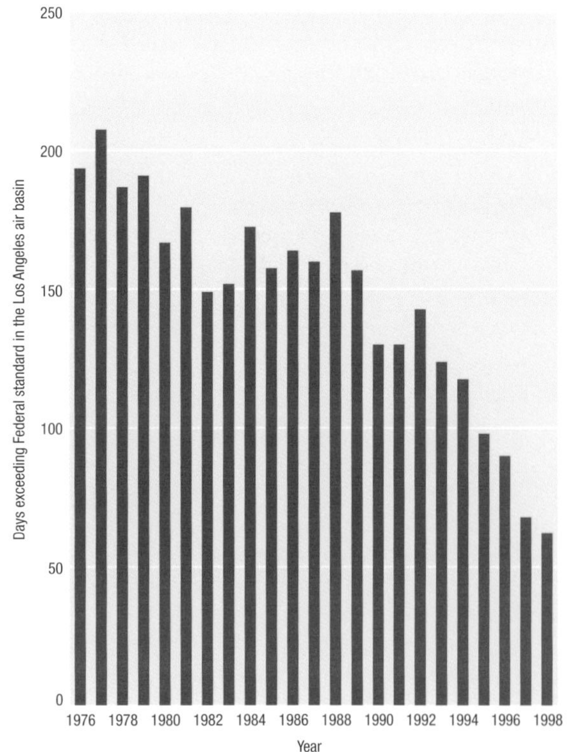

stagnates, accumulating pollutants). Combined with the high ultraviolet levels in the frequent sunny conditions the region was the ideal place to generate photochemical smogs.

Federal legislation, introduced during the 1970s and requiring cars sold in the USA to meet increasingly strict emission standards, has had a dramatic impact. In the Los Angeles basin the number of days when the ozone level exceeded the Federal one-hour standard (0.12 parts per million) fell from nearly 200 days per year in the late 1970s to 41 days in 1999. Other measures of unhealthy levels of air pollution showed equally striking reductions during the same period.

The reunification of Germany in 1990 also has provided evidence of the dramatic health improvements that can be achieved with improved air quality. There has been a steep decline in particulates in the atmosphere over the former East Germany. A survey of bronchitis, colds and infections in children showed that, in the formerly most polluted areas, within three years of reunification there had been a sharp reduction in pollution levels and a pronounced decline in sickness amongst school children.

Elsewhere in the world, many cities are facing serious problems (e.g. Ankara, Athens, Mexico City, Santiago de Chile) as populations rise and the number of motorized vehicles increases even more rapidly. Although action has begun later, there are similar stories of success in curbing the worst excesses of air pollution. For example, in Mexico City, 2000 was the first year in a decade that there was no smog alert. A programme of vehicle inspections, standards, driving bans and factory improvements has cut emissions substantially. Unusually strong winds that year also helped to keep pollution levels down.

Health consequences

The scale of the health consequences of poor air quality is summed up by a World Health Organization estimate that the excess mortality worldwide is around 200 000 people each year.

Improving air quality is, however, a continual battle. In spite of the gains made in the UK, the Expert Committee on the Medical Effects of Air Pollution has demonstrated the changing nature of air pollution in a recent study. The study concluded that 24 000 people died each year in Britain as a result of air pollution. Sulphur dioxide still claims 3500 lives each year, but now ozone, created by sunlight acting on vehicle exhaust fumes, is the major killer, causing 12 500 premature deaths. Even more distressing is the statistic that fine particulates, known as PM10s, cause over 8000 premature deaths, as these are largely the product of diesel engines, which were seen as being a more energy-efficient means of powering vehicles.

Children are most vulnerable to the health effects of poor air quality and need special protection to reduce its impact.

Forecasting and managing air pollution incidents

Around the world local authorities are using a wide variety of strategies to modify public behaviour and reduce the impact of air pollution episodes. Through improved weather forecasting methods and better predictions of likely pollution levels a few days ahead, it is possible to anticipate the worst conditions and take action. Some cities, including Athens, Paris and Singapore, have used various forms of traffic control, such as only permitting cars with odd and even licence plates access on alternate days, during emergency air pollution situations.

Another response, in the event of a smog alert, is to schedule outdoor activities for morning or early evening hours to avoid the midday peak when ground-level ozone concentrations are at their highest. This is particularly important for children whose lungs are not fully developed and who run the risk of incurring permanent damage.

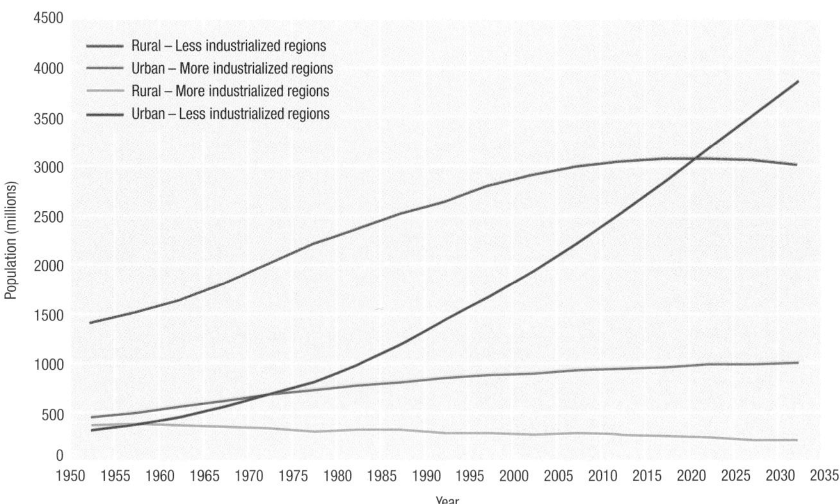

United Nations projections for population growth in the next three decades predict that the number of people living in cities in the less industrialized parts of the world will rise rapidly and the challenge of providing acceptable air quality will also grow.

3.22 Exceptional weather and climate events

Single extreme weather events have sometimes had a dramatic impact on societies. The full consequences can be hard to establish and often take a long time to become fully apparent.

It is often claimed that severe weather has played a major part in historical events. For instance, thunderstorms destroyed so much grain across northern France in the summer of 1788 that bread prices soared, which contributed to the French Revolution the following year. Such claims are the bases for historical debate, but there is no doubt that weather events, which destroy crops that cannot then be replaced until the following year, cause immense harm to food supplies. The most obvious examples of this type of damage are late spring frosts, destruction of fruit or grain by hail and torrential rain, and flooding of agricultural drylands at any time during the growing season.

Freezing weather can cause considerable damage to both trees and fruit in citrus crops. These oranges show damage following a spell of icy temperatures in Florida.

UK soft fruit disaster of 1935

Much of northern Europe is particularly prone to killing frosts in late spring because of the high latitude. These frosts can greatly damage fruit crops and vineyards. One of the most significant examples of this type of event occurred in the British Isles in mid-May 1935 when frost killed about 80 per cent of fruit production, including apples, pears and soft fruit (e.g. currants, raspberries and strawberries). At a time when far less fruit was imported from abroad and most people relied on preserves and jams made from home-grown fruit for winter supplies, the impact was enormous. Similar frosts of nearly the same severity occurred in 1938 and 1941. These frosts led to considerable comment about imminent ice ages, which was soon forgotten after a few subsequent years of abundance. These extremes underlined the importance of understanding the causes for severe frosts and of preparing for them. Such preparations range from taking action to protect crops when frost is forecast, through growing late-flowering varieties of various fruits that can better withstand late frosts, to the siting of crops to minimize the risk of frost damage.

The great Ice Storm of 1998

The storm that hit Quebec and parts of eastern Ontario in Canada and northern New England in the USA in January 1998 illustrates how even in relatively mild winters a single extreme event can have disastrous consequences. There is evidence from computer-modelling studies that even the 1997–98 El Niño event in the Pacific was a contributing factor.

For five days warm moist air was pumped up from the Gulf of Mexico and rode over static cold Arctic air over eastern Canada. During this time over 100 mm water equivalent of freezing rain fell on parts of Ontario and Quebec forming a thick coat of ice on everything it hit. This accumulation of ice was more than twice the amount of previous major storms in the area in 1942 and 1961. It disrupted electricity supplies over an area of several hundred thousand square kilometres, cutting off nearly 1.4 million homes, some of which had to survive for up to 17 days without power in the freezing weather. Approximately 35 people died as a result of the storm and the economic loss exceeded

US$ 2 billion. Included in the losses were the destruction of many magnificent trees, some up to 150 years old, in a area which provides 70 per cent of the world's supply of sugar maple syrup.

Florida frosts of December 1962

The consequences of single extreme events may take a long time to emerge. A series of severe frosts during the 1980s in Florida, USA which caused immense damage to the citrus groves, was seen as the reason for Brazil becoming a major player in the international frozen concentrate orange juice market when, in fact, it was a much earlier frost in December 1962 that killed many orange trees. For most parts of Florida it was considered the worst freeze since the beginning of the 20th century, and led to an upsurge in prices for oranges and their juice. This rise caught the attention of people in the São Paulo state in Brazil, who convinced US companies that they should invest in an alternative source of supply of oranges from a region that was much less susceptible to frost than Florida. This initiative was developed over the next two decades. The number of orange trees rose from 18 million in 1963 to 106 million in 1985, and the proportion of oranges processed for frozen concentrate juice rose from 10 to 92 per cent. This meant that, when severe frosts struck Florida in the early to mid-1980s, Brazil was well placed to exploit the US market.

It's an ill wind

Sometimes the consequences of a climatic event can produce immediate benefits for other parts of the world. The collapse of the Peruvian anchovy fisheries after the 1972–73 El Niño warm episode meant that there was a substantial shortage of fishmeal for feeding livestock. The favoured alternative for poultry and livestock feed was soy beans, and Brazil was again in a position to move into the market quickly and establish a major presence. Today, Brazil is one of the world's leading soy growers and exporters.

Brazilian coffee crisis of 1975

Frosts can affect most of Brazil south of 20°S when invasions of polar air-masses sweep northwards from Antarctica any time between May and September. They are most common in the Planato Meridional region, where widespread sharp frosts (below –5°C) occur every 5 to 10 years. Coffee production is limited to regions where frosts occur less than three times a year. Output is, however, dependent on fluctuating temperatures and cool nights that stimulate flowering and coffee is usually grown at altitudes between 500 and 2000 m in Brazil and elsewhere in the tropics. Coffee producers must therefore accept the risk of occasional frosts in return for higher average yields. They experienced severe frosts in 1953 and 1963, when up to 50 per cent of the crop was damaged. In July 1975 the frosts were even more severe and up to 80 per cent of the crop was lost.

The scale of the coffee crop disaster in Brazil in 1975 provides a good example of how public awareness of a natural disaster can have an unfortunate feedback onto the market. Not only did coffee prices rise sharply on the commodity markets around the world as the scale of the damage became all too apparent, but in addition shoppers stripped the stores bare to build up personal stocks. This additional demand raised prices in the stores and was reflected back into forward purchases on the markets, thereby further inflating the prices. The peak of over US$ 8 per kilogramme in early 1977 has not been exceeded since, in spite of droughts in the 1980s and damaging frosts in 1994.

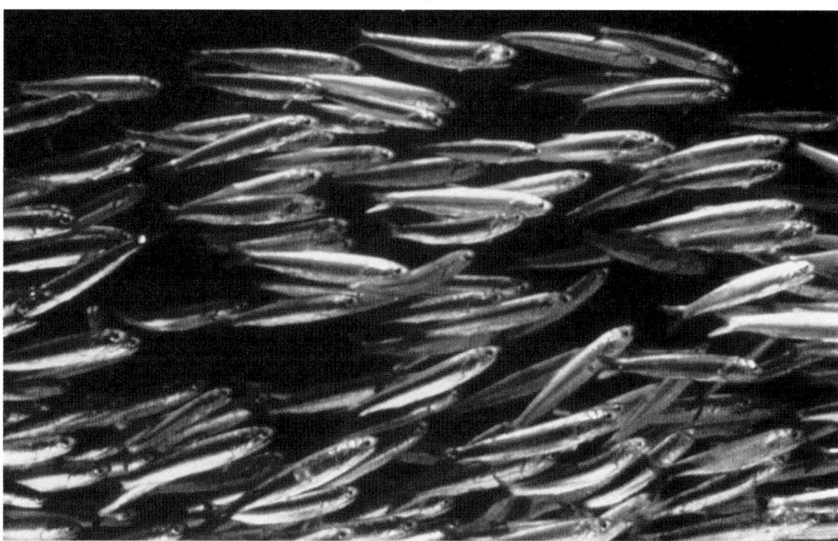

The El Niño event of 1972–73 curtailed anchovy fishing in the eastern Pacific, an industry developed to make fishmeal for animal fodder.

Freezing temperature kills all varieties of coffee. The widespread sharp frosts that hit the Brazilian coffee-growing areas in 1975 had an immense impact on the world market for coffee.

3.23 Bumper years

Ever since ancient Egypt experienced seven fat and seven lean years, farmers and the rest of the community have had to manage years of abundance and years of dearth. The abundant years are in many ways a boon, but they do also bring special challenges for both societies and the environment.

When weather conditions are right even the most difficult but fertile terrain can be used to produce bumper crops, as seen here in Mexico.

The weather conditions that bring record harvests are not necessarily the direct opposite of those that cause famine and death. One can have too much of a good thing. While abundant rainfall can be good for crops, it can be too heavy, flooding the fields, killing the crop or preventing the harvesting. Often the best outcome is a judicious mix of the right, near-average conditions. These may be barely adequate rainfall, at the right time to promote germination and rapid early growth, and thereafter only modest rainfall to maintain good growth. Dry weather is a boon at harvest time. It follows that the right conditions for abundance are not easily defined and perhaps the best way to explore our dependence on this aspect of the climate is to consider some interesting examples.

Grain production

Successful cereal growing in northwestern Europe is a fine balance between adequate moisture and reasonable warmth, especially at harvest time. The combination of sustained cool, wet weather can be as damaging as heat and drought. On balance, cool relatively wet summers produce the heaviest yields, providing the harvest can be gathered efficiently. The same applies to much of the grainlands of North America. Bumper years tend to be those with plentiful rainfall and average temperatures until the end of June, then dry and warm thereafter. These ideal conditions are rarely achieved, because they depend so much on the timing of the switch from moist to dry conditions. Nevertheless, the adaptability of agriculture means that only when certain adverse combinations of weather occur do yields fall dramatically.

Futures markets and weather derivatives

Insuring against weather disasters is a widely recognized practice. Most people who have the means will take advantage of insurance to protect against flood or wind damage, amongst other things, to their homes. The thought of taking out protection against a bumper harvest or a very mild winter seems less likely. In fact, for some producers these apparently positive outcomes can be just as disastrous. A bumper harvest can mean that prices crash and farmers cannot sell their produce, especially if their crop ripens slightly later than others. Similarly, a mild winter lowers demand for energy supplies and leaves the producers of fuel or electricity with costly stocks that may have to be carried to the next winter and hence greatly reduce their profits.

Futures markets and weather derivatives are economic instruments that enable producers to spread the risk of adverse weather consequences damaging their businesses. Agricultural producers can sell their crops 'forward' at a price agreed in advance. The purchaser in this deal estimates the likely price at the time of harvest and takes a risk that the price may fall. Similarly, the seller runs the risk that there may be a shortfall in the produce and the price may rise higher than expected.

In the case of weather derivatives the bargain is struck solely on the extent to which any agreed meteorological parameter will be higher or lower than average. For instance, an electricity producer may purchase a derivative that pays a designated amount for every degree the average temperature in a given locality is above average throughout the winter. If the winter turns out to be mild the producer collects on the derivative and uses the money to offset any loss in profits. Conversely, if the winter is colder than normal there is no payout, but then profits from sales are likely to be higher. Either way, the producer has managed to share the risk with the seller of the derivative. Unlike insurance, however, these types of arrangements do not provide full cover for disastrous losses, but merely help to spread the burden of uncertainty in managing the consequences of climatic variability.

The impact of weather fluctuations on crop yields can drive the world's commodity markets to frenzied speculation, especially as the same event can have remarkably different impacts on different sectors. In 2000, Tajikistan suffered its worst drought in 74 years and, although grain yields dropped 47 per cent, the cotton crop thrived.

When the desert blooms

Exceptional rainfall can be a blessing in arid regions. In many deserts, rare heavy rainfall can miraculously transform the landscape. Long-dormant seeds spring to life, and many flowering plants rapidly grow, flower and produce new seed which may again lie dormant for many years until the next downpour. More important for the long-term survival of flora and fauna, groundwater reserves are replenished, which subsequently feed local rivers and water holes for months or even years.

One of the most extreme examples of this phenomenon is the flooding of large parts of central Australia. Occasionally, tropical cyclones move far inland and dump huge amounts of rainfall into the dry heart of the country. Temporary inland lakes form and quickly teem with fish hatched from eggs that have lain dormant in dried-up lake beds. Other forms of aquatic life emerge from suspended animation and thrive. Seabirds and waders will fly 1500 km inland to exploit the rich pickings. Animals and birds reproduce and raise young. Once the rains come to an end, however, desert conditions soon reassert themselves and the briefly tumultuous ecosystem returns to its quieter state.

3.24 Freshwater: Life's precious resource

As the world's population increases, so does the demand for freshwater. Its availability for human activity must now be carefully managed to meet basic needs and to adapt to future shortfalls from droughts and excesses from floods. This will require difficult policy negotiations at the local, national and international levels to reach acceptable solutions for all.

Our world is the blue planet – 70 per cent covered in water. However, 97.5 per cent of all water on Earth is saltwater, leaving only 2.5 per cent as freshwater. Nearly 70 per cent of that freshwater is frozen in ice caps in Antarctica and Greenland. Less than 1 per cent of the world's freshwater is in fact available for direct human use. To meet the needs of food supply, health and welfare and industrial development, adequate supplies of freshwater are vital. Exactly how much water is available to maintain economic and population growth? One assessment states that areas are said to experience 'water stress' when the annual average water supply available to each person is below 1700 m^3, and particularly when disruptive water shortages frequently occur. These figures are likely to have risen and to rise further. Severe problems are likely in areas where the supply is less than 1000 m^3. In 1995, it was estimated that 2.3 billion people around the world lived in river basins experiencing water stress and, of these, 1.7 billion resided in highly stressed basins.

There are substantial differences in the availability of water between the countries of the world, with many already using a large proportion of their available water supplies. The greatest limitations on supplies affect countries in the subtropics of the Northern Hemisphere, especially from North Africa to Central Asia.

The world's water budget

Water moves through rivers, below the land and into the oceans. It enters and leaves the atmosphere through evaporation and precipitation and changes state between vapour, water and ice. We know the dynamics of water and the approximate quantities available on a global scale. The average precipitation on land (10 m^3 per day per person, assuming a population of 6 billion) is more than sufficient to meet our needs, but much of it falls in the wrong place or at the wrong time and is effectively inaccessible.

Each continent has a different water regime, with Australia being the driest inhabited continent and South America the wettest. These differences exert a fundamental control on the population that a continent can support. Without greatly increased recycling, and diversion of water from where it falls to where it is needed, precipitation alone will not meet the needs of a growing population over the next 50 years.

Developing a nation in a semi-arid land

Australians have forged an economy and an identity on the driest inhabited continent, and climatic variations have been a major factor in its development. In 90 per cent of years, the area receiving 400 mm or more of rainfall annually is a 200–400 km-wide strip around the east coast and in the far southwest corner of the continent; but in 10 per cent of years this amount of rain occurs up to 1000 km inland from the eastern and northern coastlines.

There is limited scope for expansion of water use as Australia's major river system lies in the southeast corner. Here, intensive agriculture already uses between 60 and 90 per cent of surface water in an average year. To combat low and variable runoff, Australians have constructed a series of large dams to divert rivers flowing eastward to irrigate the inland areas. Most of the prime sites have now been used and high evaporation in the tropical north will constrain the cost–benefit of future development. In one century, Australia has exploited its water availability – some argue overexploited because of environmental damage to the rivers and the irrigated lands. Further exploitation will be far more costly, both in monetary terms and to the environment.

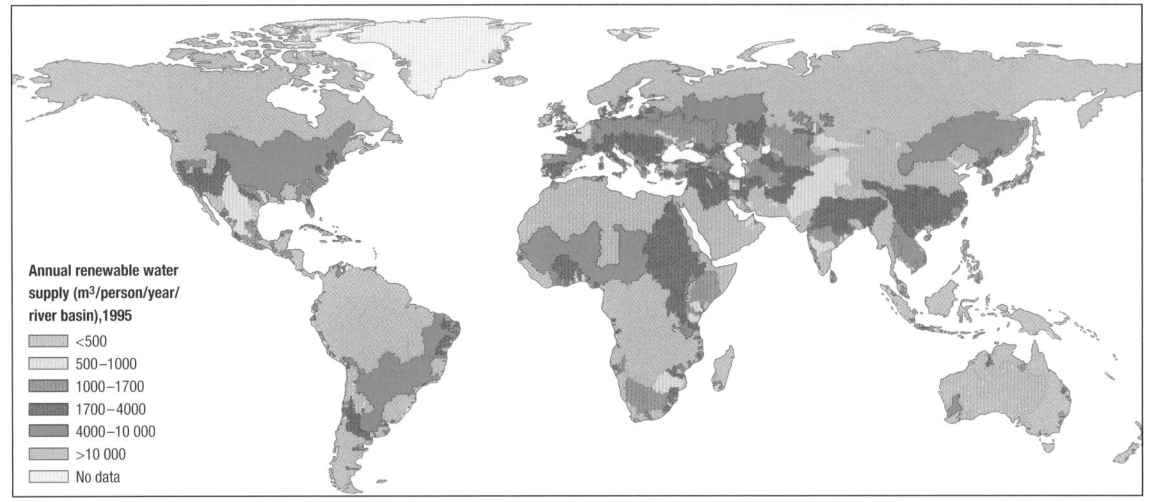

Changes along the Nile

The waters of the mighty Nile River have been the source of life for Egyptian civilizations for more than 5000 years. With the completion of the construction of the Aswan Dam in the 1960s, the management of the waters of this river was transformed. The storage capacity of the dam has eliminated the annual flood peak in September. Minimum flows have slowly risen, reflecting the increased demands downstream during low flow periods. The mean flow, however, has decreased significantly due to evaporation losses from the reservoir.

The construction of the dam was a major economic milestone for Egypt, allowing for a substantial expansion of agriculture and electricity output to help keep pace with the growing population and industrial demand. Other consequences included the disappearance of natural overbank flooding and the elimination of most of the silt downstream. Soil fertility has decreased and the river delta has been shrinking without the replacement of silt. The delta is now extremely vulnerable to storm surges and sea-level rise.

Freshwater is essential to support human life and economic growth and development. It is also critical for the healthy functioning of nature, upon which human society is built.

What is happening to the inland seas?

A number of the largest bodies of inland water have undergone dramatic changes in recent decades. Of particular interest are the differing fortunes of the Caspian and Aral Seas. The Caspian Sea is the world's largest enclosed sea and has a surface area of around 390 000 km^2 fed by two major rivers, the Ural and the Volga. Its level has varied dramatically, falling by a couple of metres since the 1930s and rising sharply by 2.5 m since 1977. These fluctuations reflect a combination of local climate variability, human action within the basin, such as reservoir and dam construction, and possibly global climate change. Dams constructed on the Ural and Volga Rivers account for much of the decline from the 1930s. Smaller fluctuations, lasting 5 to 10 years, are responses to regional variations of climate.

The fate of the Aral Sea, several hundred kilometres to the east of the Caspian, has been even more dramatic. As a result principally of withdrawal of water for irrigation from its main feeders – the Amu Darya and Syr Darya Rivers – the Aral has shrunk almost to the point of extinction. Since the 1960s the sea's level has dropped by 17 m, its surface area has more than halved to 35 000 km^2, its volume fell by more than two-thirds to 300 km^3, and its salinity tripled. The exposed seabed is a salty wasteland, while several million hectares of cotton fields have become salinized by inefficient irrigation. Furthermore, following the break-up of the former Soviet Union, management of the rivers running into the Aral is now a matter for five separate countries, which means that reversing the decline of the region has become a much more difficult challenge.

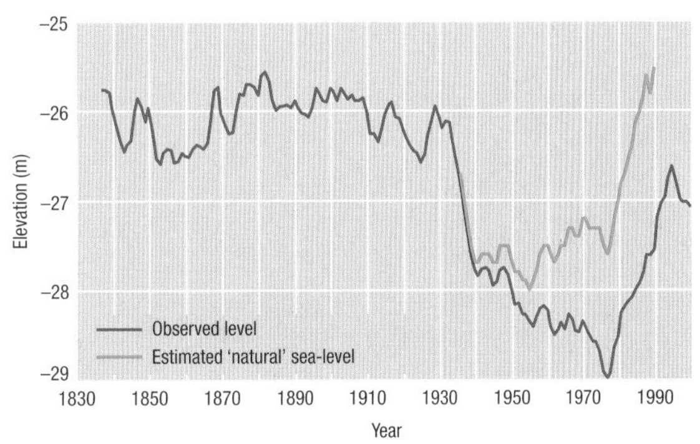

The annual mean level of the Caspian Sea fell dramatically from the 1930s to the 1970s and has since risen appreciably. Much of the change is attributed to climate factors. The model-generated sea level (1936–90), created to remove most human-induced variation in the data, shows that the sea level would have been significantly higher in recent times but for the increased extraction of water from the rivers running into the sea.

Landsat images of the Aral Sea in 1973 and 1999 show the change in area. The shrinkage is mainly attributed to withdrawal of water from its main feeder rivers for irrigation of rice and cotton plantations. The 1999 image clearly shows that Vozrozhdeniye Island (meaning 'Rebirth' in English) is no longer completely separated from the mainland.

4.1 The rise of technology

At the beginning of the 20th century scientific curiosity was driving the development of technologies to better understand the climate system. By its end, awareness of the vulnerability of society was giving a new impetus to harnessing technology in all its forms to reduce risk not only from the climate but also increasingly to the climate.

Until the advent of computers, data had to be handled and processed with mechanical calculating machines and plotted by hand on maps and charts.

During the course of the 20th century scientific research and technology fed off each other in an explosive growth of knowledge. Electronic instruments, globe-spanning satellites, fast communications and powerful computers have given us extraordinary power to measure, analyse and understand the climate and its impacts on people and society. Models of the climate system, vast databases and predictions of future climates are now being used to deliver benefits to societies worldwide. The new knowledge has also fuelled a growing concern about the environment and raised awareness of the vulnerability of societies to climate change.

The information revolution

It is easy to get carried away by the dramatic advances in technology that have transformed our view of the climate. Any analysis of these advances must be tempered by the knowledge that climate services continue to depend on the less glamorous tasks of managing the collecting, checking and archiving of data, its ongoing analysis and the publication and dissemination of meaningful information for a wide variety of applications. Technology, however, has transformed how climatologists do this work. Climatologists were early adopters of technologies for machine processing of data and have been routinely using computers for more than 30 years. Thus technology has freed climatologists from many of the traditional and often repetitive tasks associated with data quality control, processing and publishing. This has enabled them to concentrate on those tasks where human knowledge and judgement are most needed – developing applications and improving client services.

The nature of technology

From the relatively simple instruments for measuring the basic characteristics of climate there is now a huge variety of equipment to probe the workings of the climate system. Many of these use forms of remote sensing to detect the properties of the atmosphere, the oceans and the land from a distance. Radiometers in satellites measure the properties of the upwelling heat radiation from the Earth's surface and the atmosphere. Radar is used to bounce microwave radiation off clouds, precipitation and even turbulence in the atmosphere, and can measure wind speeds, rainfall and ocean wave characteristics. Laser systems have also been developed to make similar observations. Photospectrometers are used to measure the changes to sunlight as it passes through the atmosphere and is reflected back into space. These measurements reveal information about the concentrations of various trace gases and particulates in the atmosphere, or the concentration of plankton in the sea and the amount of vegetation on land.

At the same time, new and complex electronic systems have transformed routine data handling by National Meteorological Services. Observations are now transmitted instantly around the world, using satellite and high-speed cable services. In major operational centres the

Information technology

Any business, which is as dependent on both collecting and handling data as climatology, is bound to have benefited from the developments of computer technology. These advances are not simply piggybacking on the success of a new industry, but are a major factor in its development. The demands of weather and climate forecasting have been an important driving force in creating a market for supercomputers and the development of parallel processing. In the early 1950s the first computers used for numerical weather prediction were capable of performing about 100 Floating Point Operations per Second (FLOPS – each operation is a separate arithmetic manipulation). By the year 2000 this figure had risen by a factor of 10 billion to a trillion FLOPS (1 teraflop). Climate analysis and prediction centres, both those involved in operations and in research, require direct access to global data sets of climate information for the analysis of variability over space and time. These centres have been at the forefront in the development of databases holding billions of individual observations. An estimate of the speed which might have been achieved by Richardson's 'People Computer' (see 4.19) is also shown.

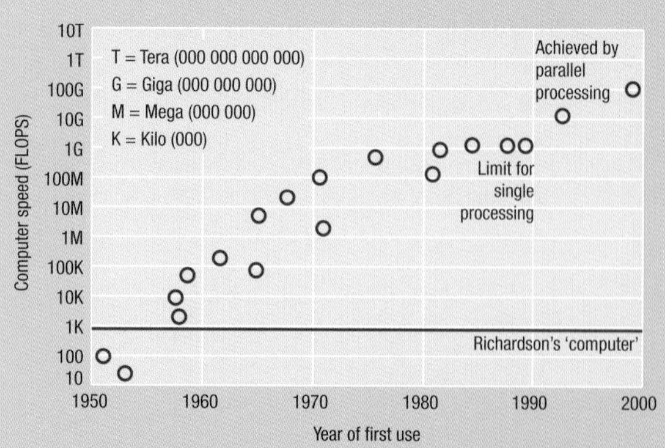

data are collected and processed automatically and fed into supercomputers to provide an instantaneous picture of the motions and forces of weather and climate.

The pace of these developments over the 20th century, particularly in the later decades, has allowed scientists to collect and process vastly more data, far more quickly and with higher accuracy than seemed previously possible. It also means that detailed images of the extent of current events (e.g. floods and droughts) and the factors influencing them (e.g. tropical sea surface temperatures) are routinely available through National Meteorological and Hydrological Services, major research organizations and, increasingly, through major commercial organizations in the private sector. Only a few decades ago most of these products were not available or only available weeks or months after the event.

Assisting development of these new capabilities has been the speed and size of computers that have increased immensely since they were first used to prepare experimental weather forecasts in the 1950s. In addition, through the Internet, it is increasingly possible for public and private sector managers in climate-sensitive industries to directly access information that can improve their decision-making, thus directly improving human welfare. All these developments have allowed scientists to address in much greater detail the difficult questions of how weather and climate affect our lives and how we might harness climate knowledge to cope better with the variability, extremes and possible change of the climate system.

Weather and climate on the Internet

During the 1990s the Internet transformed the nature of public access to many weather and climate products. It is now possible to log on to a wide variety of government, university and commercial services to find out about anything ranging from local weather warnings and radar images of severe rainfall to experimental forecasts of conditions in the tropical Pacific in a year's time. In addition, vast amounts of climate data on current conditions and changes that have occurred over the centuries are available to guide those who have an interest in knowing how the climate affects their commercial operations, the design of systems or just their day-to-day lives.

The development of weather satellites

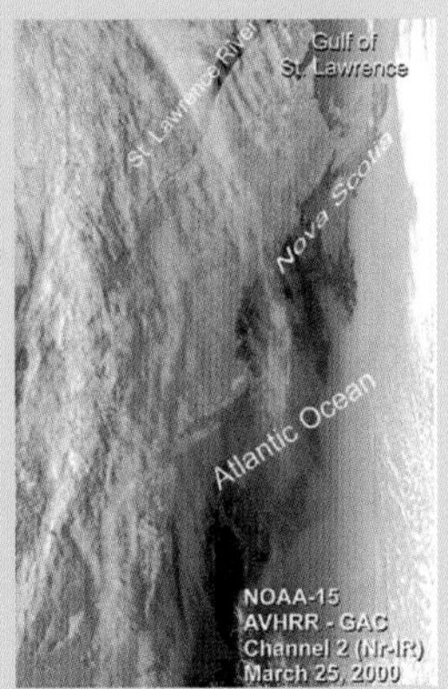

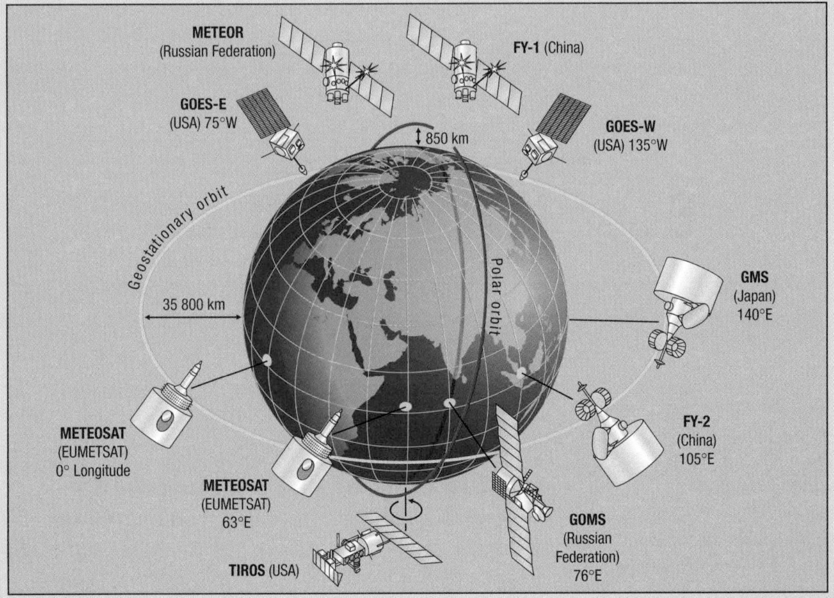

Weather satellites were one of the first products of the space age. Within three years of the launch of the first satellite (Sputnik) by the Soviet Union in 1957, the Americans had launched TIROS (Television Infra Red Observational Satellite). Its first picture (left) was a far cry from the images 40 years later on (right), but nevertheless, these early pictures provided the first view of Earth's weather from space, and within five years, polar-orbiting satellites were routinely producing regular pictures of the global atmosphere.

Professor Vern Suomi of the University of Wisconsin conceived a system to exploit, for meteorological purposes, the spinning geostationary satellite technology that was being developed for communications. His television camera scanned the visible face of the Earth and was installed in the first experimental geostationary satellite that was launched in December 1966. At an altitude of 36 000 km above the equator, this system formed images of that part of the globe in its view every half hour whereas the polar-orbiting satellites only provided images of each part of the globe every 12 hours.

A ring of geostationary satellites now sits above the equator to provide regular images of the globe between latitudes 65°N and 65°S, nominally at 30-minute intervals. The space-based constellation also includes polar-orbiting satellites such as METEOR (Russian Federation), FY-1 (China) and the National Oceanic and Atmospheric Administration's TIROS Series (USA). This array of operational satellites provides almost continuous observations of many aspects of the Earth's weather.

An astonishing fact is that despite the wonders of technology, our knowledge of the climate and its fluctuations rests heavily on the measurements made each and every day in the past by committed observers, including volunteers such as farmers, post office workers and scientists. Using simple instruments, they have over the decades not only recorded their own climate but also inexorably built up quality measurements that when put together benefit the whole world.

Standard Stevenson screens, like this one in Malaysia, are the basic foundation of the measurement of surface temperature conditions around the world.

Effective monitoring of the global climate still continues to depend in large part on the dedicated efforts of a large number of individuals, each with the responsibility for making systematic observations at his or her locality. The first meteorological networks were established through scientific curiosity. Today, National Meteorological and Hydrological Services are responsible for maintaining high-quality observations, handling the rising costs of maintaining systems and introducing changes in equipment, monitoring the location of sites and supporting adequate networks for climatological studies.

Reliable temperature measurement

Climate studies often have to detect over long periods trends in mean temperature of a few tenths of a degree a decade, in contrast to the weather systems that can have changes in instantaneous temperatures of several degrees in hours. Measuring temperature is then much more difficult than we might first think. A properly calibrated thermometer can provide an accurate measurement, but ensuring that measurements are always made consistently is less easy. The first stage in getting reliable measurements of the climate system was to ensure that we were measuring the same thing wherever instruments were kept. For this reason, surface temperature observations are made at a specified height in the shade, over natural ground cover with any grass clipped to no more than 20 cm. This specification ensures that what is measured is the climatologically important parameter – the temperature of the air – and not the capacity of a thermometer to absorb sunlight or the potential of the underlying ground to heat up locally and so influence what is observed.

It was only at the beginning of the 20th century that a standard mounting of thermometers in a well-ventilated, louvered, white timber shelter, which prevents either direct sunlight or terrestrial radiation from the ground reaching the instruments, became the accepted norm. This standard design, generally known as a Stevenson screen, is now found in meteorological observing sites around the world. Temperature measurements in Stevenson screens are directly comparable, for example in analysing climate change. Earlier records and those from non-standard shelters have to be adjusted for the different exposure characteristics.

The water content of snow can be determined by weighing or melting the snowfall caught in instruments such as the Nipher gauge (a manual gauge, left) and the Fischer and Porter automatic weighing gauge with a large Nipher shield (right).

Precipitation measurements

There are also many challenges in the measurement of precipitation. Wind eddies near rain gauges are capable of seriously distorting the amount of rain falling into the receptacle. These problems are compounded by growth of vegetation and building developments around the site that distort the airflow. Many early measurements underestimated rainfall amounts because of the siting of gauges. Snowfall, which is a form of precipitation, is particularly difficult to measure because it can be blown around after deposition. Early measurements of snowfall also generally underestimated the amounts. However, improved designs of instruments and rules for siting the gauges have improved both accuracy and consistency. An improved snow-gauge design was introduced as recently as the 1960s in what is now the Russian Federation.

Changes over time

The limitations with early measurements do not mean that these observations do not have value. There are several European locations where it has been possible to derive comparable series back to the 1750s for temperature, precipitation and pressure. Work is continuing to recreate the instruments and conditions under which early observations were made just to be able to compare observations over time. We also now know much more about how to produce internally consistent series using modern statistical tests and powerful computers. This means that in many countries there is plenty of information to be gleaned from the early records made before National Meteorological Services were set up and standards adopted.

While the siting of instruments can be checked, changes in land use over longer periods can also exert important influences on local temperatures. In particular, many early observations were made in the vicinity of towns, and urbanization has had a profound impact on the local climate.

Telecommunications

An efficient, reliable and dedicated data collection system is a crucial part of the climate business. Early weather forecasting efforts were built on the new-found ability to transmit data over the telegraph network, but the systems were prone to errors and data loss. The collecting of field-books in which the observations were recorded by hand has been an integral part of climate data management. With the introduction of machine processing, and later computers, transcribing of records from the field-books became an essential task. The introduction of automated observing systems and the growing volumes of data have further increased the need for reliable, dedicated telecommunications systems for national data collection and international data exchange. The Global Telecommunication System, a part of the World Meteorological Organization's World Weather Watch, ensures that dedicated communications channels automatically and reliably exchange data between National Meteorological and Hydrological Services for mutual benefit.

Weather balloons and radiosondes

Manned balloons involved some of the most heroic investigations of the upper atmosphere. Frenchmen Gustave Hermite and Georges Besançon conducted the first successful balloon ascent with a self-recording apparatus in 1892, reaching an altitude of 7.6 km. Subsequently, a bigger inextensible balloon reached an altitude of 16 km and obtained useful results to a height of 12 km. It was the successful introduction in the USA, by Richard Assmann in 1901, of elastic rubber balloons, which could reach an altitude of 20 km, that laid the foundation for regular upper-air observations nearly 50 years later. The modern radiosonde is a lightweight instrument package, carried aloft by hydrogen- or helium-filled balloons to an altitude of between 20 and 40 km that transmits measurements of temperature, pressure and humidity to a receiving station. Radar or the Global Positioning System (GPS) satellite system is used to track the radiosonde to provide a measure of wind speed throughout the ascent. The balloons may travel as far as 200 km downwind during the ascent.

Receiving the first weather bulletin sent by radio, on a farm in France in 1922.

Improvements to radiosondes over the last 50 years have led to more accurate observations. For example, the amount of solar energy absorbed by the instrument package is a source of error in the temperature observed. When new systems have been introduced there have been step changes in the measurements, sometimes amounting to as much as 3°C, for which adjustments have subsequently had to be made to make the data suitable for climate-related purposes.

Other sondes

Constant-level balloon sondes are designed to carry a standard instrument package at a given height in the upper atmosphere for several weeks, and provide information on the wave patterns in the global circulation. Drop sondes are launched from aircraft and parachuted into regions where severe weather is developing and no ground measurements are possible. On 12 October 1979, a drop sonde measured the lowest surface pressure on record, 870 hPa, in the centre of 'supertyphoon' *Tip* in the northwestern Pacific Ocean.

The most intense tropical storm on record, Typhoon *Tip* (T 7920, the storm to the right) in the northwest Pacific, had gale-force winds extending out for 1100 km on 12 October 1979, making it the largest tropical cyclone on record.

4.3 Essential observing networks

A major challenge in monitoring the climate system and improving understanding of the causes of change is developing and maintaining observing programmes. Now, some essential ground-based observing networks and data exchange arrangements are coming under threat.

Although historical observing networks and recent technological advances have provided climatologists with many new insights into the climate system and its processes, the progress in building adequate global observation networks has not been such good news. As long ago as 1899, in the introduction to *The Atlas of Meteorology*, the celebrated Scottish meteorologist Alexander Buchan said, "it is evident that meteorology will confer greater benefits than is now possible when a network of stations is spread over regions where there are none, or so few as to give no adequate representation of the meteorology of those regions". Now, at the beginning of the 21st century, many National Meteorological Services are facing difficulties in meeting this challenge and clearly the open oceans demand an international effort.

Competing needs of weather forecasting and climate services

A major factor limiting the effectiveness of climate monitoring is that the observing systems were designed principally to meet the needs of weather forecasting. The symbiosis that provided many benefits, including the huge store of historical data documenting the global climate, also has weaknesses. In particular, the accuracy, continuity and homogeneity of data required for weather analysis alone are not generally as demanding as are needed for climate studies. Even more significant, in many regions of the globe the number of observing stations has declined over recent decades. This is particularly true of the radiosonde network because the cost of the disposable instruments is an increasing burden.

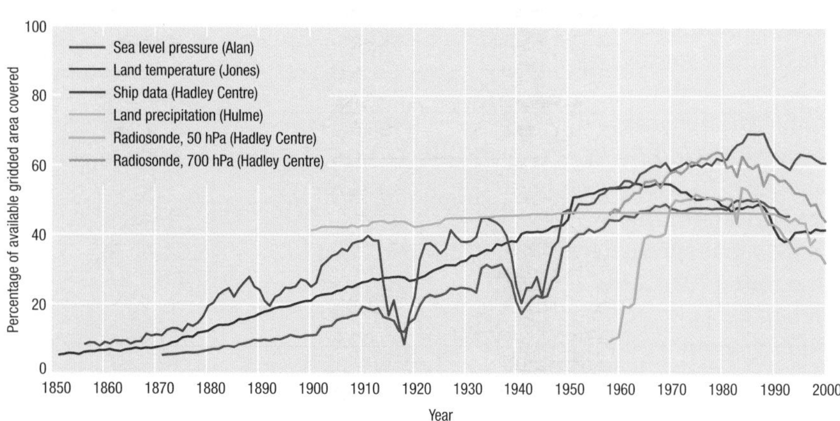

The number of meteorological observing sites on land and at sea has varied over the last 150 years for a wide variety of reasons, including economic factors and times of war and peace. For many key elements (below), the percentage of available gridded area covered by observations has decreased since the 1980s. Government commitment is essential to ensure that the necessary data are available for planning and early warning against weather and climate extremes, and for research and better management within a varying climate.

Measurements at sea

Obtaining reliable observations over the oceans requires organized arrangements to make standard measurements on ships. The impetus for the collection and archiving of data from ships came from American naval Lieutenant Matthew Fontaine Maury in the 1850s. Since then, both national and international efforts have built up a variety of impressive monitoring systems, which are the source of much of our knowledge of the climate of the world's oceans. Data from the World Meteorological Organization Voluntary Observing Ships Programme underpin better forecasting capability for maritime operations and input to research on the climate system. Approximately 6700 merchant vessels participate in the scheme.

There are also serious gaps in this upper-air network, especially in the Southern Hemisphere and over the oceans. Since accurate measurement of the rate of global warming throughout the atmosphere is essential for characterizing seasonal and regional differences, limitations to the networks have major implications for the development of sound public policies.

Land surface measurements

The longest-serving part of the observation network is the standardized ground station network that measures temperature, precipitation, surface pressure, wind speed and direction, humidity, etc. This network is comprehensive in many parts of the world, but there are still substantial gaps. Civil unrest, regional warfare and natural disasters such as flooding add to these gaps, at least over periods of months and years. Also, there are naturally very few measurements made over sparsely populated areas such as deserts and polar regions.

Can we create adequate networks soon?

To retrieve a sufficient distribution of high-quality climate data, networks must be designed to incorporate standardized measurements and to take into account how representative an observation is in a given location. The implementation of adequate national and global networks clearly requires the commitment of governments. For surface measurements the number of stations for an adequate network depends on the geography of the region — the more complex the terrain, the more stations required.

It is easier to define what is needed for the free atmosphere. In the tropics, where temperatures vary only slowly with distance, a single well-calibrated radiosonde station can provide observations that are representative for a very large area. At higher latitudes, where temperatures vary strongly in space and time, a single radiosonde ascent is representative of a much smaller area.

Even where there is a reasonably comprehensive network, there is the challenge of ensuring that the stations operate to high standards at all times. In remote locations, for all sorts of reasons, it may not be possible to make measurements as regularly as required. Sometimes, when observations are made they are not reported or transmitted immediately for wider usage.

The Global Climate Observing System (GCOS) was established in 1992 with the objective of developing an operational system capable of providing the comprehensive observations required for monitoring the climate system. The GCOS Surface Network (GSN) and the GCOS Upper Air Network (GUAN), for monitoring global-scale changes, are essential for detecting and attributing climate change, for assessing the impacts of climate change, and for supporting research toward the improved understanding, modelling and prediction of the climate system.

The GCOS networks are based on the World Weather Watch but in some parts of the world the necessary high-quality observing stations do not exist or have proven difficult for the governments to maintain. Performance monitoring at the global level is a critical part in the management of the GSN and the GUAN.

The Global Climate Observing System Surface Network (GSN) (top) and Upper Air Network (GUAN) (bottom) provide essential global coverage needed to research the climate system and understand climate change.

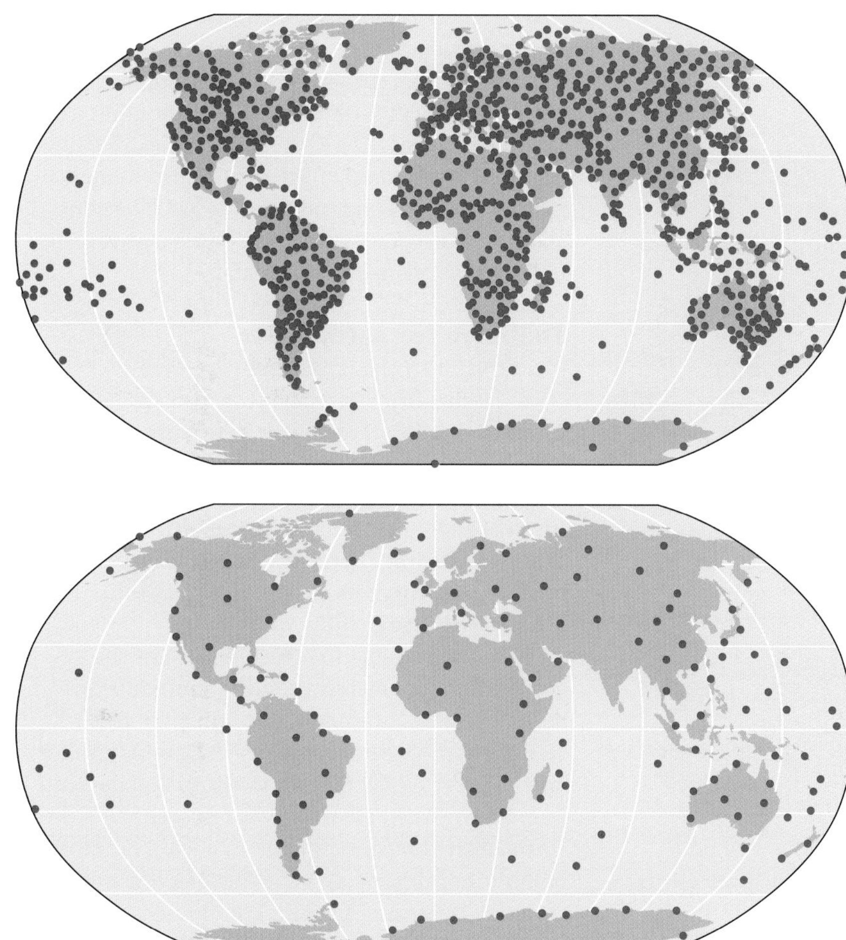

Global Atmosphere Watch (GAW)

Knowledge of changes in the composition of the atmosphere is of great importance for developing understanding of the relationships between changing atmospheric composition and changes of global and regional climate. In 1989 the World Meteorological Organization established a system that built on earlier research and monitoring activities in the field of the atmospheric environment, including the Background Air Pollution Monitoring Network (BaPMON). Known now as the Global Atmosphere Watch (GAW), its main objective is to provide data and other information on the changing chemical composition and related physical characteristics of the atmosphere.

The GAW system consists of global and regional monitoring stations. There are 22 global stations that are located in remote areas where no significant changes in land-use are expected for the coming decades within a reasonable distance (30–50 km) in all directions. The measurement programme for these stations includes water vapour and other greenhouse gases (CO_2, O_3, N_2O, CH_4, CFCs), solar radiation, ultraviolet, atmospheric turbidity, total aerosol load, reactive gas species (SO_2, NO_X, CO), chemical composition of rain and atmospheric particles, radionuclides and meteorological parameters.

At present the GAW system of regional stations consists of over 300 stations. The data are regularly published and are available to organizations, scientific institutions and individual scientists. Data from the GAW system provided the scientific basis for and stimulated the adoption of a number of international conventions and declarations aimed at protecting the atmosphere and the environment as a whole.

Modern electronic sensors and data loggers enable us to obtain measurements in difficult places, such as on mountain tops, remote islands and on the oceans. We can make routine observations through the night, when most people are asleep, but new equipment brings new challenges.

Robust and accurate automatic weather stations, like this Australian model, are becoming an increasingly important part of the climate measurement system.

The introduction of advanced sensors and microelectronics has provided many opportunities for developing more effective automatic equipment. This progress has led to systems for measuring new elements and traditional elements of the climate in new ways, but has also posed awkward questions about how to combine the new observations with the conventionally recorded data.

The drive for automation

It is difficult to ensure regular high-quality observations from sparsely-populated and inhospitable locations. Even in populated areas, however, it is sometimes difficult to find people willing to make regular observations at night and on weekends. Thus, manually recorded observations have become expensive and there has been strong pressure to develop automated observing systems.

It is now possible to measure most of the conventional climate variables (temperature, wind, pressure, humidity, rainfall, etc.) using sensors that convert a characteristic response to an electrical signal. Electrical measuring devices (e.g. thermistors) are now accurate, reliable and sufficiently robust to operate in a wide range of climatic conditions. In general, they are ideally suited to observing the climate in remote locations and can provide observations at a relatively high frequency. In fact, once the instrument and communications packages are installed the additional cost of higher-frequency observations is often very small. The switch to automation has significantly increased the amount of data being collected from many sites.

Remote automatic weather stations have to be sufficiently robust to withstand all of nature's hazards, including perhaps the interest of a family of polar bears.

Comparing the measurements

The downside of the switch to new sensors and automated instruments has been the introduction of changes in the characteristics of the quantities measured. For example, a sensitive thermistor is more responsive to the small changes of temperature associated with wind eddies than is the conventional liquid in glass thermometer. The automated system, because of its responsiveness, may generate slightly higher daytime maximum temperatures and slightly lower overnight minimum temperatures. There is also the risk that a tiny thermistor in a small plastic housing may be exposed to a different microenvironment of radiation and wind flow, and will measure something slightly different to a standard thermometer in a Stevenson screen. These differences may be no more than a few tenths of a degree Celsius – not enough to cause any problems with day-to-day weather forecasting – but they pose real difficulties for climate studies where changes of a tenth of a degree per decade are significant.

It is a challenge to produce automated sensors that measure the weather in the same way as standard historical instruments. In addition, when left untended for long periods of time the performance of many automatic sensors can slowly decline or 'drift' as pollution, grime and dust coat components of the equipment. Then there are the more direct effects such as the attentions of wildlife (e.g. nesting birds and even inquisitive bears in some locations). In spite of these limitations there often is no viable alternative. On land there are many places where it is far too expensive for human observers to be paid to conduct routine measurements every day, year in year out. This explains why, these days, so many climate observation sites are located at airports or at agricultural and other allied research stations. Similarly, voluntary sites are often on farms where the farmer benefits directly from the observations.

Inhospitable spots

For climatically inhospitable parts of the world, automatic systems are the only answer. Automatic observing stations have been developed for the icy regions of the poles and for high mountain locations. For instance, the highest stations are now two satellite-linked systems in the Bolivian Andes at altitudes of over 6500 m. This equipment is not only making measurements of current weather, but will in due course provide new information about how precipitation varies at these altitudes with El Niño/Southern Oscillation (ENSO) changes. The data will enable climatologists to calibrate ice core data from the region to make more accurate assessments of past ENSO behaviour.

Radar (RAdio Detection And Ranging)

The principle of weather radar is simple. It sends out short, sharp pulses of microwave energy and measures the signal reflected from the target, in exactly the same way as it is used in aircraft traffic control. Indeed, the first use of radar in meteorology was to track balloons to measure the speed and direction of upper atmosphere winds. Now radar is invaluable in monitoring severe storms. The amount of signal reflected by different parts of a thunderstorm is a measure of the number of droplets and ice particles present. The time taken for the signal to return to the radar system measures how far away they are. These two features can be used to build up a three-dimensional picture of the rainfall in the storm cloud. More advanced systems can also measure any frequency shift in the returning signal (Doppler radar) and from this it is possible to estimate the winds in the storm.

Since 1979, radar systems have also been flown in satellites to observe many aspects of the Earth's surface, especially over the oceans where they can now be used to measure surface wind speeds and direction, wave heights and currents, thus enabling us to build up a better climatology of the oceans, including changes in sea level.

Anticipating tornadoes and flash floods

Weather radar comes into its own for identifying and assessing the dangerous conditions associated with severe thunderstorms. Measuring the intensity of rainfall and vertical and horizontal wind speeds at a distance can give early warning of tornado developments. The most reliable indicator of a supercell is a hook-like echo in a radar image; clear evidence of the potential that severe hail or tornadoes may form. Improved warning systems have reduced but not eliminated tragic loss of life from tornadoes that nevertheless still cause terrible local damage.

Radar is particularly valuable for assessing precipitation over a river catchment because it can be used to estimate the amount of rain falling over an area. This is a crucial input to hydrological models for flood forecasting. Emergency services can then anticipate how dangerous amounts of water are likely to drain through a river basin and issue accurate alerts for flooding downstream.

The 'hook echo', as seen in radar imagery, has been associated with tornado-producing thunderstorms since the 1950s. This image is one of the supercells of the violent tornado outbreak that swept through Oklahoma City, USA on 3 May 1999. The red areas in the map generally represent the most intense rain and hail, although at the tornado location (marked) the red spot is the signature of debris being swept skyward.

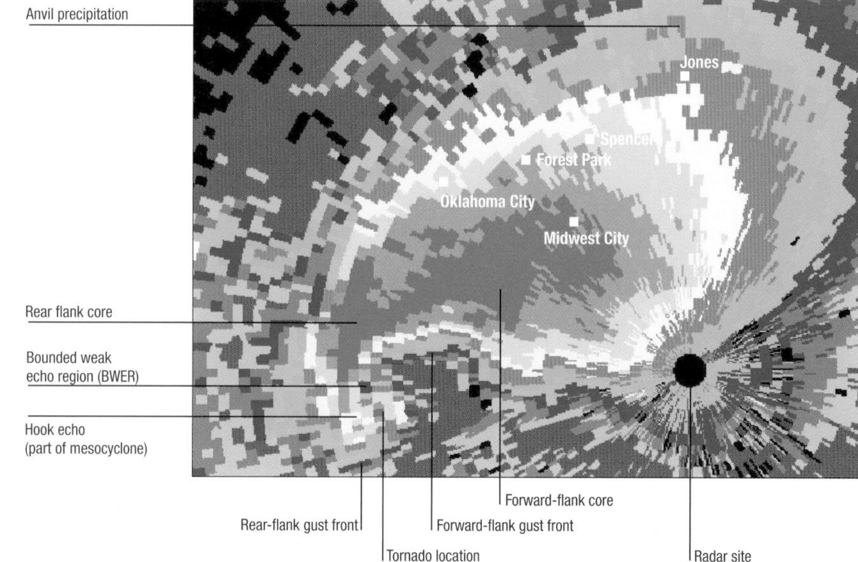

Go fly a kite

The earliest serious attempt to measure the free atmosphere was made in 1749 by Dr Alexander Wilson of Glasgow, Scotland, who used kites to carry thermometers to considerable heights. It was not until the end of the 19th century, however, that observatories in a number of countries started making regular upper-air measurements. These studies exploited the box kite, invented in 1893 by Lawrence Hargrave, of Sydney, Australia. The use of steel piano wire enabled kites to be flown to greater heights. Temperature, pressure and humidity units were designed for systems that involved flying a string of kites with instruments up to a height of 6 km or more, and these were used for many investigations until the advent of radiosondes. They provided meteorology with an excitement rarely equalled since, as a broken wire could have disastrous consequences. When a wire 7 km long flying eleven kites over Paris broke, it stopped a steamer on the Seine when it fouled the propeller and a train by jamming its connecting rods, and cut all telegraph communications with Brittany.

4.5 The view of climate from space

Weather forecasters were among the first to use the new views offered by the 'eye in the sky'. Satellite images of the Earth graphically portrayed the weather patterns of the entire globe. Streams of data transmitted from superbly designed and constructed satellite observing instruments have revolutionized our knowledge of the climate, its complex behaviour and its patterns of change.

Looking down from a great distance enables us to see much more of the weather. However, satellites give us not only real-time images of the weather, but also an almost continuous picture of the evolution of climate – the march of the seasons, the development of anomalies leading to drought, flood or heatwave, or even the slow trends of climate change.

The consistent monitoring of the Earth using satellites over the last decades of the 20th century generated an enormous database of information about various characteristics of the climate system. The data complement those available from conventional observations and add details on the spatial variability, as well as providing new insights into the workings of the climate system. Despite this enormous wealth of new data, conventional data remain essential for the calibration of many satellite instruments.

Shown is a scan from the TRMM satellite of the rainfall associated with supertyphoon *Paka* in the western Pacific in December 1997. The intense rainfall, in the central eye wall and the outer rainbands, is a characteristic feature of tropical cyclones.

Satellite images

The amount of detail provided by different satellites and sensors varies. The standard polar orbiting satellites of the US National Oceanic and Atmospheric Administration (NOAA), for example, orbiting at an altitude of about 850 km, discriminate the emitted radiation at approximately 1-km resolution and produce high-resolution pictures. Geostationary satellites, which are situated at around 36 000 km out in space, currently have a resolution of approximately 3 km in the visible and 8 km in the infrared.

Rainfall

A variety of methods have been used to estimate rainfall from satellites. Early work concentrated on correlating the brightness of clouds and, from the outgoing long-wave radiation, the temperature of their tops. This statistical approach gives reasonable indications of the overall amount of rainfall over wide areas and relatively lengthy periods (weeks and months), especially in the tropics. Later, more accurate measurements have been obtained using microwave radiometers. These devices can detect a different signal because the rainfall droplets range over sizes that give them distinctive emission characteristics.

A number of satellite instruments have been used to measure rainfall, again notably in the tropics. The latest of these is the Tropical Rainfall Measuring Mission (TRMM) run by the US National Aeronautics and Space Administration (NASA) and the Japanese National Space Development Agency (NASDA), which started in the autumn of 1997.

Radiometers

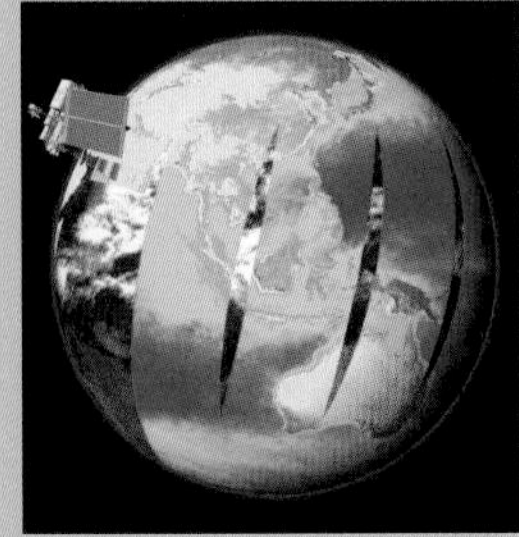

A radiometer is an instrument that measures the amount of heat radiation coming from a body. By combining the radiometer with a telescope, it is possible to look at a small area beneath the satellite and accurately measure the total amount of energy coming from it or the energy in selected wavelength bands. The telescope sweeps back and forth, building up a picture from the segment in each strip viewed. These measurements, when combined with reference observations, can be transformed into useful data.

Measurements of the total heat are used in monitoring clouds and studies of the Earth's energy budget. Measurements in selected bands of the spectrum provide information about the layers of the atmosphere (e.g. temperature or humidity) or about the Earth's surface (e.g. the amount of vegetation on the land or the amount of plankton in the surface waters of the oceans).

Radiometers that measure upwelling microwave radiation from the Earth are of particular value. These sounding units on the NOAA weather satellites can see through clouds and so monitor atmospheric temperatures with great accuracy. Since the first operational unit was launched in 1979, a continuous record of global temperature has been maintained. They can also make accurate measurements of sea surface temperatures and monitor large-scale changes associated with phenomena such as El Niño and La Niña.

Combining radar and a microwave radiometer, this system has provided new insights into the climatology of tropical rainfall and the associated release of energy that helps power the global atmospheric circulation. It will be some time, however, before the equipment on follow-on satellites achieves the precision needed to provide a measure of climatic trends in rainfall.

Sea ice and snow cover

From the earliest days weather satellite images have been used to monitor sea ice and snow cover. With the development of infrared and microwave radiometers these measurements have become the standard means of keeping track of the seasonal expansion and contraction of polar ice, continental snow cover and mountain glaciers, and any trends over time.

Wave heights, wind speeds and currents

Satellite observations of wave height rely principally on a radar altimeter. This device works on the same principle as ground-based radar and measures the shape of the signal reflected back from the ocean surface. If the sea surface is flat then the returning signal has a sharp edge and a larger amount of energy is reflected back towards the satellite. If, however, the sea is rough then the reflected radiation travels different distances depending on where precisely it strikes the waves, and the bigger the waves, the more the signal is smeared out. Radar altimeters are capable of a precision of better than half a metre, which compares favourably with estimates of wave characteristics made by ships' observers. The global coverage provided by recent satellite missions has added huge amounts of data to the records laboriously built up over more than a century from shipboard observations.

Radar altimeters can also be used to make precise measurements of the height of the ocean surface. The surface water slope in the oceans is related to the speed and direction of the underlying water flow, and scientists have been able to use measurements of sea surface topography to make new assessments of the strengths of the major ocean currents. Sea level across the equatorial Pacific Ocean is affected by the El Niño/Southern Oscillation (ENSO) and satellite measurements assist in providing early warning and monitoring of the development of El Niño events.

Other measurements

Satellites are now crucial for monitoring many aspects of the environment. For example, satellite-borne spectrometers have been developed to measure the total atmospheric ozone and amount of atmospheric pollutants in a column of the atmosphere. The Total Ozone Mapping Spectrometer (TOMS) instrument compares ultraviolet radiation scattered back from the Earth's atmosphere with the direct emission from the Sun to map, in detail, the global ozone distribution. These data are the basis for monitoring the extent and intensity of the ozone hole that develops over Antarctica each austral spring. TOMS also measures sulphur dioxide released in volcanic eruptions and detects other atmospheric pollutants.

Meteorological satellites are used to measure the state of vegetation cover around the world through the seasons. The proportion of reflected visible light and emitted infrared radiation varies with the vigour of plant growth and differs appreciably from bare ground. Below-normal growth activity may identify the early stages of drought and, later, the severity of the problem being experienced.

The SeaWinds satellite-borne scatterometer, a microwave radar, measures ocean near-surface wind speed and direction under all weather and cloud conditions over Earth's oceans. This information is vital to weather forecasting, storm detection and ship routing, to studies of the ENSO phenomenon and to studies of air–sea interaction. On 1 August 1999, Typhoon *Olga* can be seen in the China Sea. The three groups of intense winter storms near Antarctica are associated with the season of maximum sea-ice in that region.

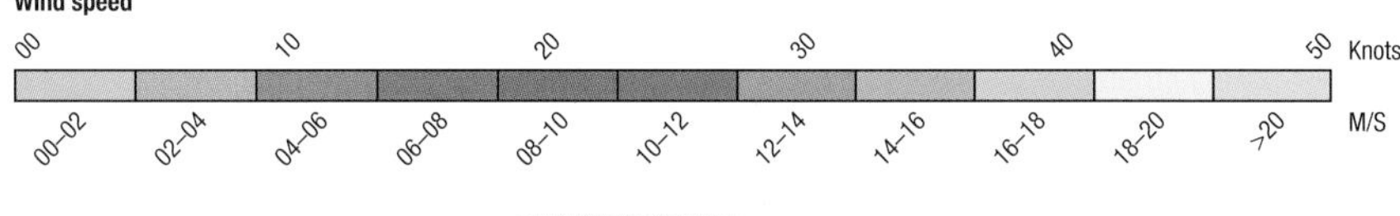

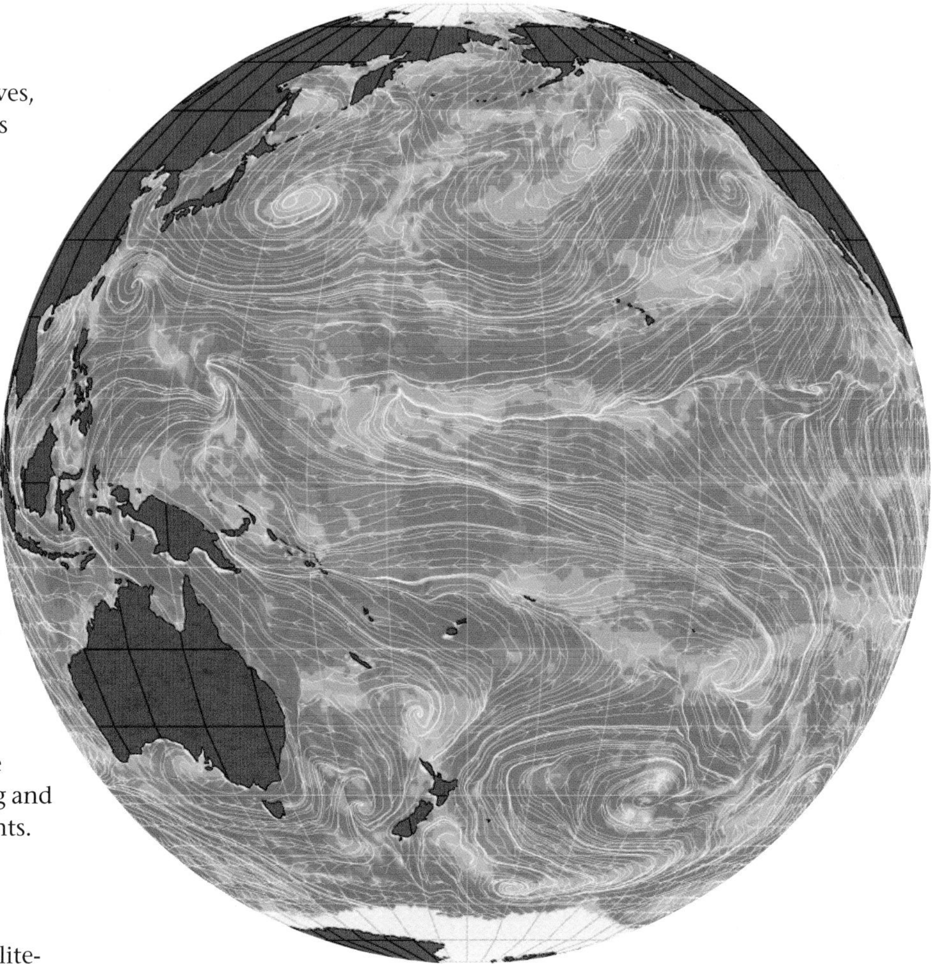

4.6 Climate knows no boundaries

The forces that drive the climate care nothing about territorial borders. To deal with problems such as flooding, drought, air pollution, desertification and climate change requires information spanning countries and continents. These requirements place modern meteorological services at the forefront of international cooperation.

The First International Meteorological Congress, held in Vienna in 1873, led to the creation of the International Meteorological Organization (IMO), the precursor to the World Meteorological Organization (WMO). The IMO provided the framework for international cooperation in meteorological data collection, research and service provision until the end of World War II.

No country can address its climate problems on the basis of its own networks and data alone; to obtain the overall picture requires putting one's own data into the global context, which requires a concerted and an international effort. International data exchange underpins world meteorology, and meteorological services provide an intergovernmental mechanism to facilitate the necessary cooperation. Every country that submits its observational data for the mutual benefit of all is entering into this spirit of global sharing.

A history of international cooperation

International cooperation in meteorology was active in the 18th century. Under the guidance of US naval officer Lieutenant Matthew Maury, the first International Meteorological Conference was held in Brussels in 1853. It established principles for standardization of observing instruments and for recording the data. Naval ships first took up the task of maritime observations and then enlisted the cooperation of commercial ships to make routine observations.

In 1873, representatives of 20 governments met in Vienna and agreed to set up the International Meteorological Organization (IMO). Although this was a non-governmental organization, it was to have a profound influence over the management of meteorological services during the next three-quarters of a century.

Major activities of the IMO in the early years were directed towards devising telegraphic codes, fixing observing standards, and establishing times for the international exchange of observations by telegraph. The IMO was also active in promoting scientific research and publication of climate information. In 1905 a conference recognized the importance of wireless telegraphy for the collection of weather reports from ships at sea, and so began an essential element of maritime meteorology. After a period of enforced dormancy during World War I, the IMO was closely involved in the exploitation of rapid advances in aviation during the 1920s.

By 1939 it had become clear that a more formal intergovernmental organization was needed to handle the growing international nature of meteorology, and a draft convention was drawn up. Following World War II, a revised version of this draft was signed in 1947, and led to establishment of the World Meteorological Organization (WMO) on 23 March 1950.

The World Meteorological Organization (WMO)

The WMO Convention came into force on 23 March 1950 and by Resolution in 1951 the United Nations General Assembly recognized WMO as a specialized Agency of the United Nations system. At the Thirteenth World Meteorological Congress (1999) there were 185 Member States and Territories.

The purposes of WMO are set out in the Convention and are generally directed towards:

a) worldwide cooperation in the establishment of meteorological observing networks;
b) systems for the rapid exchange of meteorological information;
c) standardization of methods of observations and uniform publication of observations and statistics;
d) application of meteorology to aviation, shipping, agriculture and other human activities;
e) close cooperation between National Meteorological and Hydrological Services; and
f) encouragement of meteorological research and training.

Global Observing System

Instruments deployed on land, at sea, in the air and in near outer space make up the Global Observing System. The backbone of the surface-based subsystem continues to be some 10 000 stations on land making observations at or near the Earth's surface every three hours, although data from less than 1500 of these are exchanged for global climate purposes. More than 1000 stations make observations of the upper air using radiosondes while more than 300 commercial aircraft automatically provide in-flight observations at regular intervals. Across the oceans about 6700 ships, 1000 drifting buoys, 300 moored buoys and 600 fixed platforms are instrumented to provide meteorological and some oceanographic observations. The space-based subsystem is made up of both near-polar orbiting and geostationary environmental observation satellites that augment the observations of the surface-based systems to provide global coverage.

The Global Observing System is currently being modernized to meet the requirements of the 21st century.

World Weather Watch (WWW)

With the emerging capabilities for modelling the atmosphere using fast computers, and the dawning of the space age, WMO recognized the need for rapid access to data from around the world. In 1963 WMO established the concept of a World Weather Watch and the programme came into operation on 1 January 1968. It marked a symbolic step forward, with the clear purpose of addressing the global nature of meteorology, building on the guiding principle of international cooperation. The World Weather Watch coordinates a burgeoning number of global programmes to facilitate the international exchange of data for the benefit of humankind. Built on the three pillars of the Global Observing System, the Global Data-processing System and the Global Telecommunication System, the World Weather Watch has since branched out to underpin the major WMO programmes that have provided much of our growing knowledge of the climate in recent decades.

World Climate Programme (WCP)

The First World Climate Conference of 1979 drew attention to climate as an issue of international importance, recognizing the impacts of climate extremes and the potential for climate change. As a consequence, in 1980 the World Climate Programme was established as an inter-agency, interdisciplinary effort of the United Nations and the International Council for Science to address the major climate issues. Following the 1992 United Nations Conference on Environment and Development (The Earth Summit of Rio de Janeiro) the United Nations agencies with climate activities agreed to coordinate them through the Climate Agenda. The four main thrusts of the Climate Agenda are:

- new frontiers in climate science and prediction;
- climate services for sustainable development;
- dedicated observations of the climate system; and
- studies of climate impact assessments and response strategies to reduce vulnerabilities.

WHYCOS

The wise management of freshwater resources is essential to the future well-being of humankind. In 1993 WMO, in association with the World Bank, launched the World Hydrological Cycle Observing System (WHYCOS). The system is designed to promote regional and intraregional cooperation to improve knowledge at the regional and global levels on the status and trend of water resources. Combining both the analysis of historical data and the provision of real-time data collection and transmission via satellite, the system is designed to meet the particular needs of different regions of the world. By 2000, projects had been launched in three regions (Mediterranean (MED), Western and Central Africa (AOC) and Southern Africa (SADC)) and good progress was being made in others, such as in Eastern Africa (IGAD).

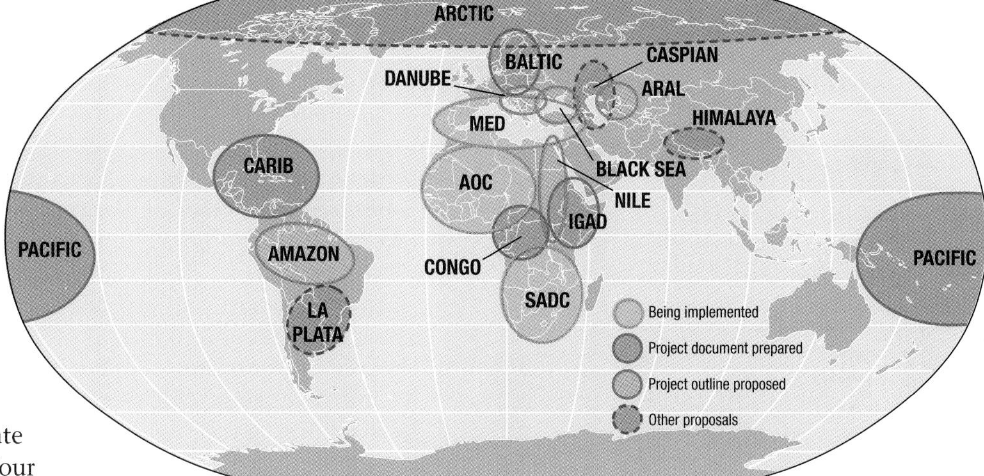

The World Hydrological Cycle Observing System (WHYCOS) is the global 'umbrella' for independent regional HYCOSs. It is designed to assist WMO Members in capturing, using and disseminating water-related data, for better understanding of the global hydrological cycle.

4.7 Serving society's needs

Technological progress and international collaboration are only part of the developments that are sweeping the climatologist's world. As understanding of the climate system grows and society becomes more aware of the potential opportunities from this knowledge, demand develops for new and better climate services.

The use of climate data and information has changed over time as the science of climatology has evolved. Early practical climate studies were based on the recording of phenological events about flora life cycles, faunal migrations and other significant events related to the seasons. With the introduction of instruments and the spread of observing stations around the world, scientists began to use quantitative measurements to explain the distribution of global life zones and soil types. They wrote eloquent descriptions of regional climates, relying heavily on observations and the availability of expanding climatological archives. Statistics became a primary tool as the climates of different places were compared and the differences highlighted.

The modern climate service

At the heart of a modern climate service are the infrastructure and trained personnel for collecting, storing and processing national climate records. Statistical and graphical presentations define the national climate, its year-to-year variability and any change over time. In addition, national services are linked into the international infrastructure for exchange of data and generation of regional and global monitoring and prediction products.

The success of a national service comes from its ability to meet the many demands from the communities it serves. There is hardly a human activity that is not weather and climate sensitive to some extent, despite the adaptation capabilities that have enabled communities to exist in climate regimes from the equator to the poles. The public benefit from national weather, climate and hydrological services comes from their contributions to safety of life and property in the event of extreme meteorological and hydrological events, and from their contributions to future development. A strong foundation of climate science is essential to ensure that development policies adequately reflect climatic realities.

Making effective use of national meteorological records, such as these in Viet Nam, is an essential part of developing more efficient climate services. Many National Meteorological and Hydrological Services have converted their manuscript records to computer-based climate archives so that the data are more accessible and can be readily analysed in different ways to meet community needs.

The climate Normals

It has long been recognized that climatological data should be processed over agreed uniform periods. This ensures a common base to compare climates all over the world and to provide a common long-term reference value or 'normal' with which shorter-term (e.g. monthly) data can be compared.

Climatological Normals are averages computed for a uniform 30-year period. The climatological standard Normals relate to the periods 1901–1930, 1931–1960, 1961–1990, etc. Data for the current climatological standard Normals (1961–1990) have been provided by Member countries of the World Meteorological Organization (WMO) for nearly 4000 stations and published by the National Climate Data Center of the USA on behalf of WMO.

Climate science at various levels underpins food and fibre production, transportation, safe housing and public infrastructures, planning and utilization of freshwater resources and the enjoyment of many leisure activities. Future policies for the generation and utilization of energy must be taken in full knowledge of the sensitivity of the global climate system, especially to increasing concentrations of greenhouse gases.

Effective climate services must bridge disciplines and combine the expertise needed to understand the totality of any problem and to analyse how to use the information wisely. This requires frequent contact between providers of climate services and those who use the information. It also requires information to be readily accessible, presented in a useful format and timely for specific needs. Whether the information is broadcast, published or accessible on the Internet, it is essential that there are minimal barriers to the exchange of the basic climate data from which the information is derived.

Sustainable development

Climate services linked to planning and early warning contribute to safety of life and minimization of property loss in the event of extreme meteorological and hydrological events. Also, a strong foundation of climate science is essential for promoting sustainable public policies, because the potentially disastrous impacts from severe events, including loss of life and destruction of property, set back development. Furthermore, the threat of climate change is highlighting just how sensitive environmental, social and economic systems are to climate.

In many sectors of our society quantified and specifically tailored weather, climate and hydrological services enhance decision-making. Often these applications require the development of special decision systems that recognize critical times in the activity cycle, or critical values of vital climate variables. For example, the performance of power stations is improved when the management of generating capacity takes into account the weather sensitivity of community power usage.

Communities can be seriously disrupted when exceptional weather events occur. Well-designed public infrastructures such as this bridge over the Una River in Barreiros, Brazil can better withstand extreme phenomena, reducing the strains on economic and social activities in stricken areas.

Long-term decisions

Decisions to establish tree plantations or build bridges are based on estimates of what the climate will be like decades in the future. But there is evidence that the Earth is warming and the climate is changing, so the premise that 'the past is the key to the future' may no longer be adequate. Any advice for the long term must address the nature of future risk, and the changing probabilities of exceeding critical extremes.

It may be relatively straightforward, for example, to give advice on harbour installations or sea-wall defences in connection with the likely sea-level rise over the next 100 years. By comparison, providing projections for planning decisions about how offshore installations should be designed to withstand more severe storms and bigger waves in the future is a more complex task.

Risk and design data

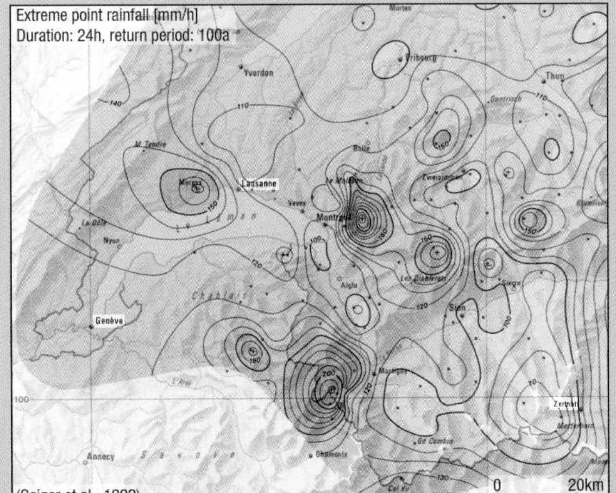

Climate is more than the average weather; it encompasses the complex variability and extreme events associated with successive weather systems. Assessing the risk of occurrence of many events is reasonably straightforward where a long series of data exists. In many countries records often go back over a century, and even longer records can be deduced from proxy data, such as that given by tree rings, lake sediments and ice cores. It is from the historical data that we calculate the probability or risk that a particular extreme event will occur. Of course, the value of such predictions depends on our knowledge of the climate's stability or regularity of change.

Data from climatological archives have been statistically analysed and widely disseminated in climatological atlases and other forms of reference material. The analyses are available for use in a variety of ways, including strategic planning, long-term investment, engineering design and policy formulation for weather-sensitive activities.

The map of once in a 100 years extreme point rainfall (mm/h) over a 24-hour duration shown here for western Switzerland is one example of a hydrological analysis. Knowledge of peak rainfall intensity and accumulated volumes over various durations are used in planning and design of water management and flood-control systems that need to cope with peak flows. Other climate analyses are used for the siting and design of recreational facilities, the selection of a best-suited cultivar for local agricultural conditions, and the building of airports and alignment of runways.

Structures built to last for decades to centuries must be able to withstand the strongest climate extreme, no matter how rare, with safety over the period. For example, because of the severe danger to occupants and passers-by, the wind loading and fixing strength for panes of glass in a high-rise building must be such as to prevent any chance of implosion or explosion in high winds. Equally, structures such as bridges and roads must be able to withstand weather extremes to maintain vital communications.

4.8 Tracking climate

Weather and climate variations have economic, social and environmental impacts at the national and regional levels and affect the course of many global markets. Managing weather- and climate-sensitive enterprises is enhanced through access to critical climate information from the past and the present, and through anticipation of future climate.

Many countries have developed a climate monitoring capability within their National Meteorological Service and others benefit from cooperative regional arrangements. Regular bulletins, advisories, and publications containing news about significant national and regional climate events, and more recently climatic outlooks and forecasts, are primary outputs from the national and regional centres. Events beyond national borders require a global climate monitoring system. Also, because of globalization of markets, governments and industry sectors need to know how climate events in other parts of the world are affecting their competitors, suppliers and potential sales.

The fusion of weather and climate information

The separation between what is recognized as weather and what is climate has become increasingly blurred in recent decades. In the past, weather covered current data, while climate described the past in terms of monthly, seasonal and annual statistics, particularly averages, totals and extreme events. Weather information was for real-time decision-making and climate information for long-term planning and design.

Data covering the most recent weeks to months have long been used in the assessment of climate events, such as drought, having severe socio-economic impacts. Until recently, statistics on these climate events took weeks or months to compile, and so the first awareness of a slowly emerging event was not through data analysis but through the impact of the event itself. Moreover, because of limited understanding of the climate system the belief was that once the seed was sown, there wasn't much one could do about a prolonged event.

Better understanding of the climate system has pointed to the value from ongoing monitoring of its variability. As a consequence, the technological infrastructure developed for weather forecasting, including the global data collection, is being harnessed and built upon for climate purposes. The daily data for weather analysis are fed directly into climate databases and are available for rapidly generating climatological maps of monthly averages and anomalies, and for analysing various indices for monitoring climate phenomena such as the El Niño/Southern Oscillation (ENSO), the North Pacific Oscillation and the North Atlantic Oscillation. Also, the data are being used to initiate climate forecast models that are showing potential to identify and map the evolution of anomalies several months in advance.

Exploiting the latest information

Recent weather and climate data have proved to be a powerful and essential information source

Exploiting marine observations

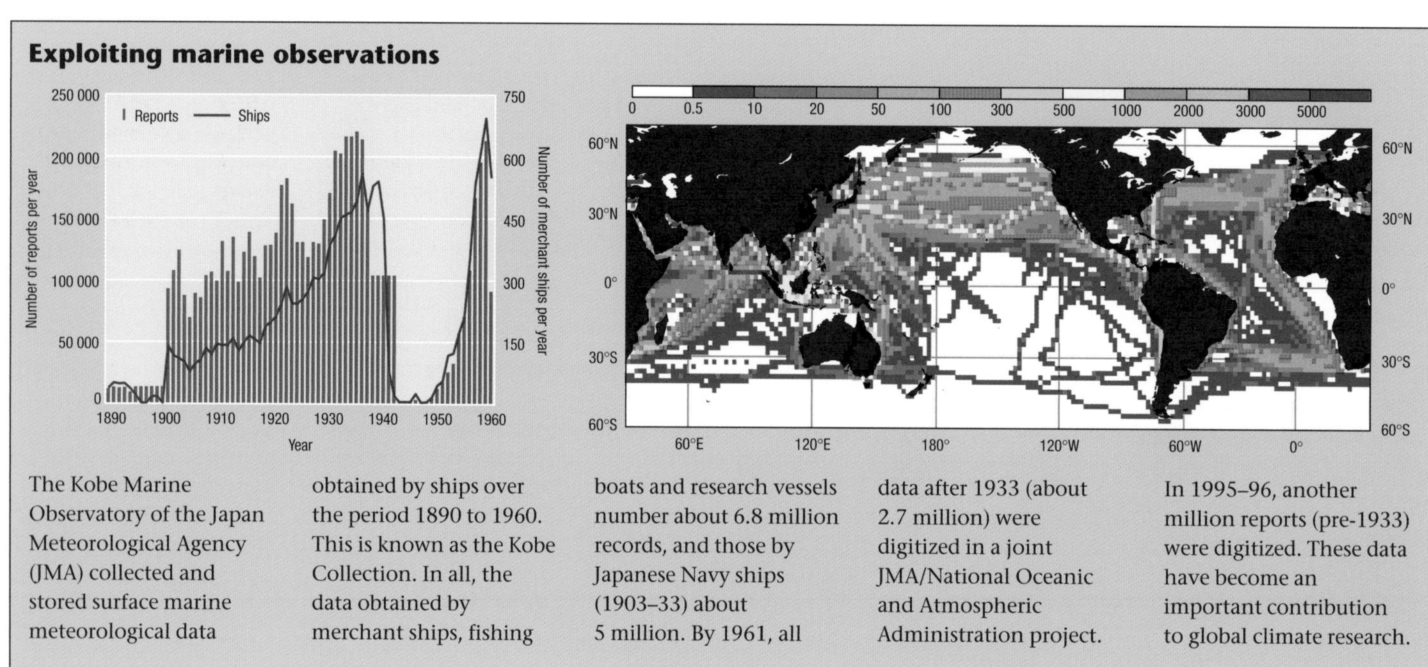

The Kobe Marine Observatory of the Japan Meteorological Agency (JMA) collected and stored surface marine meteorological data obtained by ships over the period 1890 to 1960. This is known as the Kobe Collection. In all, the data obtained by merchant ships, fishing boats and research vessels number about 6.8 million records, and those by Japanese Navy ships (1903–33) about 5 million. By 1961, all data after 1933 (about 2.7 million) were digitized in a joint JMA/National Oceanic and Atmospheric Administration project. In 1995–96, another million reports (pre-1933) were digitized. These data have become an important contribution to global climate research.

for users concerned with day-to-day activities, or with processes that are happening now or that may happen in the near future. Climate input to operational decisions is found in many activities, including irrigation scheduling, pesticide application, forest fire surveillance, fuel deliveries, food production forecasts, water flow regulating, commodities trading and electrical power generation and distribution. These routine activities are generally concerned with variations on timescales ranging from hours through days to seasons. Climate monitoring, based on a comparison of the integrated weather over the past month or season, is now routinely used for early warning of emerging hazards, such as drought or conditions favourable to diseases or pests (e.g. locusts) that attack crops.

In many countries effective climate services providing early warning are built upon infrastructures providing access to the relevant climate information, and interpretation of the information in the context of the hazard of concern. For example, drought monitoring requires a rainfall reporting network and an historical archive of information that defines what is the 'normal' expectation for a particular month or season of the year. Without the historical data, reasons for a developing water shortage and/or soil degradation may be misinterpreted as due to increased resource usage rather than reduced rainfall, and an incorrect analysis could lead to implementation of wrong policy responses.

The value of climate-monitoring information is enhanced by its timeliness and the ability of decision makers to respond positively to a changing situation. The infrastructure of a National Meteorological and Hydrological Service must be capable of quickly collecting and analysing data over relevant timescales, such as months or seasons, and integrating these with regional and global analyses of the climate system that are increasingly becoming available from major meteorological centres. Putting information into the hands of decision makers without delay is thus critical, either to minimize losses and damage or to exploit opportunities.

Mining the archives

Computers and the Internet have revolutionized the management and use of climate data. Where there are on-line electronic databases it is now possible to manipulate raw data for the many applications, and to produce on demand tabular, graphical and chart-form summaries of meteorological data. It is possible to immediately relate current weather events with historical records and to determine, for example, whether today's heavy rainfall is an all-time record.

Routinely, computers are being used to produce averages, extremes and probabilities of critical events. Computers are also being used to run a variety of applications incorporating climate data, such as estimating crop-yields under changing thermal and moisture conditions, calculating water budgets for estimating lake level and river flow, and undertaking climatic impact assessments in sensitive industries.

More and more, climate data and summaries are becoming accessible on the Internet such that climate services are available beyond the government offices or public libraries to the field, factory and boardroom. Unfortunately, worldwide many valuable historical climate records that form the basis for these systems and capabilities remain in manuscript form, deteriorating and in danger of being lost forever. A major international effort is required to identify and digitize this vital memory of past climates.

Anticipating when the climatic conditions are particularly favourable for the breeding of swarms of locusts is an essential part of reducing the impact of this scourge.

The global reanalysis project

Since the earliest days of using computers for weather analysis the charts have been archived in digital form and used as the basis for building regional and global climatologies. However, many observations were delayed in the communications systems and so were included in neither the weather analysis at the time nor the subsequent climatology. Also, with time, big improvements in the automatic data analysis systems were introduced. By the late 1980s it was recognized that the missing data and the changed analysis methods contributed to significant deficiencies, particularly in the early years of analysis, that had the apparent effect of altering the climate. Even fundamental quantities, such as the strength of the Hadley cell, appeared to vary over the years but were in reality the result of the changes in the data assimilation systems.

Several reanalyses of the global data sets have recently been carried out using all available data and standard data assimilation and analysis methods. In addition to the missed observations, more comprehensive observations (e.g. satellite radiation measurements, better sea surface-temperature analyses, wind fields obtained from geostationary satellites and the results of the latest field experiments) were used. Improved quality control and better vertical resolution (e.g. in the stratosphere) have also contributed to form a much more comprehensive picture of the state of the atmosphere over the years.

Reanalysis data sets independently compiled by the European Centre for Medium Range Weather Forecasts (ECMWF), the Japan Meteorological Agency and, in a joint endeavour, the National Centers for Environmental Prediction (NCEP) and National Center for Atmospheric Research (NCAR) of the USA are benchmarks for climate studies. They provide the most comprehensive analyses of the climate of the second half of the 20th century and are particularly valuable for climate variability and predictability studies.

4.9 Predicting the seasons

Piecing together the puzzle of the El Niño/Southern Oscillation (ENSO) phenomenon has led to the prospect of useful seasonal climate forecasts. The next step is to learn how to apply these forecasts to reduce global ENSO impacts, in such areas as fisheries failures, floods in Peru and droughts in East Africa, Australia and India.

Predicting the behaviour of ENSO-related phenomena over the months and years ahead offers the best prospect of seasonal forecasting for many parts of the world. In other parts, such as Europe and West Africa, regional sea surface temperatures seem to be the important factor. The success of seasonal forecasts depends on a detailed knowledge of how the atmosphere and ocean interact. As our understanding of the relevant processes has evolved, increasingly complex models have been produced to exploit improved measurements of conditions and provide improved seasonal forecasts.

The demand for predictions

The prospect of forecasts on seasonal and longer timescales, for example about whether rainfall or temperature will be above or below average, and by how much, shows enormous potential benefits. These forecasts are generally the products most eagerly sought for longer-term decisions and early warning of potential hazards but, as for any forecast, credibility depends on a proven track record. To be accepted, climate predictions will have to demonstrate skill beyond the climatological experience that is traditionally used to make decisions.

Empirical methods

The simplest way to forecast departures from the normal climate months to years ahead is to establish statistical rules linking future patterns to current climatic anomalies. Large-scale, slowly varying anomalies, in sea surface temperature for example, that can persist for many months, may force changes in atmospheric circulation patterns and hence departures from normal of local climate cycles. At first, this approach was not a great success, but now the growing understanding of what drives ENSO and other forcing patterns has brought some improvements.

Empirical methods using, for instance, sea surface temperatures assume that local climate will be affected in roughly the same way each time there is a similar large-scale forcing. Their advantage is that they are relatively easy to apply, as they rely entirely on climate statistics and use modest computer resources. There are, however, limitations because the statistical models generally attempt to use a linear approach to predict complex interactions without any specific links to the underlying physical and dynamical processes. This means that they work best when large-scale developments are well and truly under way, but they have difficulty anticipating shifts from, say, warming to cooling or vice versa. As a consequence, they generally do not predict sudden developments well.

Computer models

A more physically based approach to seasonal forecasts uses computerized general circulation models. In one form of this approach the first step is to predict the development of sea surface temperatures in the tropical Pacific. These predictions may be based on a regional model that considers developments in the tropical Pacific in isolation. Once this model has made forecasts of how the Pacific may behave up to a year ahead, the forecast sea surface temperature patterns are used to drive an atmospheric general circulation model to predict how the weather around the globe might respond. These predictions of seasonal weather have produced promising results, especially in the tropics. The improvement of these forecasts has been built on both advances in the models and better observations from the equatorial Pacific Ocean. Significant developments are now under way to build fully coupled systems in which the ocean, atmosphere and land surface components of the computer model continually interact with each other to produce a forecast up to several months ahead.

Making the forecasts relevant

A challenge to forecasters is to ensure that the forecasts are both timely and understandable to users. Since the most successful forecast methods to date have been mostly associated with ENSO events, it is the lesser-developed tropical countries that stand to gain most from these developments.

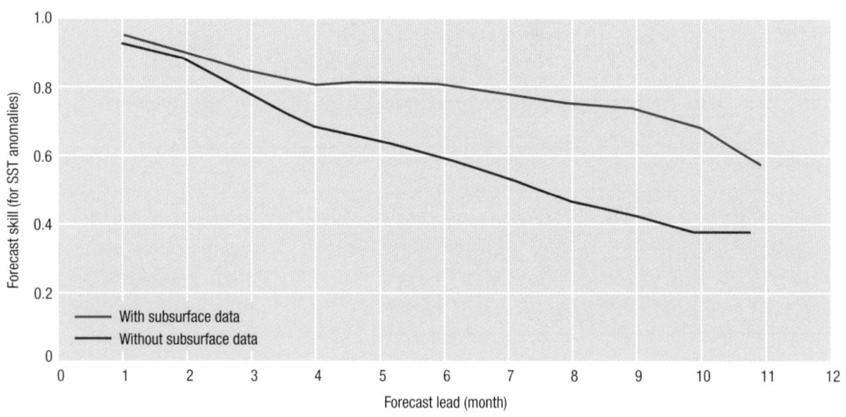

Forecast 'skill' from a coupled ocean–atmosphere model can be expressed in terms of correlation between forecasts and observations of ocean sea surface temperatures (SSTs) for various lead times. Skill is generally higher for forecasts that used subsurface data than for those that did not, particularly at lead times longer than three months. The TOGA-TAO moored buoy array in the equatorial Pacific provides the vital subsurface data which underpin improvements in forecasting ENSO events.

It is important, then, that the predictions are presented in ways that farmers or fishermen can readily use; useful predictions cannot rely on computer graphics or statistical arguments, but on statements that can be broadcast over the radio or published in newspapers. In addition, they must be available in time for people to make decisions about what to sow, and in a form that is relevant to making such important decisions. This involves understanding local agricultural practice and variations in climate, and integrating both into methodologies for better decision-making.

Seasonal forecasts: the acid test

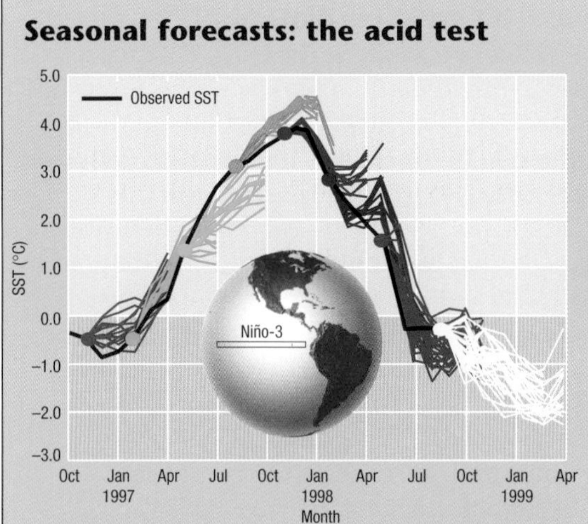

The first real test of modern seasonal forecasts was during the 1997–98 El Niño event. During 1996 many computer models had been predicting that a warming (i.e. El Niño event) was on the way. The strength of the evolution, however, was predicted well only after the event was under way. Several centres produced and published their forecasts including the National Centers for Environmental Prediction (NCEP) of the USA and the European Centre for Medium Range Weather Forecasts (ECMWF). The 'plume' diagram represents the evolution of sea surface temperatures from forecast ensembles for the eastern equatorial Pacific (see 4.19). The model from October 1997 clearly predicted warming. Although the magnitude was underestimated for May to July 1998, the forecast was excellent from then until the following March.

Subsequent forecasts had difficulty both with predicting the sudden onset of the 1998 La Niña event and with its continuation into 2001. The models show least reliability in predicting an onset of or switch to either a warm or cool event, as do empirical forecasts. Representing forecasts of the sea surface temperature in the eastern equatorial Pacific, the 'plume' diagrams plot the ensemble predictions from each forecast point on the ECMWF model.

Probability forecasts

There is always uncertainty about the future, and a challenge for forecasters is how to represent this uncertainty. One way is to assign a specific probability, for example to each of dry, near average and wet conditions. At a location, the climatological data define the distribution of past rainfalls (e.g. seasonal totals) and the expectation, or climatological probability, of any of these totals in the future. One method, as used by the International Research Institute for Climate Prediction based in the USA and shown in the sets of boxes of the upper map, is to predict for a season the probability of totals in each of the climatological terciles (or thirds of the distribution). The lower tercile (left box) represents dry conditions, the middle tercile (middle box) is near-average rainfall, and the upper tercile (right box) represents wet conditions. In the map, a number that is different from 33 per cent indicates the forecast shift from the climatological distribution for the period. A number that is much higher than 33 per cent in the left box indicates an increased chance of drier conditions, while a number much higher than 33 per cent in the right box indicates an increased chance of wetter conditions for the region represented. In the lower map the distribution of observed rainfall, expressed as a percentage of average, gives a visual pattern of the overall success of the forecast.

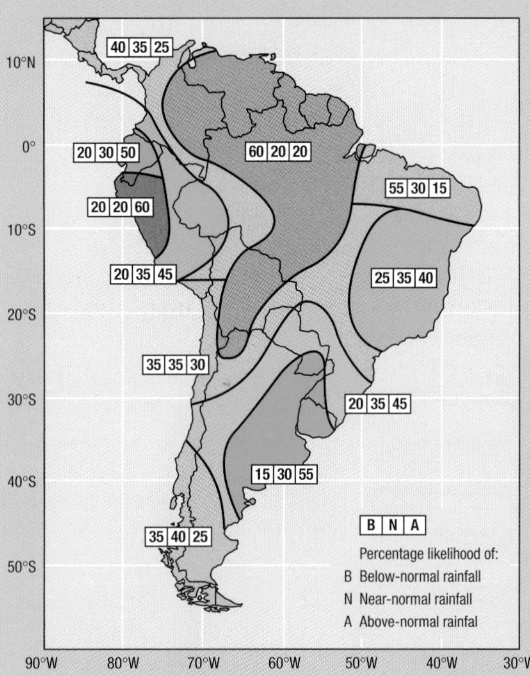

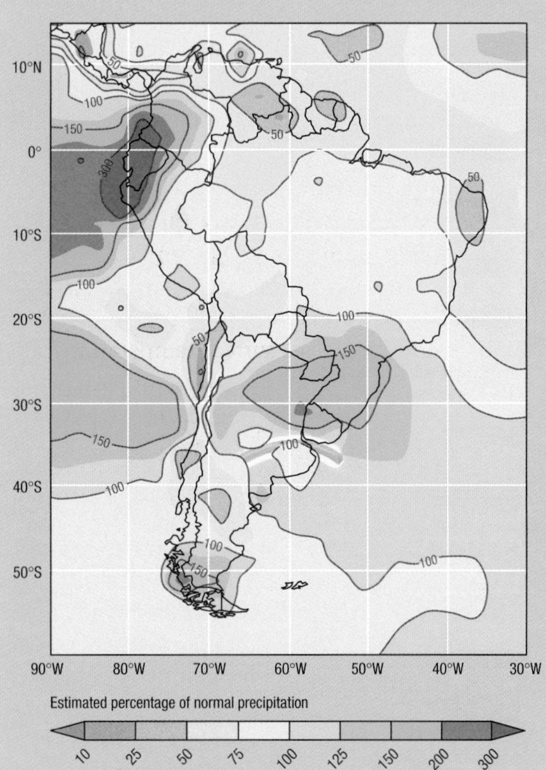

4.10 Forecasting years ahead

The ability to predict changes in the climate on the timescale of a few years to several decades would have a profound effect on how we manage our lives. Confirming the existence of quasi-periodic cycles in the range 2 to 100 years could be of great value in making investment decisions and planning economic activities.

The winds in the equatorial stratosphere swing from westerly to easterly and back again every 27 months or so. This regular behaviour, called the quasi-biennial oscillation, or QBO, may be linked with other less well-defined periodic fluctuations in the climate.

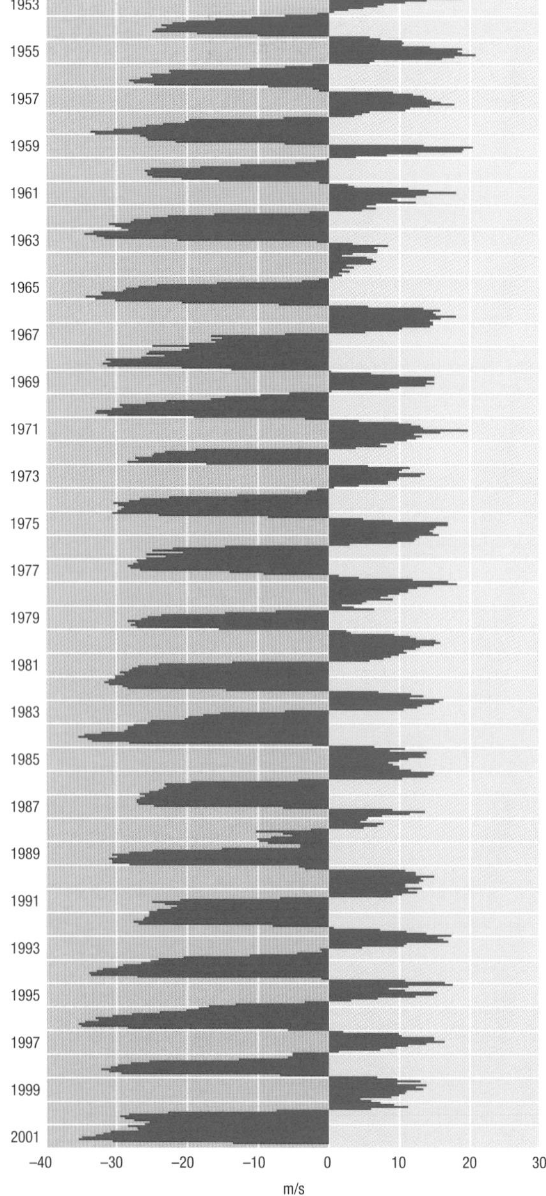

The progress made in understanding the interactions between the atmosphere and the oceans that have provided insights on the El Niño/Southern Oscillation (ENSO) and other climatic 'oscillations' raises the question of whether there is any prospect of extending predictions of natural climatic variability years and decades into the future. Central to this endeavour is discovering whether we can establish physical explanations for any of the claimed 'climate cycles' and shifts and so use them for forecasting. This search is also of great importance in establishing the extent to which the consequences of current global warming, attributed to human activities, can be predicted in the future.

What are we looking for?

Seasonal forecasts are built on a growing understanding of how the atmosphere interacts both with the oceans and with other more slowly varying components of the global climate (e.g. snow cover, pack ice and soil moisture). This offers the possibility of predicting the probabilities of periods of anomalous weather occurring months or even a year or so ahead.

Weather or climate cycles have fascinated many meteorologists for more than a century. It has been a long and largely fruitless search for clear evidence for predictable cycles. There are, however, a few periodic features that scientists continue to examine closely. In particular, there is a quasi-biennial oscillation (QBO) observed in winds in the stratosphere, with periodicities around 3 to 5 years associated with ENSO, and there are also 11, 18–22, and 80–90-year cycles which may be linked with solar activity, or possibly lunar tides in the case of the 18–22-year cycle.

Quasi-biennial oscillation

There is no doubt about the QBO. This regular reversal of the winds in the stratosphere over the equator has been closely studied since it was first discovered in the early 1950s. In recent years work has focused on possible links between this stratospheric periodicity and tropospheric weather. The QBO has been linked with the fluctuations in the ENSO and its characteristic rainfall patterns, and also with the incidence of hurricanes in the tropical Atlantic. The QBO has also been linked with the 11-year variation in solar activity associated with the incidence of sunspots.

During the 1980s there was considerable interest in the fact that the north polar stratosphere during winter tended to be colder during the westerly wind phase of the QBO than during its easterly wind phase. Also, at the maximum in solar activity it was unusually warm if the QBO was in its west phase. By sorting out the east- and west-phase winters it was possible to show a strong correlation with solar activity. In terms of tropospheric weather, examination of this relationship between 1956 and 1988 suggests that, during a west phase at times of high solar activity, cold winters were much more likely down the east coast of the USA. When it was used to forecast that the winters of 1988–89 and 1990–91 would be cold, however, it failed. Since then seasonal forecasts based on the QBO–solar activity connection have attracted much less interest.

Solar and lunar cycles

In many meteorological records there is evidence of a periodicity of around 20 years. This has been widely attributed to both the 22-year double sunspot (Hale) cycle and to the 18.6-year lunar tidal cycle. The periodicity could also be linked to quasi-periodic changes in the oceans, notably an interdecadal oscillation in the Pacific. The most significant evidence of this periodic behaviour occurs in the analysis of western and central USA rainfall records, central England temperature records, global marine temperature records and

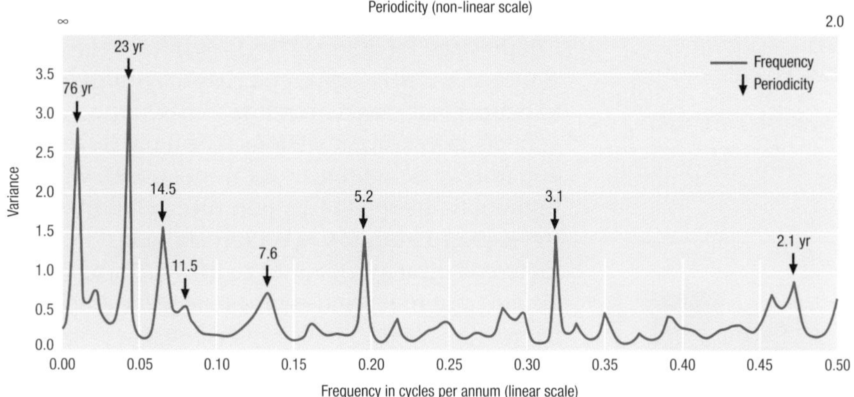

Many meteorological series, like the temperature record for central England, can be analysed for evidence of periodic behaviour. Here the 'cycles' at 76 and 23 years are interesting, but are not strong enough to be used for forecasting.

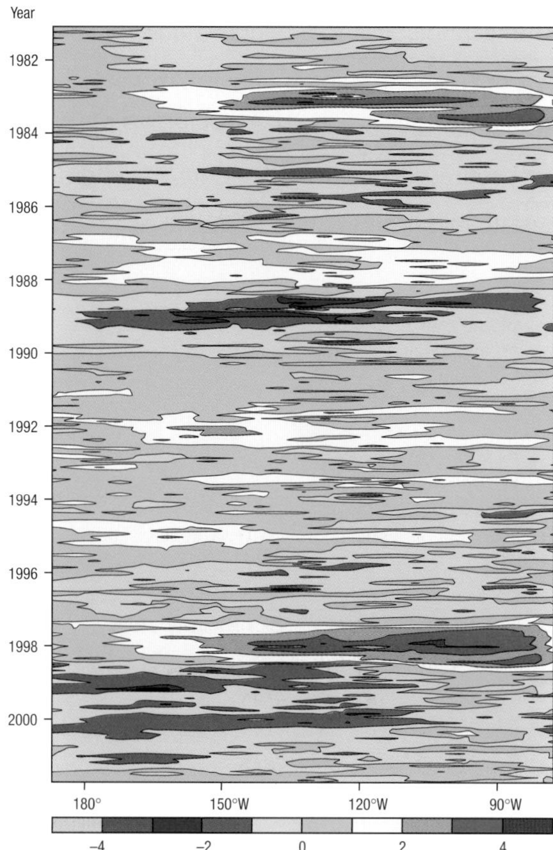

The variability with time of the anomaly of monthly sea surface temperature for an equatorial strip right across the Pacific can be used to build up a marvellous image of the swings from warm to cool conditions over the years. The major El Niño events of 1982–83 and 1997–98 stand out strongly as warm bands.

Greenland ice-core records. The debate about the origin of this ubiquitous signal in climatic records has yet to be resolved. However, in the absence of any plausible physical explanation for these statistical relationships, most scientists are highly sceptical of any links between long-term climate variability and solar or lunar activity.

The 80–90-year variation in solar activity shows some symmetry with global temperature trends. Although satellite measurements show that solar output does vary with solar activity it is far too small to explain directly the observed global trends. It is possible that other more subtle variations in the nature of the solar output could have stronger effects on the Earth's climate. These include such factors as changes in the length of the 11-year cycle and, more importantly, variations in the solar magnetic field and its effect on cosmic rays reaching the Earth.

These physical proposals have yet to be adequately explored.

Prospects for the next decade or two

The limited evidence for consistent weather cycles and the complex processes underlying quasi-periodic variations in the climate suggest that their prospects for long-term predictions are not good. The search for a consistent climate signal from these cycles has made little progress, even in the case of the 20-year cycle, which can be found in many records. Here, recent evidence suggests that solar activity can influence the climate but the physical connections between solar and climatic behaviour have not been established.

As for the various atmosphere–ocean 'oscillations', the prospects for success, if any, are still largely in a research phase. In the case of ENSO, although warm events have occurred every four years or so since the late 1960s, the strength and duration of different events have varied considerably. The varieties of responses that have occurred around the world with each event underscore the complexity of the problem. Further, the depths of the oceans and their changing characteristics have been poorly monitored and there is still much to learn. It is expected that new ocean observing systems being deployed will add to the knowledge of the oceans and to our ability to advance climate predictability.

J. Murray Mitchell (1928–90)

Murray Mitchell made a major contribution to the detection of global warming. At a time when the subject was not regarded as scientifically significant, he published seminal papers on the impact of city growth on temperature measurement and then in 1961 produced a highly influential estimate of changes in world and hemispheric temperatures since 1880. This widely quoted paper set high standards for much of the subsequent work on measuring global warming and identifying its causes. He also participated with Charles Meko and Dave Stockton in the most thorough analysis of tree-ring data to measure variations in rainfall in the western USA. This provided the most convincing evidence for a 20-year cycle of drought in this region.

4.11 Technology transfer – bridging the gap

The technological gap between developed and developing countries has been a constraining factor in monitoring the global climate system and advancing our understanding. Bridging this gap will be essential for progressing knowledge about the entire globe and for underpinning sustainable development everywhere.

WMO assists developing countries through education and training programmes in meteorology and operational hydrology.

Since its inception in 1950, the World Meteorological Organization (WMO) has promoted collaborative efforts to enhance the capabilities of its Members for the benefit of the global community through socio-economic development and protection of the environment.

In the 1960s, WMO initiated a Voluntary Cooperation Programme (VCP) that has helped National Meteorological and Hydrological Services to benefit from and contribute to observing systems, telecommunications, satellite receivers and human development across the world. VCP funding is also used to support the operation of two drought monitoring centres and one meteorological applications centre in Africa, and also to fund numerous educational fellowships. During the 1990s, WMO implemented activities worth US$ 200 million in support of technical and regional development projects that have contributed to bridging the technological gap between developing and developed nations.

To secure the necessary funding, WMO has worked in partnership with agencies including the United Nations Environment Programme, the United Nations Development Programme, the World Bank and regional development banks. Funding and technical support have also been provided through multilateral or bilateral cooperation, and in cooperation with non-governmental organizations and the private sector.

Building capacity through enhanced knowledge

Scientists from the developed world have, to a large extent, led the management and implementation of most international climate-related activities. This is understandable because many developing nations have been unable to generate the necessary scientific, institutional,

Rescuing climate data

In many countries the establishment of a National Meteorological and Hydrological Service and its responsibility for managing the climate data archive has been a relatively recent development. However, particularly in those countries with a colonial past, there are often accumulated climate records in manuscript form available from earlier periods that are not readily accessible. These documents, some dating back to the late 19th century, are deteriorating with age. There is a danger that the unique information could be lost forever, and with it valuable input to national development plans. Also, the data can provide important insights into local climate change.

The WMO Data Rescue (DARE) project aims to assist countries in the management, preservation and use of climatic data over their own territories. The DARE project in Africa, funded primarily by Belgium until 1997, dates back to 1979. It resulted in more than 5 million documents from more than 30 countries being saved on microfilm. In 1995, a DARE project began in the Caribbean with funding support from Canada and has served as a testing ground for the use of more advanced DARE technology such as digital scanners and digital cameras. A similar project involving hydro-climatic data has recently been launched in Africa.

DARE is now committed to transform the data currently in paper records and in various image formats to digital form so that they can be readily used in computer-based climate studies.

financial and technical resources in the face of more immediate developmental challenges. Helping to address this imbalance is the Global Change System for Analysis, Research and Training (START), which aims to evaluate the impact of global change and enhance scientific capacity in developing countries. START was a response to a 1989 United Nations General Assembly resolution that governments "increase their activities in support of the World Climate Programme and the International Geosphere-Biosphere Programme". It further recommended "The international community support efforts by developing countries to participate in these scientific activities".

One of the major capacity-building activities of WMO is to promote training in meteorology and hydrology and to coordinate the international aspects of such training. Its Education and Training Programme activities are supported by a worldwide network of 22 Regional Meteorological Training Centres, organizing regular training programmes and specialized courses in response to the needs of different regions. WMO was active in encouraging the Conference of the Parties to the United Nations Framework Convention on Climate Change, at its fifth session in late 1999, to formally recognize the capacity-building needs related to developing country participation in the systematic observation of the climate system.

The CLICOM story

The development of computer database management systems has revolutionized the ability to process and apply climate information. The World Meteorological Organization recognized early that the meteorological services of developing countries were not getting access to this important technology and in 1985 instigated the CLImate COMputing (CLICOM) Project. Through its Technical Cooperation Programme, WMO has assisted more than 130 countries with implementation, training, maintenance and upgrading of a standard climate data management system for computer-based applications.

The CLICOM project has led to the establishment of electronic databases from which climate applications services can readily be provided for research on climate variability and change, and for the development of climate information and prediction services in the national interest. Additionally, the project has provided a valuable opportunity to bring computer technology and expertise to hundreds of staff members of National Meteorological and Hydrological Services in developing countries.

Climate Information and Prediction Services (CLIPS)

The decade-long (1985–95) Tropical Ocean Global Atmosphere (or TOGA) project of the World Climate Research Programme identified the potential of using new knowledge of the El Niño/Southern Oscillation (ENSO) phenomenon for prediction on seasonal to interannual timescales, at least in some regions of the globe. New capabilities for managing and processing national observations, for monitoring climate on a global scale and for prediction all provided the opportunities for a paradigm shift in the provision of climate services.

Set up by WMO in 1995, CLIPS builds on past decades of atmospheric and oceanographic research and recognizes climate as both a resource and a hazard. It is designed to assist countries, especially developing countries, to access regional and global climate monitoring and prediction products and, with their own local information, provide services in sectors such as agriculture, tourism, disaster preparedness, water resources management and other social and economic activities.

WMO is also working towards the establishment of regional climate centres, each as a focus for regional analysis and prediction, and for the development of climate expertise. A major objective of CLIPS is to develop national and regional capabilities for early warning and mitigation of natural disasters related to extreme meteorological and hydrological events, such as those that were associated with the 1997–98 El Niño event.

4.12 Food and fibre production

Food and fibre production is central to human life, and agriculture and pastoralism figure prominently in the economies of most countries. Climatic data, monitoring and predictions are important ingredients to farming success.

Even with the impressive technological advances of recent decades, climate remains a critical factor in the success or failure of any form of food or natural fibre production. Also, the planned introduction of exotic species that either have higher production or better meet market requirements, must be carefully planned to ensure their suitability for the target climate.

The seasonal cycle controls the various stages of crop growth but to be successful, farmers must minimize the risks from weather and climate hazards. Drought, frost, high wind, saturation of soils, hail and excessive humidity that influence diseases and insect infestations, are examples of weather and climate factors that can limit plant growth, and may even cause crop ruin. Farmers can exploit weather and climate services to minimize the impact of these hazards, either by planning to avoid the risk in the first place or by taking precautionary measures when there is warning that a hazard threatens. Meteorological services for agriculture make a significant contribution at the farm level and benefit national economies.

The widespread introduction of canola (oil-seed rape) to temperate latitudes has transformed many landscapes in spring and summer.

Building on experience

The fact that many crops have been grown for centuries in certain parts of the world is testament to their climatic suitability. However, in relatively recent times, exotic species have been introduced to many parts of the world with varying success. Early in every summer seas of yellow flowers now emerge across the prairies of western Canada and the fields of northwestern Europe. This is proof that warm sunny days and cool nights are ideal for growing canola (oil-seed rape). Farther south, in Minnesota or southern Europe, the oil yield drops off and the crop is uneconomic. Failure to recognize climate limitations, especially the impact of recurring climatic extremes, has caused costly losses through expansion into unsuitable lands or the introduction of unsuitable species (see the East African groundnut scheme below).

A climate that is suitable for a commercial crop will supply not only the appropriate growing conditions but also will allow a whole host of farming operations at the appropriate times. Such seasonal operations include ploughing, seeding, irrigating, spraying, cultivating, harvesting, storing and all the associated scheduling of labour and equipment. In addition to the provision of data and simple statistics (e.g. degree-days and drying indices) support for farmers covers input to complex crop-yield models. For example, a national early warning system that monitors the effects of weather on crop growth and for planning purposes was developed for Zimbabwe. Food crops can be monitored to anticipate and mitigate damage caused by adverse weather conditions, diseases and pests (such as army worms and locusts). Locust plagues, for example, take several months to build up and the insects only swarm under certain meteorological conditions that, if carefully monitored, give windows of opportunity for effective control.

Crop forecasting is big business now. Governments and businesses use climate monitoring and prediction information to formulate crop market strategies and policies. In an age of rapid transportation and global markets, assessing crop health at home and abroad has become part of the marketing process.

Forestry

In the forestry sector, climatic anomalies and severe weather are of greatest importance in forest fire control and prevention. Monitoring fire-weather parameters such as moisture conditions of the soil and ground litter, wind speed and direction, temperature, humidity and lightning strikes are crucial for preparedness

Monitoring the conditions that increase the risk of forest and grassland fires enables early action to be taken, such as clearing land for a firebreak.

before a blaze ignites, and for early action before it gets out of control.

In addition to using current and projected weather conditions, foresters use historical climate data to develop strategic plans from planting to harvesting. These decisions cover practices such as zoning land for commercial forestry based on climate suitability, site preparation, regeneration, thinning and fertilizing. Information on potential climate change is equally important. If the climate changes as the trees grow the yield could be significantly different to what was expected.

Environmentally friendly bananas

Forecasting the optimum times for carrying out fieldwork operations and estimating crop development phases are crucial management decisions. Scheduling irrigation based on daily rainfall and estimates of potential evaporation is a simple example of tactical decision-making. Many pests and diseases also respond to weather and climate and their management is better achieved by including meteorological information and predictions in decision-making. In Guadeloupe, for example, banana plantations are treated with fungicide to maintain fruit quality. Twenty-five applications per year were the norm before research established how the rate of development of the fungus was related to meteorological conditions. Armed with this information, reliable weather observations and short-term forecasts, the number of applications has been reduced to as few as six per year, saving US$ 2.8 million for a 3500 ha area – a very large cost–benefit ratio. Using weather-based scheduling of pesticides and insecticides not only reduces the number of applications but also minimizes spraying costs and protects the environment.

Fishing

The success of a fishery in any year depends on the presence or absence of climatic anomalies during the fish growth and harvest seasons. Wind, waves and swell all affect fishery operations, as does ice formation and break-up in some waters. Knowledge of the distribution and migration of many fish species is also essential for their successful commercial management. Ocean currents, sea surface temperatures, environmental degradation and salinity can all influence the size of stocks, the migration patterns of different species and the spread of disease. Monitoring of monthly and seasonal variations in sea surface temperatures is enhancing the tactical planning of operations for ocean fishing fleets (e.g. by minimizing time and costs in locating and travelling to fish) and could improve the overall management of the world's fisheries.

The sustainable management of many fisheries around the world depends, in part, on an improved understanding of the climatic conditions that influence fish stocks, the migration of species and successful regeneration following a period of climatic disruption.

The East African groundnut scheme

The British government launched a scheme in 1946 to grow groundnuts (peanuts) in Tanganyika (now the United Republic of Tanzania). Designed to meet the shortages of margarine, cooking fats and soap in Britain, and to develop agriculture in Africa, this ambitious scheme, which planned to cover 3.2 million acres, turned into an economic disaster. While there were many social, economic and agricultural reasons why this scheme would have difficulty meeting its objectives, the one factor that doomed it to failure was ignorance of the inadequate normal rainfall and the range of its variability. This meant that the optimistic production figures could never be achieved with an acceptable level of consistency.

The first small-scale field experiments were carried out for a single and, as so often happens, unrepresentative wet season in 1947. Full-scale operation was embarked upon in 1948. In 1949 unfavourable weather combined with other troubles made it clear that the project could not succeed. The yields fell far below those the indigenous population had attained with local crops. Eventually, after six years of desperate efforts, the experiment was abandoned. In total, the cultivated area never exceeded 75 000 acres. The total capital cost was estimated at US$ 100 million, a huge expenditure at the time.

4.13 Transport and trade

Technological advances have ensured that maritime, aviation and land transport systems are generally robust in the face of extreme weather events. However, when infrastructure failures occur they have severe impacts on dependent communities and their industries.

Keeping shipping lanes open during winter in ice-bound seas uses not only standard weather information but also satellite and radar measurements of the extent of ice.

High winds and waves can make towing a supply barge extremely hazardous.

Extreme weather events historically have been a hazard to transport and trade. The safety of naval fleets was the impetus for organizing the meteorological services of several countries in the mid-19th century, but the benefits were quickly taken up more broadly and cooperative international arrangements established for international maritime trade. Organized weather services have also been crucial in the expansion of aviation services during the 20th century.

The infrastructure for transport and trade consists of the ports and loading facilities for shipping, the airports for aviation, and the roads, railways and bridges, etc. for land transport. Operations can be temporarily disrupted by local weather events, including storms, fog, etc. If, however, planning of infrastructure does not take sufficient account of meteorological and hydrological extremes there may be an elevated risk of failure, loss of lives and more lasting disruption. Such breakdowns can have wider social and economic impacts if communities are denied access to markets. Weather and climate services ensure safety of operations and maximum utilization of facilities.

Maritime industries

The potentially devastating impacts of winds, waves, sea ice and storm surges on all aspects of marine operations dominate the design and operation of ships, port facilities and coastal hydraulics systems. The design considerations range from ensuring the seaworthiness of vessels to withstand defined sea states, through offshore platforms that can withstand, for example, a one-in-one hundred-year storm for a given location, to building sea defences and harbour installations to cope with similarly rare storm surges. All these activities require the fullest possible use of climatic data collected over many decades.

A good illustration of the complexity of issues, uncertainties and challenges is in ship routing. Journeys across the major oceans, which may take from five to 20 days, can utilize forecast winds, waves and swell to obtain the optimum route that balances the pressures to achieve the scheduled voyage on time against fuel and other costs while avoiding damage to the vessel and cargo. For one-off voyages, like ocean tows of offshore drilling rigs and platforms where winds above 30 knots (15 m/s) could spell disaster, routes and timing have to be planned with great precision to avoid any heavy weather. More generally, shipping lines now use forecast information to improve the efficiency of operations and reduce damage to cargoes.

Land transport

More than 2000 years ago, even before the Romans laid down an infrastructure of roads to the far corners of their empire, structures were being built to make traffic more efficient and to minimize disruptions from rain and flooding.

The 20th century saw enormous upgrading and expansion of land transport systems globally to accommodate increased trade and commerce, especially in the collection of raw materials and the distribution of finished goods. Many meteorological hazards have been 'engineered out' of land transport systems, as high bridges have been built to avoid potential flooding from rain-swollen rivers and tunnelling has avoided disruption from snow accumulation and avalanches on high mountain passes. Even so, because of the heavy reliance on these transport arteries, meteorological services that can minimize disruptions are vital, and meteorological data to ensure avoidance of infrastructure failure remain crucial.

Major bridges, such as the Second Severn Bridge in the United Kingdom, have to be designed to reduce the impact of high winds on the vehicles that cross them, otherwise they would have to be frequently closed when conditions became unsafe.

Local strong winds

In the vicinity of mountain ranges, hillsides and gullies, local winds can become very strong. They are known by local names in different parts of the world, including Chinook along the Rocky Mountains of North America and Föhn in the Alps of Europe. The dry grasslands on Olkon Island in Lake Baikal are blocked from rainfall by the mountains to the west, but are subject to the 'Sarma', a treacherous westerly gale of up to 160 km/h that hits with little warning. In southwestern Newfoundland the Wreckhouse Winds, named after the small local community, appear as gales as strong as 140 km/h that sweep down from Table Mountain across the highway and out to sea. Wreckhouse Winds are notorious for lifting freight cars off their tracks and blowing over tractor-trailers. At the slightest indication of a Wreckhouse Wind, all nearby trains are kept out of the area or are chained to their tracks until the winds abate. In Antarctica, cold, heavy air frequently sweeps down the slopes from the domed interior to the coast, where wind gusts of up to 300 km/h have been recorded. These fierce, often turbulent, winds can produce sudden, localized blizzards.

In temperate climates a particular hazard in winter is the formation of frost or ice on road surfaces (black ice). Many highway and road authorities have installed effective road weather information systems, either as sensors embedded in the road pavement or as automatic weather observing stations along the roadway. A network of such sensors can provide up-to-the-minute data on road weather conditions, alerting a traffic manager, for example, when there is a high risk of slippery roads. Electronic traffic aids can also then alert motorists about dangerous road conditions. Studies in England and Sweden show that the cost of related forecasts is recouped manyfold by just reducing the cost of road salting. Other benefits include reduction of accidents, reduced damage to the environment and reduced corrosion to vehicles.

The exploitation of mineral resources in many parts of the world has required climatological analyses for the building of special-purpose railways to carry ore to coastal ports or refinery facilities. The economic viability of the massive investment hangs on continuity of operations, so disruption of the transport system due to weather incurs a large economic cost. Even in the most arid regions, the land route must be carefully surveyed and the climatology assessed to design against occasional flooding of crossings over generally dry riverbeds. In particular, flood rains associated with tropical cyclones or their remnant storm systems have been damaging when adequate engineering had not been carried out to prevent erosion of the track and supporting structures.

Aviation

In the development of new airports, and even the expansion of existing ones, siting and design must take into consideration the meteorological factors that could disrupt operations and therefore add to the overall cost of operations. Fog, low clouds, snowfall, locally strong winds and poor visibility are a few of the important weather-related events.

The new Hong Kong International Airport opened in July 1998 provides a case study for the design of airports. Meteorological analyses related to planning and design of the airport site began in the late 1970s. These included wind shear and turbulence investigations necessitated by the hilly terrain near the site. Measurements of the prevailing easterly winds helped in deciding the alignment of the runways. Other climate information, including cloud base and visibility, contributed towards determining the projected runway usability as part of the economic evaluation. One finding from the analysis of the data gathered is that, over the past few decades, there has been a trend towards reduced visibility. Operational efficiency of the airport could be reduced in future if the trend continues.

The design of the new airport in Hong Kong, China took particular account of the local topography and weather conditions to ensure maximum safety and minimum disruption by adverse winds.

4.14 Industry and commerce

The globalization of industry and commerce has been possible through development of complex supply, manufacturing and distribution networks that can cope with the range of climates, including the variability and extremes, experienced around the world. Effective utilization of climate information has nevertheless become crucial to the success of many global enterprises.

Globally there are about six million hectares of rubber trees in plantations such as this one in Uganda. The wide distribution of sites across many regions means that, overall, production and market prices are not significantly affected by localized year-to-year climate extremes.

The characteristics of industry and commerce have changed markedly during the 20th century as expanding world trade has altered patterns of supply, manufacturing, distribution and marketing of goods. Suppliers of raw materials now rarely have a monopoly, the skills and technologies for manufacturing goods that evolved in specific localities have been exported to all parts of the globe and similar goods can be purchased in most cities. This process of 'globalization' represents, in part, a throwing off of the shackles of local climate that previously restricted peoples' lifestyles.

The diversification of supply

The introduction of exotic species native to one part of the world to new locations with favourable climates has been part of the expanding production of many goods. Long before rubber plants were smuggled out of Brazil in the late 19th century to establish a competitive industry in Malaysia, this process had been a feature of agriculture around the world. Another recent example is the kiwifruit, also known as the Chinese gooseberry or *yangtao* that has its origins in South East Asia. From a successful cultivation strategy in New Zealand, the fruit is now also grown as far afield as California and the Mediterranean region, and is widely available in global markets. This process has been aided by modern transport speeding the movement of agricultural goods and opening up markets worldwide.

The diversity of sources for agricultural goods has reduced the level of weather-related price fluctuations from previously when there were dominant regional suppliers. Frost occurrence still has a major impact on coffee production in Brazil, but the world price of coffee is no longer so sensitive because of production in other tropical and subtropical regions of Latin America, Africa and East Asia.

Major food companies monitor the global climate for early indications of regions where supplies may be interrupted or where a good season

People purchase according to the weather

Weather sensitivity extends to food purchases where there is considerable impulse buying. Brewers, ice cream parlours, air-cooling installers and swimming pool salespeople revel in hot weather because it increases demand. On rainy days, wet weather gear and umbrellas are prominently displayed in stores and in the marketplaces. One Japanese brewer linked sales data collected over 30 years to create a weather index for 15 areas across the country. This showed that on fine days in July and August beer sales increased 8 per cent if the temperature was 1°C above normal. The development of such quantitative links allows retailers to monitor their stocks and weather forecasts to ensure adequate supplies are on hand when there is an increase in demand. Monitoring the weather forecast is particularly important for manufacturers of perishable goods for which unsold stocks can become a heavy cost burden.

may produce a bumper crop and lower prices. Attention to the impacts of global weather and climate is giving many companies a competitive edge through marginal savings in the cost of supplies and ensuring a continuity of supply regardless of regional weather impacts.

Manufacturing

The export of skills and technologies for manufacturing to all regions of the world generally requires adaptation to local climatological conditions if quality is to be maintained. Many materials change their physical characteristics, including volume, surface roughness and flexibility, over ranges of humidity and temperature. Machinery developed in one climatic zone to process materials may have to be carefully adjusted to perform efficiently in a different climatic zone, or expensive air conditioning may have to be utilized.

The 'just-in-time' management philosophy is widely used in manufacturing and retail industries. It maintains minimal inventories, as this reduces operating costs by not tying up finances in stockpiles of raw materials or processed goods. Such an operating framework requires reliable supply chains, because interruptions hold up the whole manufacturing process, with loss of output and increased average costs. Therefore manufacturers must keep a close watch on the sensitivity of their supply chain to weather factors, carefully balancing inventory costs against the potential for interruptions to supplies.

Marketing

Our consumption of some goods is particularly weather-sensitive. Successful marketing of such goods depends on anticipating changing demands. Clothing is marketed according to the season and an early or late onset of seasonal conditions can have a significant financial impact on sales of fashion and sporting goods, particularly in those climates where the peak season is relatively short. For example, people generally are not enticed to buy new summer clothes if a cool and cloudy spring extends into the normal summer period, neither do they purchase sporting equipment if the weather doesn't offer an opportunity to use it. An abnormally cool summer or mild winter can be financially disastrous in the fashion industry if seasonal stocks are unsold and, by design, are out of fashion the following year.

The global marketing of goods means that products, whether computers or automobiles, have to be designed to function with minimum adjustments in all climatic extremes. Lubricating fluids, coolants and electrical components that are effective in snow and ice of polar and high altitude regions must be equally effective in hot deserts. Similarly, combustion engines that are efficient near sea level should also be easily adaptable to function effectively in the high altitudes of the South American Andes, or the Himalayan Mountains and Tibetan Plateau of Asia.

Climate extremes disrupt industry

Droughts and floods disrupt river transportation, which is still important in many parts of Europe and North America. Drought and low levels of the Mississippi River in 1988 reduced and in some places completely halted the carriage of goods. Notwithstanding the negative impact on the river barge industry the climate extreme provided opportunities for competing road and rail freight industries.

The widespread drought experienced by countries bordering the western Pacific Ocean during the 1997–98 El Niño event was also disruptive for local industries. Major mining companies in Papua New Guinea are reliant on river transport for securing supplies and taking out mineral ore. Large quantities of water are also used in the mineral extraction processes. The failure of seasonal rains was so severe during the El Niño event that operations had to be halted for several months. The national economy of Papua New Guinea was affected because it is reliant on mineral exports and the loss was a primary cause of devaluation of the national currency. Port Moresby and Manila were two cities where reduced water supplies affected hydroelectricity generation and industry. On the other side of the Pacific Ocean, evaluation of the freshwater resources of the region is vital to the efficient operation of the Panama Canal. Each ship that passes through the Panama Canal needs more than 200 000 cubic metres of water to float it through the series of locks. Reduced rainfall over Panama in the early stages of the 1997–98 El Niño event resulted in shortages of water to operate the locks and shipping capacity was reduced.

4.15 Buildings and public infrastructures

Modern cities have evolved around networks of buildings and public infrastructures that provide comfort and safety to the inhabitants and maintain business, even during extreme weather and climate events. However, the reliance on artificial heating and cooling is being challenged by the need for more energy conservation.

People, particularly urban dwellers, spend much of their lives within buildings, whether they are houses, apartment buildings, places of work or places for entertainment. They move between buildings on public roads and transport facilities and have access to a range of common utilities. Buildings and public infrastructures are essential to modern living and to maintenance of trade and commerce. Futurists have often portrayed cities evolving into completely artificial environments insulated from the local climate.

Modern cities and their urban surroundings are generally places of safety during extreme weather events and, despite some escalation of damage costs, loss of life and serious injury are declining during such events. The increased safety is due to better city planning and construction standards that take account of expected extremes, and to early warning that allows timely implementation of specific response strategies. Many low-lying coastal margins and river flats are evacuated when there is danger of inundation and flooding, while sturdy public buildings, such as schools, are often used as safe havens during storms.

Building codes

Not too long ago, building practice was almost entirely a cultural process based on tradition, with styles developed over time to suit the local climate and techniques making use of available, often local, building materials. Since World War II many countries have established research agencies to assist in the planning and design of buildings, especially to set standards of construction for safety. Weather-related events are a major cause of building failures so there is good reason for using climate services to define building standards and performance, and to integrate climate statistics into national building codes.

In most countries with a variety of climatic regimes, differences are significant enough to require regional design considerations, and not just for heating, cooling and ventilation purposes. In lower latitude coastal regions where tropical cyclones are a particular hazard the design wind loading is much higher than inland or in more temperate coastal zones. At higher latitudes the potential snow loads on roofs vary according to geographical location, climate, site exposure, shape and type of roof, etc. In many mountainous regions (e.g. Switzerland and Austria) snow loads on roofs can last all winter, creating a dangerous burden on buildings. The buildings must therefore be designed to withstand the load and permit access to the roofs so that dangerous accumulations can be cleared.

Optimum design

Insurance statistics highlight the escalating damages associated with extreme meteorological events. These include the costs of replacing not only damaged or destroyed buildings but also the cost of contents destroyed with the building. As a home is also thought of as a place of safety, particularly from extremes of weather, the rising economic losses seem at odds with the falling number of deaths. The answer lies partly in the early warning and evacuation procedures established in many places to ensure that people are removed from danger to the extent possible, and partly with more people living on vulnerable foreshore areas that have rapidly rising property values.

The basic economic question is whether the cost of building all structures to withstand hazards may be too high. Put simply, is it realistic to engineer every home and building and associated urban

Warming and thawing of permafrost on slopes in northern regions can lead to instability and landslides. Such soil movements have to be considered when designing buildings and roads.

The vulnerability of shoreline communities to storms is obvious in this picture of waves crashing on Monmouth Beach, NJ, USA during the 'Halloween Storm' of October 1991, later dubbed the 'Perfect Storm'.

infrastructure to withstand a Category 5 tropical cyclone? City planners have to focus on the safety of lives and the preservation of vital infrastructure so that, after an extremely damaging event, life can return to normal as quickly as possible – public utilities are restored, business and industry recommence, and reconstruction can get under way. What determines optimum design is linked to the purpose of the structure, the personal or community loss caused by its failure, and the availability of insurance and other financial mechanisms to fund its replacement.

Solar panels on the roof of this Colorado, USA home are part of a cost-effective water heating system. High insulation standards and use of solar or wind energy in housing design can help us reduce our consumption of fuels from fossil carbon deposits such as oil, gas and coal, and hence lower emissions of greenhouse gases.

Building for the future

Building construction in the 20th century was characterized by widespread adoption of concrete, metals and glass as basic materials. Reinforced concrete provided strength and glass permitted light to enter without the nuisance of wind, dust and insects, etc. Such construction methods are common from the tropics to the near-polar regions. The thermal limitations of these materials have largely been countered by the use of artificial air conditioning – cooling in the hotter lands and heating where it is colder. On the downside, cement manufacture and air conditioning are both major contributors to the build-up of greenhouse gas concentrations in the atmosphere, and hence to global warming.

The building industry and its design and construction methods offer significant potential for reduction in the future contribution to greenhouse gas emissions. Research is being carried out in many parts of the world to identify house designs that are more thermally efficient. In many instances, architects are reverting to indigenous designs that had been discarded in favour of imported mass-production building methods. They are also adopting principles from physics to promote comfort and aesthetics while minimizing the need for artificial lighting and heating.

Wind damage

Wind is perhaps the greatest destroyer of buildings. Therefore building codes developed in many countries are designed to ensure that structures, including homes and places of work, can withstand wind speeds that occur not more than once in 50 years. Wind, however, is seldom steady but usually consists of gusts and lulls. Hence, structures have to be built to withstand the pressures and suctions caused by a wind fluctuation that may last only a few seconds. Though estimates of the steady wind (the ten-minute average at the time of the observation) are regularly recorded with weather observations, gust speeds lasting several seconds have been routinely measured and recorded from only a few stations. It is possible, however, to estimate gust speeds from steady wind speeds, to aid in the setting of building standards.

World Meteorological Organization Headquarters

The Headquarters building of the World Meteorological Organization (WMO), which opened in 1999, utilizes both innovative and time-honoured techniques for energy conservation.

Harnessing the laws of physics to ensure efficient heat transfer through a building can take many forms. For example, the Romans used convection for warmth (hypocaust) while the Arabs use the same principle to produce cooling draughts (*burj el hawa*). The WMO Headquarters uses Canadian Wells, built into the foundations of the underground car park and linked through the supporting structures of the building to all floors, to store cool fresh air. As the air in the building warms it rises and is replaced so that the natural process of heat transfer maintains the building at a near constant temperature.

Aluminium, glass, stone and concrete are the main materials of the building and are a delicate compromise between maximum light penetration and optimal heat retention or cooling. The east–west alignment leaves the facades exposed to the bitter mountain winds from the north and the full glare of the sun to the south. The northern windows of the outer shell are therefore permanently closed, providing insulation. The southern façade is, however, made up of panels that can be opened and closed. The entire façade is specially coated to reduce ultraviolet radiation and to the south the glass is further coated to provide 40 per cent shade.

Overall, the building will use significantly less energy than the new Swiss Federation maximum of 400 MJ/m^2 per year and compares more than favourably with 800 MJ/m^2 per year energy usage of normal office buildings.

4.16 Energy and society

Harnessing energy to mitigate temperature extremes has allowed communities to survive and prosper in some of the harshest climates. Energy is at the heart of economic and social development but is also the cause of a wide range of environmental problems, including the threat of climate change.

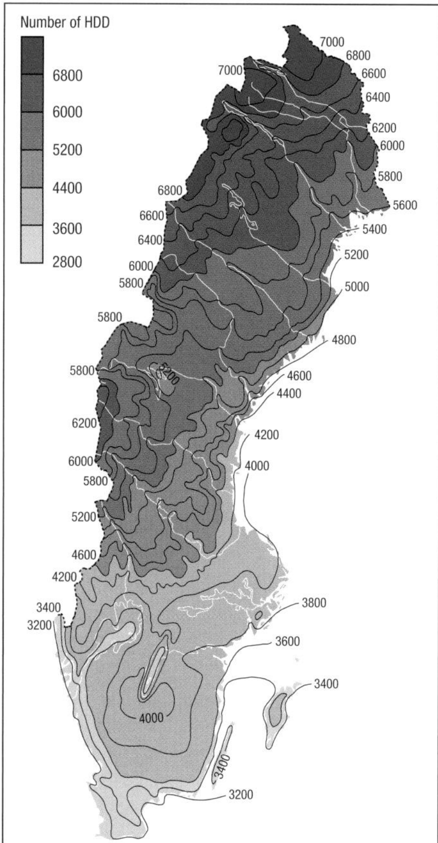

The number of heating degree-days (HDD) for space heating in different parts of Sweden is calculated for each separate day as the difference between 17°C (the base temperature) and the mean daily temperature (primarily in the colder months of the year). The data shown are annual means for the period 1961–62 through 1978–79.

Communities have developed lifestyles that have enabled them to cope with the wide range of climates experienced from equatorial to near-polar regions. Nevertheless, humans are most comfortable in a relatively narrow temperature range from about 15°C to 25°C. This comfort zone is reflected in energy usage patterns of modern cities. When air temperatures move outside this range energy demand increases; for heating as the temperature falls below about 18°C and for cooling as the temperature rises above about 22°C.

The degree-day is a useful statistic that has been developed to assist the monitoring of energy usage and prediction. The degree-days can be either the accumulated departures of daytime temperature below a specified threshold (the heating degree-days) or above a specified threshold (the cooling degree-days). Thus, if a winter is milder than normal there are fewer heating degree-days and less demand for energy for home and office heating. By contrast, a severe winter will create a greater demand for energy. Degree-days do not take account of sunshine, wind or exposure, or how well insulated buildings may be, but as a simple measure reflecting energy usage the notion works well. Energy companies use the link between climate variability and energy demand for supply planning to guard against shortages during the most critical times.

Severe winter weather not only disrupts traffic but also causes great damage to utilities. These power and telephone lines are sagging after a heavy ice storm.

Electricity generation and distribution

The ability to generate electricity centrally and to distribute it where it is needed has been a significant factor in the development of cities. Urban air quality also benefited as electricity replaced wood, coal and coal gas for domestic heating and cooking purposes, and the smoke and soot that were characteristic pollutants in many cities were no longer prevalent. Communities have, however, developed a dependence on ready access to electricity. Complex distribution networks have evolved to meet the various needs, and grids of high voltage lines criss-cross the countryside linking communities. Often the power lines that service individual buildings are hung from street poles and other support structures. The exposed wires are subject to wind loading, ice accretion, lightning strikes and storm damage. Increasingly, electricity generation is far from where it is used and the transmission distances and exposure have increased accordingly. Climate statistics are one of the key factors for planning the energy generation and distribution systems, and for ensuring that outages through extreme weather events are minimized.

Renewable energy

With apparently abundant fossil and nuclear fuels, the global electricity generating capacity has steadily increased to meet expanding industrial and societal needs. However, this strategy is now under serious review. Significant accidents at nuclear plants in the USA (Three Mile Island, March 1979) and the Ukraine (Chernobyl, April 1986), reported radiation leaks from other facilities, and no worldwide agreed strategy for safe disposal of nuclear waste material, have contributed to widespread concern over the long-term use of nuclear fuels. The implications for climate change of burning increasing amounts of fossil fuels in the future also raises difficult issues.

Renewable energy resources have been exploited for centuries, using windmills and water wheels.

1995 World electricity generation

Total electricity generation 13 204 TWh

Primary energy source	Percentage
Solid fuels	38
Oil	10
Gas	14
Nuclear	18
Hydro	19
Other renewables	<1

(source: International Energy Agency)

More recently, renewable energy technologies have been extended to utilize the enormous potential of the Sun. Direct solar conversion, harnessing of hydro, wind and wave energies and bio-fuels can all, in certain circumstances, replace significant quantities of non-renewable energy sources. Tidal and geothermal energy sources may also be tapped for electricity generation in some locations.

Hydroelectric power is extensively used in many countries. A benefit of hydro power is that efficient generation plants can be built on all scales. A disadvantage is that the damming of major rivers has environmental consequences, both to the local aquatic species and to the downstream riverine ecosystems. A major challenge for managers of hydroelectric facilities is to match energy generation to seasonal and long-term water supplies, and often to competing water demands for urban and irrigation needs. During periods of drought the demand for electricity has to be balanced against the need to conserve scarce water supplies. Long climatic records on the year-to-year variability and the duration and intensity of past drought events are essential to the design process and are crucially important in water management at times of meteorological extremes.

Wind power has traditionally been harnessed for lifting, grinding and other mechanical applications. Small wind systems were widely used early in the 20th century for domestic electricity generation and pumping water, but were displaced due to the convenience of petrol or diesel generators. Large wind generators now offer economic power supplies in climatologically favourable locations of northwestern Europe and the United States. Long records describing the diurnal and seasonal patterns of local winds are essential for planning the economics of a wind-generation project.

The most successful use of solar energy has been in direct heating of water for domestic and space heating purposes. Solar hot water heaters are increasingly common in tropical and middle latitude regions of both developed and developing countries. Generally the economics are such that they are preferred, even though electricity from the grid is readily available. In higher latitudes electric or gas boosters are often needed to provide supplementary heating during winter and this means that the economic benefits of current solar heating technologies become more marginal.

Attempts have been made to generate significant quantities of electricity from photovoltaic solar panels. Large-scale solar home systems have demonstrated the feasibility of photovoltaic conversion but currently such applications generally do not compete economically with centrally generated electricity. Photovoltaics do, however, become economical for applications in remote locations and where the requirement is for low voltage to power electronic systems. Such solar panels have found wide application for communications relay and other systems that are often operated without attention for extended periods. Solar panels are also finding general applications for sailors and people travelling in remote locations but requiring a source of power for electronics and low demand applications.

Solar home systems

More than 2 billion people worldwide live without access to basic amenities including electricity. Many of these people use candles and kerosene for lighting and wood for cooking with both having negative impacts on health and the local environment. Renewable energy could play a significant role in bringing power to communities, especially in rural areas. Commercially supported demonstration solar home system projects have been implemented with success in a number of locations in Africa and Asia.

Improved designs of cookers that use biomass fuel can greatly reduce the demand for firewood and minimize the impact on the local environment.

Protecting offshore energy facilities

Exploitation of offshore oil fields requires special climatic information. The Gulf of Mexico, the North Sea and the Timor Sea are a few of the places where giant rigs are extracting valuable oil and gas resources. These places can also be climatically inhospitable as gale-force winds whip up seas that endanger platforms. In the tropics the main threat comes from tropical cyclones while in middle to high latitudes storms are generally less intense but of longer duration.

In polar regions icebergs pose an additional threat. In the 1970s, climatological information on their size, frequency and movement off Labrador was used in the design and deployment of a US$ 250 million drilling platform. More generally, climatological information has been vital in the designing of offshore platforms to withstand winds and waves. Monitoring and prediction of storms in the area are crucial for providing essential early warning to safely close down and secure operations and even evacuate workers if the platform is threatened by extreme weather.

4.17 The Earth's freshwater resources

Half of the world's people live in cities and rely on distant sources for their water supplies. There is an escalating freshwater crisis due in large part to water shortages, pollution and ineffective sewage management.

In parallel with the growth of communities there has been a need to harness freshwater resources. On a worldwide basis, demand for freshwater has increased ninefold since 1900, forcing ever more careful use of available resources. Rapid urban expansion over recent decades has led to an emerging water crisis that is aggravating the socio-economic, health and environmental conditions of large cities. There are now several megacities with more than 10 million inhabitants and the number is growing. As a direct response to a lack of clean municipal water there is an expanding use of bottled water, particularly in developing countries. Such water is priced at 500 to 1000 times the price of municipal water and the disposal of bottles is contributing to the waste burden of cities. Consequently, there are economic and environmental imperatives to protect rivers, lakes and underground freshwater resources.

Water is naturally recycled and purified by the hydrological cycle and rainfall is the primary source of freshwater, whether collected in surface catchments or accumulated in underground aquifers. Good design of storage reservoirs and water supply systems for cities is based on measurements of how much water is available and how the supply fluctuates in space and time. Thus, every aspect of the hydrological cycle, and consequently all water in the world, is intimately connected with climate.

Human actions such as the removal of trees and other vegetation cover, expansion of paved areas, building of dams and channels, etc. modify the local hydrological cycle. Changes to the evaporation, soil infiltration and surface runoff rates all modify the amount of rainfall available for collection and utilization. Also, global warming is expected to lead to changed precipitation patterns. The current trends in water availability and quality are therefore a serious concern for the future.

The Red River Floodway around Winnipeg in Canada provided vital relief when the exceptional floods struck in the spring of 1997.

Major reservoirs

Climate data and information underpin the planning and management of surface freshwater supplies and mitigation of damage from high and low water flows. Long records of catchment rainfall provide the basis of planning for sustainable freshwater harvesting but it is the hydrological extremes of flood and drought that pose significant problems for water resource managers. It is vital that major reservoirs handle peak flood flows without danger of exceeding the spillway capacity. Statistics based on long records of catchment rainfall are essential for calculating quantities such as the Probable Maximum Precipitation for given periods and frequency of heavy rainfall events. Water managers use rainfall forecasts, especially early warnings of heavy rainfall events, to regulate outflows and maintain safety.

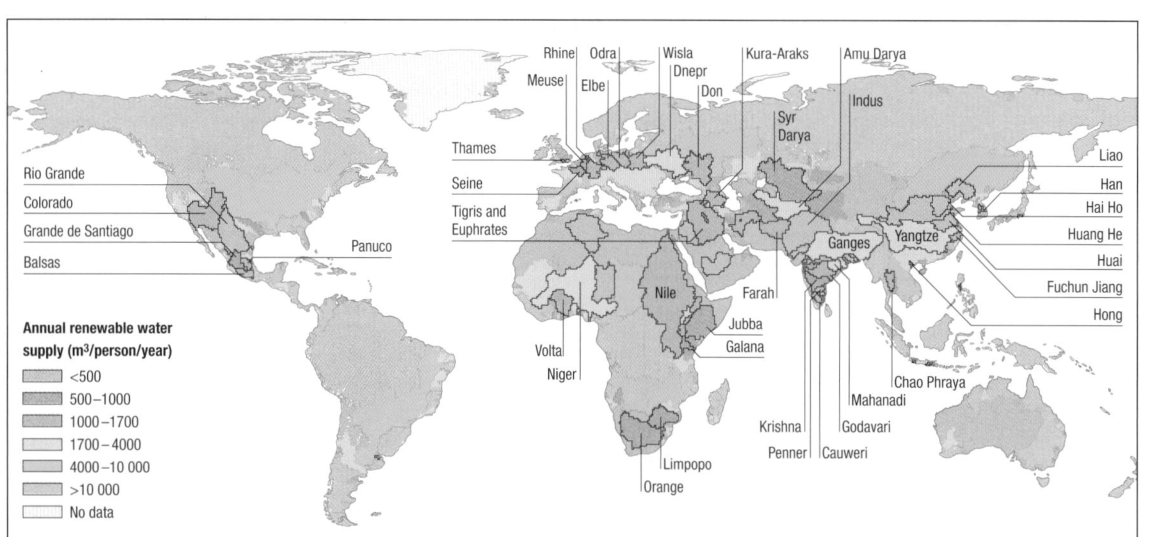

The future availability of renewable freshwater around the world will show greater disparity in supplies between countries as demand rises in many places. In this estimate of annual renewable water supply per person per river basin for 2025, outlined basins are projected to have a population of 10 million people or more, and less than 2500 m^3 of water per person per year – a scenario for water scarcity.

Where urban demand has increased over many years during which the rainfall has been average to above average, the onset of a drought period that results in water rationing may have serious social and economic effects. Many reservoirs are multipurpose, serving urban water supply, crop irrigation, hydroelectric generation and/or recreational purposes. At times of rationing, managers have to make difficult decisions in terms of the competing demands. In recent years there has also been growing recognition of the need to allow for reservoir outflows for downstream environmental health and biodiversity.

Groundwater

About 30 per cent of global freshwater resources are in the form of groundwater and provide the major source for a large part of the world's population. Water levels in rivers, lakes and reservoirs respond rapidly to rainfall events but groundwater has a much longer natural accumulation and discharge time. Groundwater is buffered against short-term weather and climate processes. However, because of its long residence time in aquifers, groundwater is highly vulnerable to pollution and overexploitation. Large scale and intensive tapping of these aquifers is virtually equivalent to a process of non-renewable mining for water.

The most vulnerable groundwater systems are the freshwater lenses typical of coral atolls that are vital for vegetation cover and the people. These water resources are fragile and generally under increasing threats from overexploitation. Knowledge of the recharge rate is essential for managing the sustainable extraction of potable water; overexploitation will lead to salinization that might take many years to rectify. In most smaller coral atolls the freshwater is at a depth of only 2 to 5 m and the lens thickness is generally only about 20 m. Leaching of surface pollutants and wastes readily contaminates the groundwater.

Climate variability and change

Rainfall at any location is highly variable from year to year and sustainable exploitation of rainfall requires careful planning. Also, the year-to-year variability is generally highest in those areas where the annual average rainfall is lowest. Communities in semi-arid regions that must rely on locally generated water resources will inevitably experience prolonged dry periods, often lasting years. When the rains do come they can be disruptive, or even destructive, as the sudden surface-water flow picks up soil and other debris to be carried into storage reservoirs.

Patterns of rainfall and its variability are expected to change with global warming. Heavy rainfall events are expected to increase in frequency and intensity. The generally warmer temperatures and higher evaporation rates will lead to greater soil moisture loss during dry periods in some areas. Such changes could have a direct impact on freshwater supplies and their management. As already noted, of particular concern are the many tropical coral atolls, where drawdown of the freshwater lenses would increase if there were more dry spells and greater demands from higher populations. Here the potential for contamination is further increased because of the threat of sea-level rise that would change the hydraulic pressures and encourage salt-water intrusion. It is estimated that a typical freshwater lens could lose 40 cm of thickness for every 1 cm rise in sea level.

Harvesting clouds of their water

An ingenious example of the application of meteorological information and technology is the capturing of water from fog in areas where potable water is scarce or non-existent. This idea is not new. For hundreds of years it has been known that trees growing in heavy fog areas get their water needs supplied by drifting fog (as do certain animals, such as spiders in the Namib). Scientists working in small isolated communities in the Andes of Chile have demonstrated the feasibility of wringing water from fog for drinking and irrigation. The key is finding these locations facing acute water shortages where there is a steady horizontal drift of moisture-bearing air.

These water-extraction schemes are not meant to provide water to millions but are invaluable for settlements like Chungungo, Chile, where water brought in by truck is expensive. Here, vertical panels of polypropylene mesh intercept the wind-blown clouds. Tiny droplets stick to the mesh and by accretion gradually form larger drops. It is estimated that 10 million fog droplets are necessary to form one drop. The drops trickle down and flow into collectors to be piped away for consumption.

Desalination plants

A number of cities extract freshwater from sea water through desalination in combination with electricity generation, particularly in the relatively dry but oil-rich countries of the Middle East. Such freshwater is suitable for most urban purposes and the price of desalinated water is comparable to importing freshwater, either by way of shipping or piped from neighbouring countries. However, the thermal energy usage from fossil fuels used in the process contributes to greenhouse gas emissions.

Early hydrological measurements involving basic observation techniques contribute to the assessment of likely fluctuations in river flow and their potential to flood.

4.18 Recreation and leisure

Increased prosperity worldwide has led to greater community focus on recreation and leisure. Travelling to far-away places involves preparation for a climate regime that may be significantly different from home. Many adventure activities challenge the natural elements, so safety demands understanding the climate and paying close attention to prevailing and forecast weather conditions.

Travel is a major thread linking the history of humankind. Exploration, trade, cultural curiosity and therapy have all taken people away from their homes to foreign lands and climates. During the 20th century, with the evolution of mass air transport, tourism literally 'took off'. Regular flights now take people halfway around the world in less than 24 hours, a journey that would have earlier taken many months by sailing ship and even several weeks by steamer.

One of the characteristics of rapid aircraft travel is that passengers can board a plane on a hot summer's day and alight hours later on a freezing winter's night, or vice versa. The human body may have to suddenly adjust to temperature changes of 30°C and more. Such metabolic shocks make a mockery of the concept of travel to exotic places for rest and recuperation, unless there is careful planning and time for adaptation.

Notwithstanding the climate sensitivity often associated with long-distance travel, tourism is one of the fastest-growing economic sectors in the global economy. The World Tourism Authority projects that international tourist arrivals will top one billion by the year 2010. Many communities, even some countries, are dependent on the income from tourism and recreational activities for their prosperity.

The rapid development of resorts in favourable climatic regions, like Cancun in Mexico, is part of the burgeoning of international tourism. The development places heavy demands on the local environment and exposes many people to more challenging climatic factors than they might normally experience.

Safety and tourism

Recreation and leisure cover many concepts, be it a chance to relax, to discover exotic lands and their cultures, to attend a specific event or to participate in a special activity. Having enticed travellers to far-away locations there is then a necessity to ensure their safety. Avalanches on ski runs or tropical cyclones on resort islands illustrate the extremes, but in between there are a host of weather events that can lead to tragedy if not carefully planned for. Hikers, rafters and fishing people are all involved in the sort of recreational activities where they can be caught out and their lives endangered by a sudden weather change. Many resorts make arrangements to receive the latest weather information and post it prominently to assist in planning daily recreational activities.

Knowing what to expect

Being prepared is an important part of international travel, and this can mean the difference between an enjoyable, rewarding and relaxing experience, and a period of pain or even serious illness. Sunburn, dehydration and heat stress are a few of the dangers common to tropical resorts or desert regions in summer that can easily be avoided. Amazingly, tourists frequently have succumbed to these afflictions because they have not taken precautions or not recognized the symptoms at the onset.

Sudden ascent to higher elevations can also cause unexpectedly unpleasant experiences without adequate preparation. Many holidaymakers in Hawaii, for example, go to see the sunrise over Halealaka. Within half an hour they have gone from sea level to an altitude greater than 3000 m, the temperature has fallen from 30°C to near zero, and the air pressure is such that they are only getting 70 per cent of their usual oxygen intake with each breath!

Scaling mountains like Kilimanjaro in the United Republic of Tanzania requires climbers to pass through a wide range of climatic zones in a short time.

Around the world in a balloon – high adventure

On 20 March 1999, when the Breitling-Orbiter 3 with Bertrand Piccard and Brian Jones accomplished an around-the-world balloon voyage after 19 days, 1 hour and 49 minutes, the last great aeronautical adventure of the 20th century was completed. Such a feat was possible because of the technology of the craft and the planning and skill of the crew. The pressurized capsule had provisions for 20 days and the balloon could fly to altitudes of 12 000 m. However, the timing of the flight was based upon identifying the time of year with the most favourable climatological conditions, and the flight was managed using the most advanced weather forecasting models and navigational software. Under guidance from the supporting forecast team the balloon altitude ranged from 5000 m to 11 500 m during the flight to take advantage of the optimum wind direction and speed. While travelling with the subtropical jet stream, speeds of 220 km/h were reached.

Tourists often go to places and engage in activities, which, if taken leisurely, should be well within their capabilities. However, with limited time and budgets, every moment counts, and stress is put on the body, compounded by not having acclimatized or not having developed sufficient physical fitness for the sustained exertions in the unfamiliar climate. People who visit desert regions and then hike to experience nature's wonders or who explore the cultural ruins in the high mountains of Central and South America are putting strains on their bodies that may be compounded by the unfamiliar climatic conditions.

The safety of people participating in adventure recreation, such as white-water rafting in Uzbekistan, requires constant monitoring of local weather and hydrological conditions, and early warning of severe weather that might prove hazardous.

Adventure recreation

Increasing numbers of tourists participate in adventure recreation, where they develop skills in mountaineering, rock climbing or canoeing, etc. to tackle nature's many challenges. Many adventurers develop high levels of physical stamina and carry out planning to a high degree of detail. Although they are challenging nature, through careful planning they are also minimizing their overall risk. Experienced mountaineers tackling high peaks plan their ascent for the period of the year when climatologically there is the most favourable weather. The weather on high peaks can change quickly with sudden onset of high winds and cloud that significantly lower comfort levels and increase the danger. Paying close attention to the prevailing weather and forecasts is vital for safety.

Many adventure activities, while requiring physical stamina, do not require a high degree of skill in the particular pursuit. Such activities as rafting in fast-flowing rivers and streams are often carried out within groups under the guidance of experienced tour operators. Without a level of excitement, even to the extent of a dunking in an upturned raft, such activities would not be 'adventurous'! However, there is always the possibility that the weather will suddenly change for the worse, putting the participants in real danger.

The expansion of adventure recreation worldwide has benefited from access to climatological information for planning and to regular weather forecasts for managing activities and ensuring safety.

The solo sailor – not alone on the sea!

Many ships and their crews have circumnavigated the world since Ferdinand Magellan's surviving ship *Victoria* and 18 crew completed the voyage in September 1522 (of the five ships and about 250 crew that set out in September 1519). Since Sir Francis Chichester did it solo in the *Gypsy Moth V* in 1967, many more lone yachtsmen have followed. However, Vinny Lauwers was the first paraplegic person to circumnavigate the world solo in a yacht. Like adventurers before him, Vinny's success and safety were due to planning, including studying the likely weather patterns (the climatology). Vinny was not alone on the sea, because he was in daily e-mail contact with a weather forecaster who provided guidance on the best course under the prevailing and predicted weather conditions.

4.19 Computer modelling comes of age

The accumulating knowledge of the climate system, enhanced by streams of data from satellite systems supporting integrated modelling studies, bore fruit in the 1990s. Computer models truly came of age when they simulated atmospheric processes so accurately that they could be used for both weather and climate forecasts.

A representation of Richardson's 'fantasy factory' where he envisaged that 64 000 mathematicians would be needed to make calculations to forecast the weather for the whole globe.

Weather forecasting has now become largely a problem of mathematical physics. The potential to use mathematical equations representing well-established physical laws to predict the behaviour of the atmosphere was first proposed by the British meteorologist Lewis Richardson in the early 1920s. At the time it was an impractical proposal. As Richardson himself noted, he 'played with a fantasy' of a forecast factory of 64 000 mathematicians, which he estimated would be the number needed to keep pace with the weather. This dream did not begin to become a practical reality until the 1950s with the advent of high-speed digital computers. Long before these advances had occurred, however, weather forecasting, based on detailed analysis and empirical rules, had become a vital part of our lives.

Numerical weather predictions

Computer models used for weather forecasting calculate the birth, growth, movement and decay of large-scale weather systems. The models incorporate the principles of conservation of momentum, mass, energy and water in all its phases; the Newtonian equations of motion applied to air masses; the laws of thermodynamics and radiation for incoming solar energy and outgoing infrared radiation; and the equations of state for atmospheric gases. Factors specified in advance include the size, rotation, geography and topography of the Earth; incoming solar radiation and its diurnal and seasonal variations; radiative and heat-conductive properties of the land surface; and the surface temperature of the oceans.

The ENIAC computer, which filled a room measuring approximately 10 m by 15 m, was capable of 300 multiplications per second and was used to produce the first successful numerical weather forecasting experiments in early 1950.

Ensemble forecasts

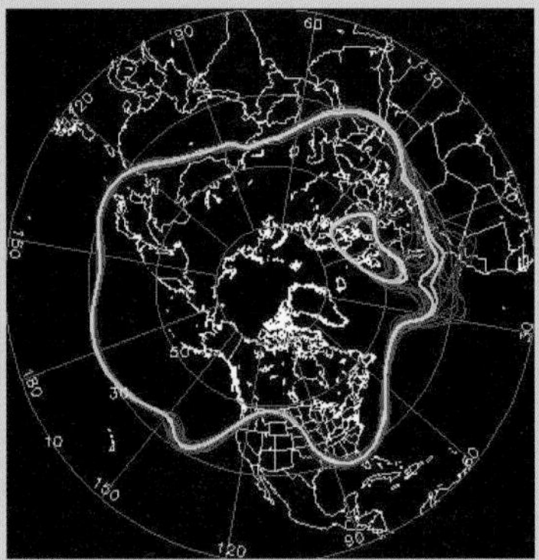

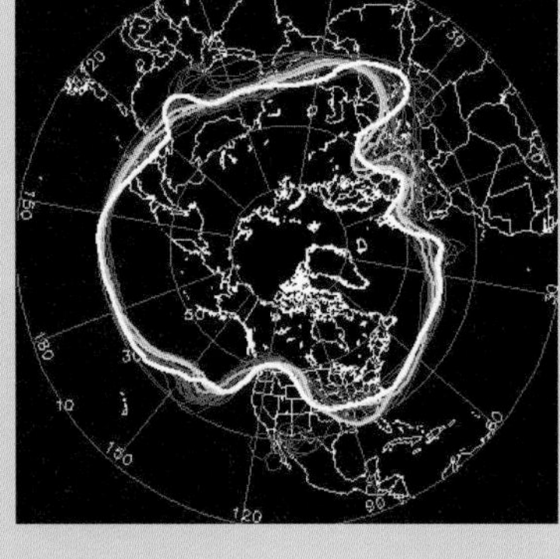

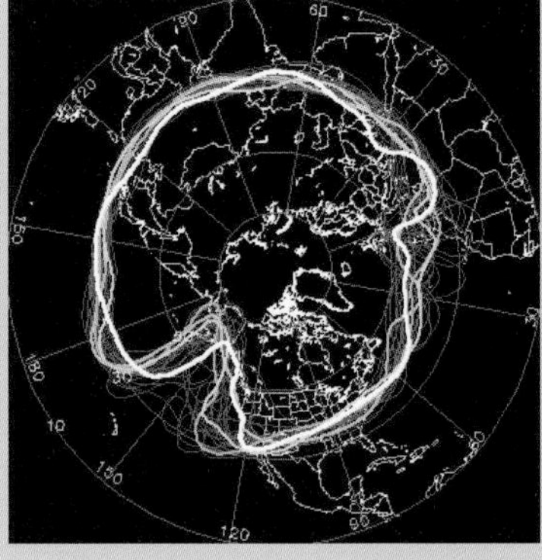

By running prediction models several times with slightly different starting conditions, it is possible to identify the more reliable and the more uncertain aspects in a forecast period by observing how the different forecasts remain similar or diverge rapidly. With the growing power and speed of modern computers, this ensemble approach is now a central part of weather and climate forecasting. The coloured lines in this example from the US National Centers for Environmental Prediction each represent one solution in an ensemble at the initial time, and 4 and 9 days into the future. Minor differences in the configuration of the 5640 m contour of the 500 hPa geopotential height at initial time (left) are associated with the uncertainty in determining the exact state of the atmosphere due to observational and other sources of errors. These initially minor differences amplify due to the chaotic nature of the atmosphere.
At four days lead time (middle) some of the subsequently observed features (heavy white line) are covered well by the ensemble, but others less well, notably over Europe. At nine days lead time (right), none of the main features has been predicted that well. The high pressure ridge over the northeast Pacific is predicted by most ensemble members but its exact location and amplitude is uncertain. Other aspects of the flow are completely unpredictable. Over western Europe, some members predict a low, while others predict a high pressure system. With this level of uncertainty, the predictability limit has clearly been exceeded in these regions.

The physical state of the atmosphere is updated continually, using observations from surface land stations, ships, buoys, aircraft, balloons and satellites to establish the initial conditions for the model prediction. The number and accuracy of the observations are fundamentally important since any deficiencies in this initial representation will expand as errors in the forecast.

Each of the points in the 'model atmosphere' is assigned new values of temperature, pressure, wind and humidity with each step forward in time (15-minute time-steps are typical) to provide forecasts up to ten days ahead – an enormous number of calculations. The final products are predictions of atmospheric pressure, temperature, wind, humidity, vertical motion, rainfall and other meteorological parameters, which are used to generate a variety of forecasts for different users.

Forecasts have shown significant improvements in the last 20 years or so in terms of both accuracy and lead time. Nonetheless, they still only provide consistently useful forecasts out to around six days. Progress in making single explicit weather forecasts beyond this limit is constrained by the chaotic nature of the atmosphere that restricts how far ahead its detailed evolution can be predicted.

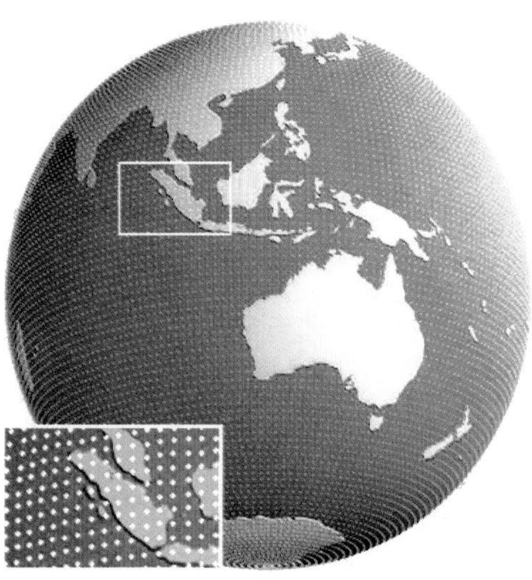

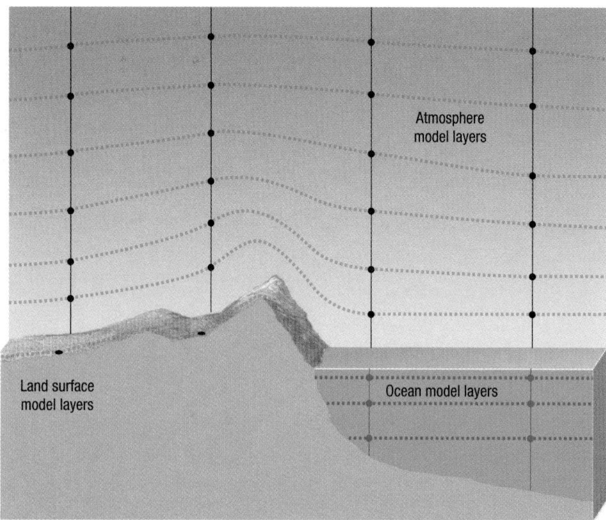

Computer models of the coupled climate system represent the global fluids of both the ocean and atmosphere by a set of grid points at defined layers from deep in the oceans to high in the atmosphere. For global-scale models, the atmosphere is typically divided into 30 or more layers between the ground and around 30 km and, in the more advanced models, each level is divided up into a grid of points about 50 km apart – some 4 million points in all (and far in excess of Richardson's vision).

Satellite temperature measurements

It took a long time to fully exploit the potential of satellites for weather forecasting. The earliest instruments relied on measuring reflected sunlight, and observations were limited to the daytime. Also, it was impossible to see below the clouds or over the polar regions in winter. The development of infrared radiometers that measured the upwelling heat radiation from the Earth (bottom figure) made it possible to measure continually the temperature of the Earth's surface, the tops of clouds and, by measuring the radiation in selected wavelength bands, to estimate the temperature of layers in the cloud-free atmosphere (top figure). The first infrared radiometer that could measure the temperature of successive layers and build up a temperature profile throughout the atmosphere was launched in 1969.

The better measurement of the temperature profile of the atmosphere is an essential part of improving weather forecasts. Orbiting satellites now provide assessments of the temperature of the atmosphere from the surface to high in the stratosphere.

Early atmospheric temperature measurements from satellites proved of limited value in weather forecasting, but did produce significant benefits in the Southern Hemisphere in regions where measurements are few and far between. The advent of operational microwave sounding units in 1979 that can see through clouds then gave global coverage of data. Now, numerical weather forecasts routinely use satellite data as part of the continual process of defining the current state of the atmosphere and producing better predictions.

4.20 Modelling the global climate

Just as an architect might build a scale model of a building to understand and predict its behaviour, so too the climate scientist can build a computer-based model of the climate system to understand and predict its behaviour. Climate models incorporate the physics of the atmosphere and the oceans and aim to answer questions such as how a current El Niño might evolve, and what might happen if greenhouse gas concentrations double.

The challenge for weather and climate models is to run forward in time much faster than the real atmosphere and oceans. To do this, they must make a large number of simplifying assumptions, and perform prodigious numbers of calculations. Various types of model are used to analyse different aspects of the climate. They can be relatively simple one-, two- or three-dimensional, and can be applied to a single physical feature of climatic relevance, or they may contain fully interactive, three-dimensional processes in all three domains: atmosphere, ocean and land surface.

Flux adjustment

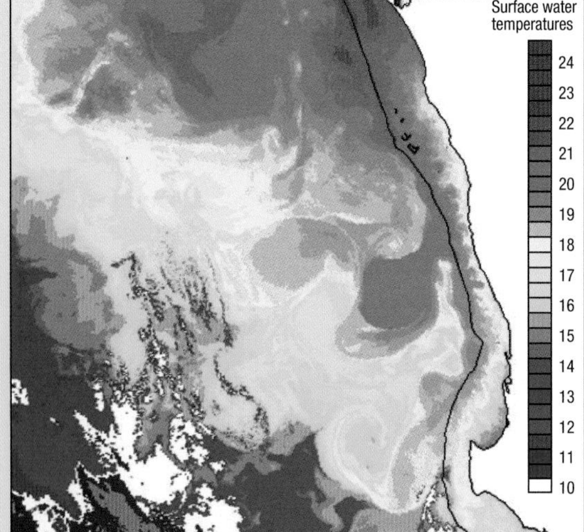

One of the most challenging features in the development of coupled GCMs has been how they deal with the exchange of heat, momentum and freshwater between the oceans and the atmosphere. The rates of exchange (or fluxes) vary dramatically throughout the year and from place to place. For instance, the oceans will provide large amounts of heat to the atmosphere in the winter hemisphere and absorb solar energy in the summer hemisphere. Over some regions the heat exchanges are huge and vary dramatically between warm water currents and adjacent cold water, as with the strong temperature gradient between the warm tropical waters (red/orange) of the Leeuwin current and the colder waters (green/blue) off Western Australia (the thin black line is the edge of the continental shelf).

Almost inevitably, small errors build up over time. To prevent this, pragmatic adjustments, known as 'flux adjustments', have been made in some models to keep them balanced. Although this approach has worked rather well, it was seen as a temporary fix to a problem. The latest high-resolution coupled models appear to show more success in managing climate simulations without using flux adjustments.

Matching the model to the problem

Where complicated processes vary according to a wide variety of factors, initially it may be best to explore the processes in one dimension. For example, when looking at chemical reactions that vary with the physical conditions through the depth of the atmosphere, one approach is to look at the reactions at each level from the ground to the top of the atmosphere using average climatic values appropriate to each level. One-dimensional models were initially used for energy-related studies of the climate system. As greater confidence is obtained in the way a simple model handles a particular process, the ideas are then incorporated into more complex two- , three- and four (time)-dimensional representations which incorporate the dynamics of the climate system.

Weather forecasting models must handle the properties of the atmosphere in three dimensions, and work with current analyses of the ocean surface temperatures and at least some basic land surface processes. These models have come to be known as atmospheric general circulation models (GCMs). In parallel, studies of the oceans can concentrate on three-dimensional properties of the oceans and are generally known as ocean GCMs. When it comes to simulating the general behaviour of the climate system over lengthy periods, however, it is essential to use models that represent, and where necessary conserve, the important properties of the atmosphere, land surface and the oceans in three dimensions. At the interfaces, the atmosphere is coupled to the land and oceans through exchanges of heat, moisture and momentum. These models of the climate system are usually known as coupled GCMs.

Global models for climate

Climate models have been developed from weather forecasting models but for the present use bigger grid spacing and longer time-steps so that they can be run further ahead in time for a given amount of computer time. Without more powerful computers, simulation of the climate with the same detail as in weather forecasts would take far too long, especially if we wanted to explore many different scenarios of the future. Nevertheless, there is increasing convergence between weather forecasting and climate models, especially for predictions in the range out to months and seasons.

The values of the predicted variables, such as surface pressure, wind, temperature, humidity and rainfall are calculated at each grid point over time, to predict their future values. The time step (the interval between one set of solutions and the next) is a function of the grid size: the finer the resolution the shorter the interval between each computation.

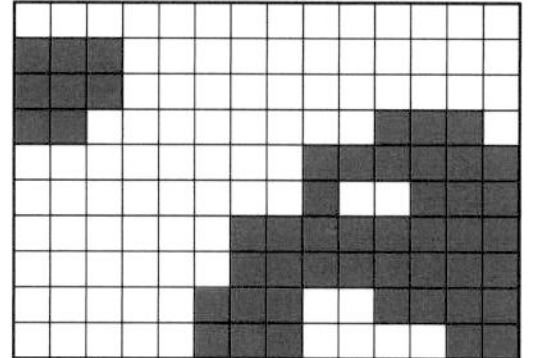

Model T21
horizontal resolution of 500 km

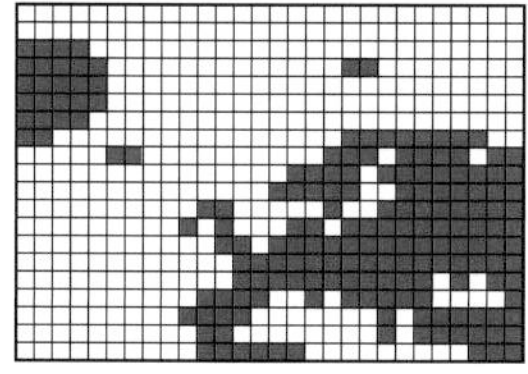

Model T42
horizontal resolution of 250 km

Model T63
horizontal resolution of 180 km

Model T106
horizontal resolution of 110 km

As the resolution in global climate models is increased it is possible to provide more detailed representations of atmospheric processes and important phenomena. Of particular importance are those sensitive to the underlying topography and coastal boundaries.

A coupled GCM with a 100 km horizontal resolution and 20 vertical levels, for example, would typically use a time-step of 10–20 minutes. A one-year simulation with this configuration would need to process the data for each of the 2.5 million grid-points more than 27 000 times – hence the necessity for supercomputers.

Even then, there are certain physical processes that act at a scale much smaller than the characteristic grid interval (e.g. clouds and turbulence). Additionally, the complete physics of, for example, clouds would, if computed explicitly at each time-step and at every grid-point, swamp the computer. These processes cannot be eliminated, so simplifying equations are developed to represent the gross effect of the many small-scale processes within a grid cell as accurately as possible. This approach is called parameterization and much research work continues to be directed at devising better and more efficient ways for incorporating these sub-grid-scale processes into climate models.

Coupled model systems

Coupling the ocean processes to atmospheric GCMs is a major challenge. The thermal capacity of the oceans is massive compared with the atmosphere and can provide to, or extract from, the atmosphere massive amounts of latent and thermal heat. Representing the heat storage, and the absorption of greenhouse gases by the oceans, in long-term simulations of climate requires a full three-dimensional ocean model, which simulates even the deep currents. Changes in the intensity and location of deep-water currents can ultimately have profound effects on the atmosphere. In the past, changes in the circulation of the oceans have produced major atmospheric responses. The models must also be able to handle shorter-term fluctuations such as those associated with the El Niño/Southern Oscillation.

Recent developments in climate modelling, which take into account not only surface processes at the ocean–atmosphere interface but also those acting at depth, have produced considerable improvement in the quality of climate model results. An oceanic GCM typically requires very high spatial resolution to capture eddy processes associated with the major currents, bottom topography and basin geometry. High-resolution ocean models are therefore at least as costly in computer time as are atmospheric GCMs. Further coupling of other climate system component models, especially the cryosphere and the biosphere, are also necessary to obtain more realistic simulations of climate on decadal and longer timescales.

Regional climate models (RCMs)

Simulating climate change at the regional and national levels is essential for policy-making. Only by assessing what the real impact will be on different countries will it be possible to justify difficult social and economic policies to avert a dangerous deterioration in the global climate. Furthermore, understanding processes on the regional scale is a crucial part of global research. Processes acting on local or regional scales, such as mountain ranges blocking the flow or dust clouds interacting with radiation will ultimately have impacts at the global level.

One technique used to overcome the coarse spatial resolution of coupled GCMs is that of nested modelling. This involves linking models of different scales within a global model to provide increasingly detailed analysis of local conditions while using the general analysis of the global output as a driving force for the higher resolution model. Results for a particular region from a coupled GCM are used as initial and boundary conditions for the RCM, which operates at much higher resolution and often with more detailed topography and physical parameterizations. This enables the RCM to be used to enhance the detailed regional model climatology and this downscaling can be extended to even finer detail in local models. This procedure is particularly attractive for mountain regions and coastal zones, as their complexity is unresolved by the coarse structure of a coupled GCM grid.

4.21 Models versus reality

While weather forecasting and climate models are very complex and use very advanced mathematics and computer techniques, as well as masses of atmospheric and oceanic data, they are mere surrogates compared with the complexity of the climate itself. Current climate models are known to have characteristic limitations and to exhibit systematic errors, yet it is quite remarkable how well they do perform and provide valuable insights and useful predictions for the future.

The first test of a model is to see how well it can simulate various features of the current global climate. The performance of models has been kept under close review by the Intergovernmental Panel on Climate Change (IPCC). Its Second Assessment Report (SAR) in 1995 concluded that "large-scale features of the current climate are well simulated on average by current coupled models". In tackling the immense complexity of the global climate the modellers have adopted a variety of ingenious approaches to simplify their computational task. This multiplicity of approaches provides insight into which processes make a difference to how the models perform, and where the models have the greatest difficulty simulating the real world. Also, the scatter in their results is a measure of confidence that can be given to their performance.

In its Third Assessment Report (TAR) in 2001 the IPCC noted that modelling work has continued to make progress (see 4.22). The simplest measures of performance are the global values of temperature and precipitation that they predict. The notable feature of the temperature results (top figure) is that, while on average there is reasonable agreement with observed values, the scatter between different models is several degrees Celsius, which is large compared with both past climate change and predicted future changes. The biggest discrepancies are at high latitudes and over the continents. By contrast, differences over the ocean are rather small.

The simulation of precipitation (centre figure) produces a comparable scatter. The broad features of peaks associated with tropical rainfall and the middle latitude storm tracks appear in all the models. The discrepancies are related to the intensity of rainfall and are largest in the tropics. These discrepancies are important as the precipitation rate is a measure of the intensity of the hydrological cycle. More generally, the models with higher average temperature had the larger precipitation rates, as might be expected, as this should intensify the hydrological cycle.

The seasonal average mean sea level pressure (bottom figure) was represented rather well except, again, at high latitudes. When it comes to simulation of snow and sea ice cover the models do, however, have major unresolved difficulties. Some models had an excess of winter snow cover, which persisted into summer, while others without flux adjustment (see 4.20) lost all their sea ice in the Arctic and almost all their sea ice in the Southern Hemisphere throughout the year. Clearly, the failures to simulate accurately these essential features of the climate are serious deficiencies that need to be addressed.

Until recently, model representations of the scale of the general circulation of the oceans had fundamental limitations. However, the latest high-resolution models have produced more realistic figures for the amount of heat transported towards polar regions by the oceans. This is important both to establish greater confidence in the models and to examine the possible impact of global warming on the future circulation patterns of the oceans.

Handling clouds remains a substantial source of uncertainty in most models. This is a critical test because clouds exert such a strong influence on both incoming solar radiation and outgoing terrestrial radiation. The treatment of cloud type and distribution has a major impact on the

How each computer model represents the Earth's basic climatology is a fundamental characteristic. The IPCC, as part of its review for the Third Assessment Report, compared characteristics of a large number of climate models.

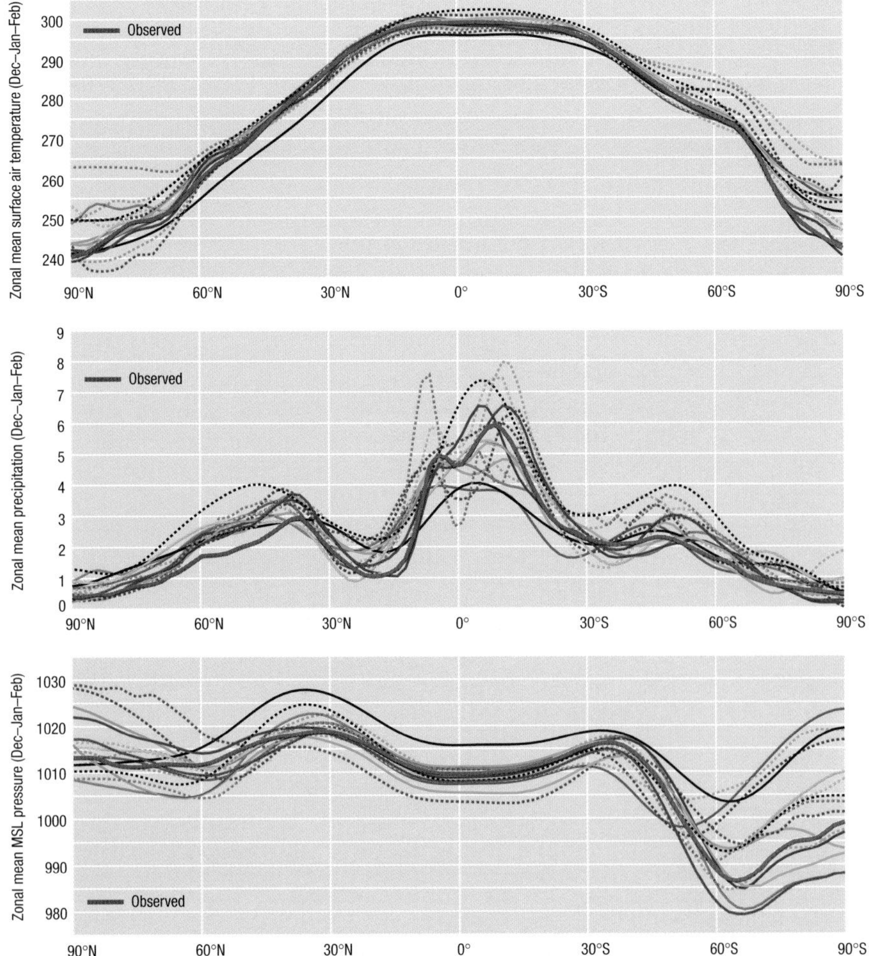

climatology provided by a model. Although the latest models represent clouds better, these processes continue to account for most of the uncertainty in predicting human-induced climate change. Moreover, the models do at best a modest job of portraying the latitudinal and seasonal distribution of global cloudiness. They systematically underestimate the cloudiness in low and middle latitudes in both winter and summer. In higher latitudes they overestimate cloudiness, especially over Antarctica, even allowing for the uncertainty of satellite measurements in discriminating between clouds and underlying snow and ice in these parts of the world.

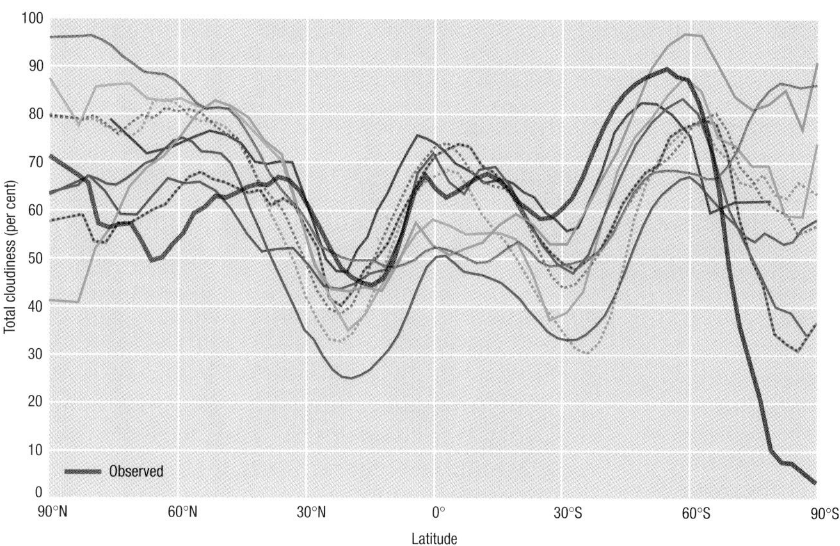

The representations of the amount and radiative properties of clouds are critical to models and their ability to realistically represent the Earth's climatology and how it might change. The cloud climatology generated by each of the various models reviewed within the IPCC Third Assessment Report show significant departures from what is observed.

Simulation of climatic variability

The ability to model natural variability is a further important test in assessing whether a given climate model might perform well in predicting climate change. By running general circulation models (GCMs) for the equivalent of many centuries of the global climate we can see how much they vary of their own accord. On the shortest timescales they do reasonably well, reflecting the variability of weather systems in the climate. They do, however, underestimate the variability of the regions of maximum storminess, and the differences between the models are substantial. The models also tend to underestimate the amount of blocking in the atmosphere.

Comparisons of the output of coupled models averaged over timescales of a month to a year suggest that the variability of the near-surface temperature is in reasonable accord with observed values. The simulated values in the tropics were generally larger than those observed but smaller in high southern latitudes.

Interannual variability is a more complex matter. Some of the coupled GCMs reproduce variability that resembles some aspects of the El Niño/Southern Oscillation (ENSO), although the magnitude is usually underestimated. Also, although many models produce realistic modes of interannual variability, the differences between them are considerable. While coupled models are simulating important aspects of large-scale atmosphere–ocean interactions, they cannot yet handle the seasonal cycle of the oceans well and the associated exchanges of heat, moisture and momentum.

The challenge faced in simulating decadal and longer fluctuations is to establish what constitutes the real level of natural variability. The combination of the limited geographical coverage of lengthy instrumental measurements and doubts about whether proxy records accurately reflect the scale of longer-term variability means that the true level of decadal and longer fluctuations has probably been underestimated. Nevertheless, comparing the results from coupled GCMs that have simulated the climate for hundreds of years suggests that the models reasonably reproduce interannual and decadal variability. They do not, however, succeed in producing a peak at around three to five years, which corresponds to the ENSO, and some of them seem to appreciably underestimate the magnitude of longer-term variability.

Reproducing the past

A popular test of models is to see how well they can reproduce the extreme conditions of the past. Two particular favourites are the Holocene climatic optimum 6000 years ago, and 21 000 years ago, during the depths of the last Ice Age (the last glacial maximum). The models meet the challenge of 6000 years ago quite well. They are able to simulate several well-established large-scale features of the Holocene climate but they underestimate the magnitude of these changes. In particular, they show that ocean and vegetation processes introduce important feedbacks that are necessary to explain the stronger monsoon at the time. The results show that, in studying the potential impacts of future climate change, more research is needed on the role of changing vegetation on future climate.

The simulations of the last glacial maximum are not fully consistent with the standard view of sea surface temperatures at the time. Some coupled models produce realistic results, especially in the tropics, and enhance our confidence in the estimates of climate sensitivity used in future climate change studies. They do not, however, include an adequate treatment of ocean heat transport, and need to be refined to raise further confidence in their capability.

4.22 Seeking answers on climate change

During the past two centuries our success as a species has created a first major global environmental problem. It was no-one's fault, but now we understand many of the causes. Importantly, people want to know whether characteristic features of the climate where they live, and to which they have become adapted, are going to change, and whether there is anything they can do about it.

During the 1970s there was a spirited debate about whether the climate was about to cool rapidly or start to warm appreciably. We now know that the global climate has markedly warmed since then.

"It is possible that we are on the brink of a several-decade-long period of rapid warming," observes Dr. Wallace S. Broecker of Colombia University's Lamont-Dohery Geological Observatory. "If the natural cooling trend bottoms out ... global temperature would begin a dramatic rise ... this warming would, by the year 2000, bring average global temperatures beyond the range experienced during the past 1,000 years."

Judging from the record of the past interglacial ages, the present time of high temperatures should be drawing to an end ... leading into the next glacial age....

NATIONAL SCIENCE BOARD, 1972

Were the cooling trend to reverse ... the earth could warm relatively rapidly, with potentially catastrophic effect.

NATIONAL SCIENCE FOUNDATION, 1975

Early predictions of climate trends were based on studies having only a few decades of data from limited regions of the globe, and so were bound to have shortcomings. For some scientists, the downturn in temperatures in the Northern Hemisphere during the 1960s and early 1970s, and the simultaneous rising level of dust and particulates in the atmosphere, suggested that the global climate was about to cool. By the early 1980s we had global data and a better understanding of climate dynamics, but not the modelling tools of today.

The progress made in the 1980s and 1990s on both modelling and measuring the global climate system has altered our understanding of the impacts of human activities and has prompted a sharper, more focused effort on trying to understand climate change. How can we be sure that humans are affecting the climate system? This is the detection and attribution problem.

Detection and attribution

Measuring the rise in global mean temperature over the past century is only the start. We must demonstrate that an observed climate change is unusual in some statistical sense. This is the detection problem. For this to be successful we have to know how much climate varies naturally. Having detected a climatic change, the cause of that change has still to be established. This is the attribution problem. Can we attribute the detected change to human activities, or could it be due to an unusual combination of natural causes?

Neither detection nor attribution can ever be 'certain', but only probable in a statistical sense. The attribution problem has been addressed by comparing the temporal and spatial patterns of the observed air temperature increase with what we would expect to see on the basis of model calculations. This technique depends on how well the models predict that human activities will produce patterns of warming different from those resulting from natural variability. Unfortunately, we do not know enough yet about this natural variability.

At the surface the analysis is complicated by how the warming in the last 100 years or so has varied around the world. For example, some of the regional surface warming over Europe and Asia since 1980 relates to the upward trend in the North Atlantic Oscillation (NAO), and this accounts for about one third of the wintertime hemispheric interannual surface temperature variance over the

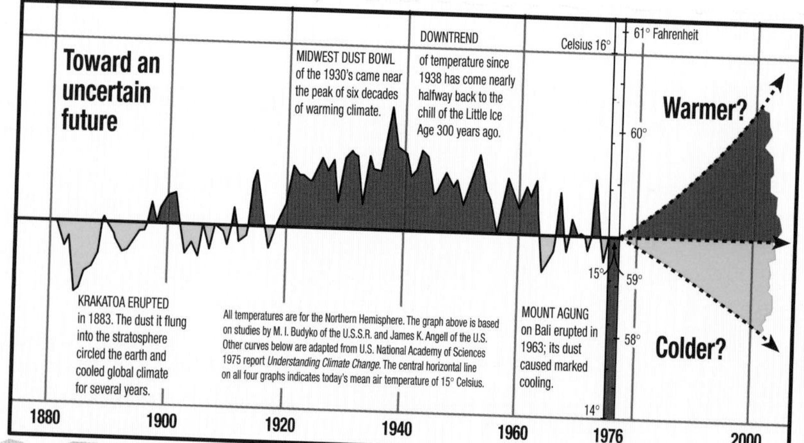

Matching surface and satellite observations

The integration of satellite observations to prepare global temperature trends has required great care to correct for drift in the satellite observations, small changes in their orbits and changes in the measuring equipment.

An issue that led to extensive scientific discussion in measuring global warming was the fact that, over the decades of the 1980s and 1990s, satellite assessments of the temperature for the lower troposphere showed less warming than surface observations. The satellite measurements also showed a lack of warming in the upper troposphere (approximately at the levels that the highest clouds reach). Although 20 years is too short a time to draw definite conclusions about long-term trends, the differences were hard to interpret. The issue was, however, partly resolved by combining the satellite data with earlier temperature measurements from radiosondes dating back to the 1950s. Over this longer period the discrepancy between the troposphere and the surface disappeared. The disparity over the shorter period may be due to different responses to short-term perturbations in the climate system, such as stratospheric ozone depletion, atmospheric aerosols and the El Niño phenomenon. Nevertheless, the differences are not fully resolved.

past 60 winters. This pattern has also led to cooling in the North Atlantic, south of Greenland. Therefore, discriminating between the effect of the build-up of greenhouse gases and that due to large-scale decadal fluctuations in the natural circulation of the atmosphere and oceans requires some careful detective work.

For example, there is an aspect of observed climate change in recent decades that helps in the attribution debate. This is the cooling of the stratosphere. The fact that this cooling has occurred while the lowest levels of the atmosphere have warmed supports the thesis that part of the observed change is attributable to the build-up of greenhouse gases in the atmosphere due to human activities.

Recent assessment

The IPCC, in its Third Assessment Report issued in 2001, concluded that the warming over the 20th century, at least in the Northern Hemisphere, was unusual when compared with the previous nine centuries for which sufficient proxy temperature estimates have now been derived. Moreover, by including naturally occurring phenomena (solar variations, volcanic activity) and anthropogenic forcing (increasing greenhouse gas concentrations, sulphate aerosols) in coupled general circulation models it was possible to simulate the global temperature variations observed since 1860. As a consequence, the IPCC concluded: "In the light of new evidence and taking into account the remaining uncertainties, most of the warming over the last 50 years is likely to have been due to the increase in greenhouse gas concentrations."

The Intergovernmental Panel on Climate Change (IPCC)

The World Meteorological Organization (WMO) and the United Nations Environmental Programme (UNEP) established the IPCC in 1988. It is open to all Members of UNEP and of WMO.

The Panel does not conduct new research, monitor climate-related data or recommend policies for governments. Its mandate is to assess the relevant research information on climate change available in peer-reviewed literature, journals and books. It provides scientific, technical and socio-economic information to the world community, particularly to the 170-plus Parties to the United Nations Framework Convention on Climate Change (UNFCCC).

For its Third Assessment Report, completed in 2001,the IPCC was organized into three working groups and a task force on national greenhouse gas inventories. Each had two co-chairmen and a technical support unit. Working Group I assessed the scientific aspects of the climate system and of climate change. Working Group II addressed the vulnerability of human and natural systems to climate change, the negative and positive consequences of climate change, and options for adapting to them. Working Group III assessed the options for limiting greenhouse gas emissions and otherwise mitigating climate change, and economic issues.

Modelling the 20th century warming

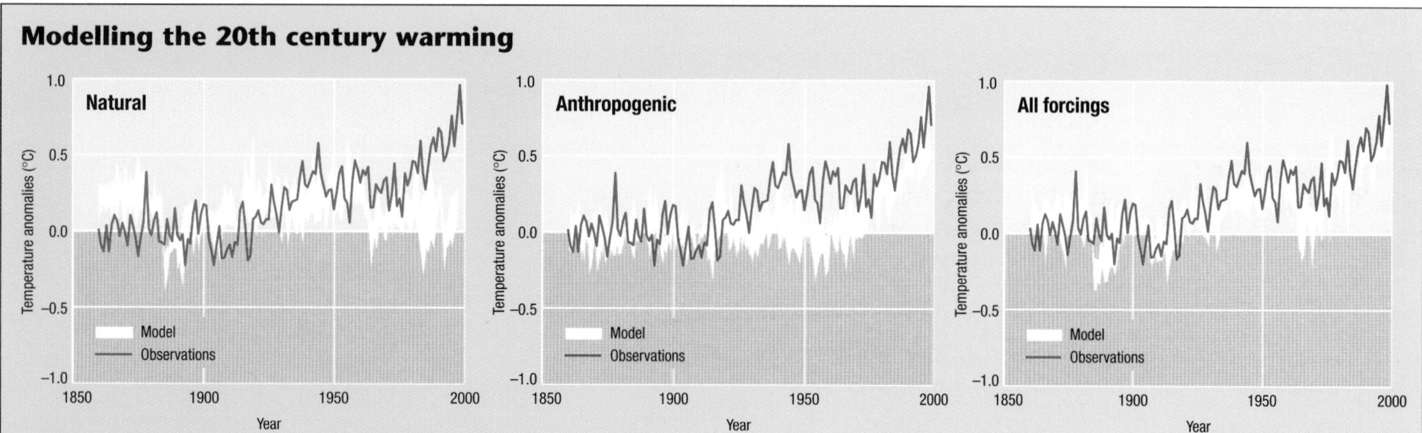

The first test of our ability to predict future climate is being able to reproduce what has happened in the past. A basic test of global climate models is whether they can simulate the changes that occurred during the 20th century. This is a matter of setting up the models to reproduce an analysis of the equilibrium climate at some time during the 19th century. This initial state is then perturbed by realistic changes in the radiative forcing due to the build-up of greenhouse gases since then. In addition, the models can be programmed to include estimates of how other factors have changed (e.g. sulphate particulates, solar and volcanic activity).

The broad conclusions that can be drawn from these simulations is that modelling exercises which include only the build-up of greenhouse gases do not simulate the interrupted course of the warming during the 20th century very well. When the models include the additional effects of the formation of sulphate particulates and natural forcing changes (e.g. solar variability and volcanic emissions) they reproduce rather well the observed large-scale changes, including the lack of warming between the 1940s and the 1970s.

5.1 Where do we stand now?

The technical progress of the last century, together with growing understanding of the climate system, will be carried forward in the 21st century. These advances will be an essential part of tackling the growing challenge of global warming and any associated changes in climatic variability.

The global average surface temperature increased by 0.6°C during the 20th century. The record shows that the rate of warming has not been uniform, with most of the increase occurring during two periods, 1910 to 1945 and 1976 to 2000. What is more, the increase in surface temperature in the 20th century is likely to have been the largest in any century of the last 1000 years. It is also likely that the 1990s were the warmest decade and 1998 the warmest year of the last millennium. Furthermore, it is the consensus of scientific opinion that there is new and stronger evidence that most of the warming over the last 50 years is attributable to human activities.

An unplanned environmental change

Although the climate changes attributed to human activities have been relatively modest so far, they are expected to become more dramatic by the middle of the 21st century. By then temperatures are likely to exceed anything seen during the past 10 000 years. Although some regions may benefit for a time, overall the alterations are expected to be disruptive. Clarification of the extent to which specific activities influence climate is essential in deciding which strategies are most likely to reduce the worst disruptions. Can we achieve this objective? We can get the answer, but only if the goal remains an international priority.

What we really need to know is how particular human activities will affect both the local and global climates. Accurate climate models are needed to provide this level of detail. This will require the technological muscle of supercomputers many times faster than those in use today. More data and research will also be needed to understand better how the climate functions. We need to disentangle the myriad

Energy demand and greenhouse gas emissions

If current rates of economic growth are sustained, global energy demand is expected to triple by the middle of the 21st century. In spite of this huge expected increase, it is possible to plan for more effective controls on how we generate and use energy, reducing our dependence on fossil fuels and switching about a third of all our energy consumption to renewable sources. Such changes will be driven by technological advances and pressures to reduce the environmental impact, including climatic, of fossil fuel use. Competition between renewables and cleaner, more efficient fossil fuel-fired equipment (e.g. high temperature gas turbines and combined heat and power systems) will produce energy generation systems that have less environmental impact.

If economic growth is sustained over the next 50 years or so, although the proportion of energy supplies met by fossil fuels will decline, at best their use will not change much, and at worst could still increase appreciably. The amount of carbon dioxide added to the atmosphere each year will therefore continue at current levels (see maps) or higher; a mid-range estimate is that emissions to the atmosphere will rise at about one per cent per year. Even with considerable success in shifting to alternative non-fossil fuel energy sources, this rate of rise will only slow gradually, and hence, the absolute level will continue to build up for much of the 21st century.

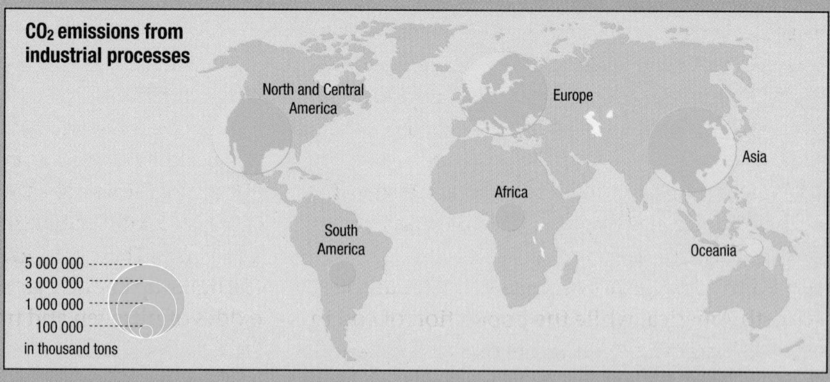

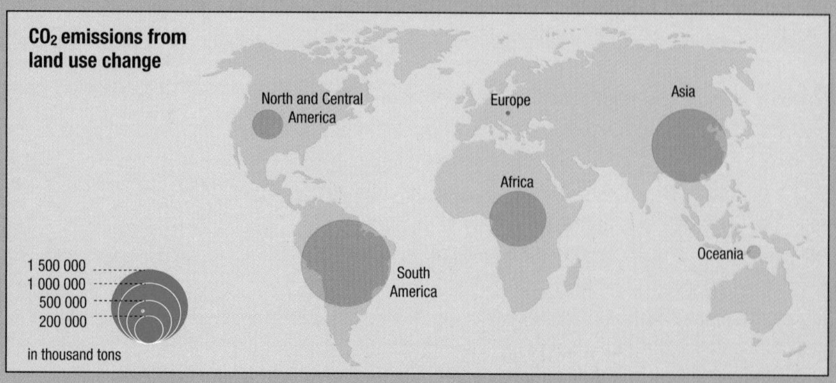

Climate change in the Arctic is already threatening many species that depend on the past ecological balance.

interactions among the oceans, atmosphere and biosphere to decide exactly which variables and which processes matter most.

Consequences of 20th century warming

The effects of global warming in the last 100 years are evident all around us. Perhaps the most striking evidence is in the changes in flora and fauna that have occurred in recent decades. Across Europe, observations of leaf unfolding in spring and colour changes in autumn show that the growing season has lengthened, in some locations by nearly 11 days since the early 1960s. Here, leaves are now emerging six days earlier in the spring and lasting nearly five days longer in autumn. These observations are supported by satellite measurements of vegetation activity across the Northern Hemisphere, which show even bigger changes in the growing season since the early 1980s, especially across Eurasia. Some of these changes in growing season are linked to milder winters that are part of the overall pattern of global warming since around 1970.

Other changes in Europe include plants growing at higher altitudes in the Alps, and birds laying their eggs earlier in the spring. Butterflies have extended their range northwards in both Europe and North America, while the population of cod in the North Sea has crashed where overfishing has combined with the water becoming too warm for the fish to spawn. In the tropical oceans the rise in sea surface temperature, which is part of the recent warming pattern, is suggested as the cause of widespread coral bleaching and raises the fear that many reefs may be wiped out by continued warming. In polar regions, the decline in ice cover in the Arctic is posing serious threats to the walrus, and is reducing the time available for polar bears to hunt, with disastrous consequences for their survival. In the Southern Ocean, vegetation is thriving on the most southerly islands and is expanding on the Antarctic Peninsula.

Climate change and climatic variability

The whole question of how much further the climate will change in the 21st century and whether it will become more variable will continue to be a major environmental issue. In terms of the human consequences of climate change, a substantial shift in variability could have a significant impact on our lives, and it is essential that the uncertainties are resolved.

The build-up of greenhouse gases

The level of carbon dioxide in the atmosphere rose from around 280 parts per million (ppm) in preindustrial times to around 370 ppm at the end of the 20th century. At the same time, other greenhouse gases also rose appreciably. A consequence of current energy consumption patterns is that the amount of carbon dioxide added to the atmosphere (as modified by absorption within natural sinks) is rising at a rate of about one half of one per cent per year.

The actual rate of growth will depend not only on the amount of fossil fuels burnt in the coming decades, but also on how much carbon dioxide remains in the atmosphere. There have been considerable fluctuations in the rate of growth of carbon dioxide levels since the mid-1970s. This raises fundamental questions about the rate of uptake of carbon dioxide by the biosphere, and how this will vary with climate change. Also, there are uncertainties about the future rate of growth of the global economy and how carbon dioxide emissions will be linked to this growth.

Current estimates of carbon dioxide concentrations in the atmosphere by the year 2100 range from 540 to 970 ppm, providing that climate change does not itself affect the ability of the biosphere to absorb carbon dioxide. Uncertainties about the future ability of the biosphere to absorb carbon dioxide widen this estimate to 490 to 1260 ppm. There are equally large uncertainties about the build-up of other greenhouse gases (e.g. methane, oxides of nitrogen and tropospheric ozone).

Measurements of carbon dioxide levels in air trapped in the polar ice sheets, together with atmospheric observations made since the 1950s, show that the concentration of this gas has risen by about a third since the early 19th century.

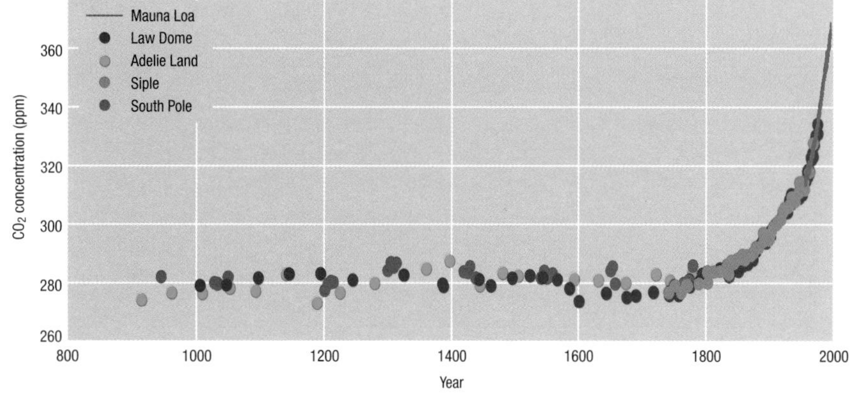

5.2 Global warming in the 21st century

Our estimates of what the global climate will be like at the end of the 21st century depend in large measure on computer models and the assumptions made in their formulation. The ability of the most advanced models to simulate the warming during the 20th century gives us some confidence in how much the temperature will rise. More detailed aspects of future climate change are less certain.

Scientists have refined the expectation of further global warming over the 21st century. They have drawn the many strands of previous observations, studies and computer modelling together under the auspices of the Intergovernmental Panel on Climate Change (IPCC). The IPCC reports represent the most comprehensive and thorough analysis of past and possible future climate change. The Third Assessment Report (TAR) of the IPCC, completed in 2001, provides several detailed scenarios for the future.

Predicted changes

The performance of the models in simulating global warming during the 20th century is reassuring. Their formulation includes the important contribution of natural forcing factors and the results give reasonable confidence that the various simulations of impact of human activities on the climate of the 21st century are plausible. A number of broad conclusions have been drawn from the various models, based on a range of scenarios of future greenhouse gas concentrations in the atmosphere.

The globally averaged surface air temperature is estimated to increase from 1990 to 2100 by between 1.4°C and 5.8°C. The magnitude of this increase over such a time period would be without precedent during the last 10 000 years. By 2100, the spread of surface air temperature rises predicted by the various models for any given scenario is large. This is because underlying assumptions and the different representations of complex processes mean that some models are much more sensitive to changes in concentrations of greenhouse gases than others.

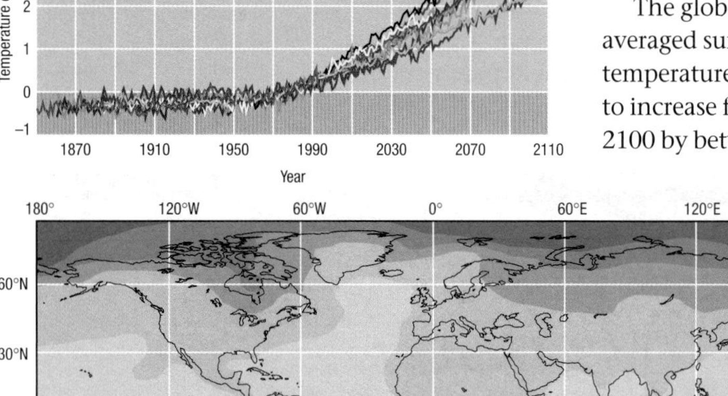

Various climate models predict that by the end of the 21st century the global mean surface air temperature will rise by about 3°C (time series). The multiple model ensemble map for the end of the 21st century predicts that most warming will occur over the Arctic and land areas, when compared with the 1961–90 Normals.

Climate models cannot yet provide a coherent picture of regional climate change, but some general characteristics can be identified. For example, it is likely that nearly all land areas, particularly those at high latitudes in the winter season, will warm more rapidly than the global average. Most notable is the warming in the northern regions of North America, and northern and central Asia. In contrast, the warming is less than the expected global-mean over South and South East Asia in summer and southern South America in winter. The surface temperature is likely to rise least in the North Atlantic and the circumpolar Southern Ocean.

Globally averaged water vapour and precipitation are projected to increase along with the warming. It is expected that precipitation will increase over northern middle and high latitudes and Antarctica in winter. At low latitudes both regional increases and decreases of rainfall over land areas are expected. Larger year-to-year variations are likely over those areas where the mean precipitation is predicted to increase.

Global mean sea level is projected to rise between 0.09 and 0.88 m above the 1990 level by 2100. This rise is due primarily to thermal expansion of the warmer oceans combined with melting of glaciers and ice sheets. The global mean temperature and sea level would continue to increase beyond 2100 because of the slow thermal response of the oceans, even if carbon dioxide concentrations in the atmosphere had stabilized by then.

At high latitudes in the Northern Hemisphere snow cover and sea-ice extent will continue to decrease. Glaciers and ice caps are projected to continue their widespread retreat. Over Greenland, the ice sheet is expected to lose mass as ice melt runoff and iceberg calving exceed the precipitation increase. Only the Antarctic ice sheet is likely to gain in mass because of greater precipitation.

What do the models say about future extremes?

Much of what climate-model studies predict about future weather and climate extremes is what we might intuitively expect. For example, warmer oceans supply more water vapour to the atmosphere. Research has shown that there is a relationship between the precipitation of individual storms and the prevailing surface humidity, so a warmer climate holding more moisture would be expected to produce more precipitation. A number of other predicted changes in weather and climate extremes have already been seen in observations in various parts of the world (e.g. decreased diurnal temperature range, warmer mean temperatures associated with more very warm days and fewer very cold days, etc.).

In spite of this progress, we must be careful not to attach too much confidence to predictions about specific extremes, as there are uncertainties about some climate processes and their inclusion in climate models. Nevertheless, most models predict that future climate change could include:

- higher maximum temperatures and more hot days in nearly all land areas;
- more intense precipitation events over many Northern Hemisphere middle to high latitude land areas;
- higher minimum temperatures and fewer cold days and frost over virtually all land areas;
- reduced diurnal temperature range across most land areas;
- summer continental drying in some areas and associated risk of drought; but,
- there is much less confidence about any increase in tropical cyclone peak wind intensities, or peak precipitation intensities.

For some other extreme phenomena, many of which may have important impacts on the environment and society, there is currently insufficient information to assess recent trends, and the confidence in models and understanding is inadequate to make firm projections. Uncertainties of what further global warming may do to the intensity of middle latitude storms is a good example. Also, the trend, after the mid-1970s, for conditions to become more El Niño-like in the tropical Pacific, is projected in some models to continue. However, confidence in such projections is tempered by shortcomings in how well El Niño/Southern Oscillation is currently simulated and how small-scale phenomena, such as thunderstorms, tornadoes and hail, are parameterized in coupled general circulation models.

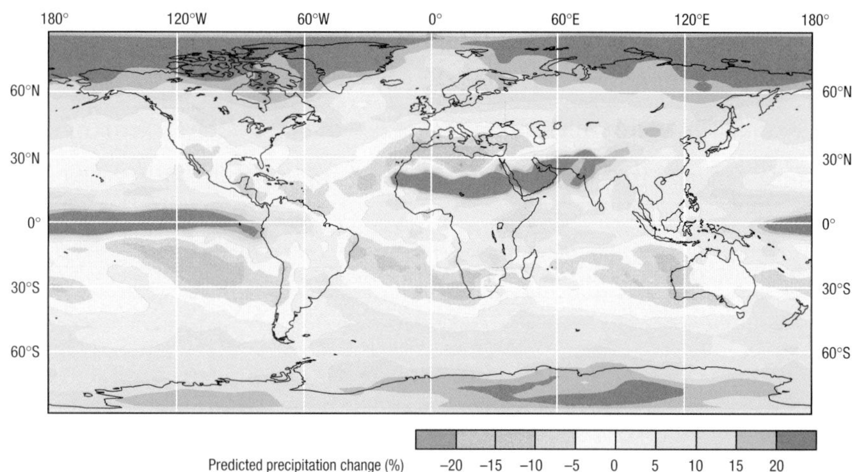

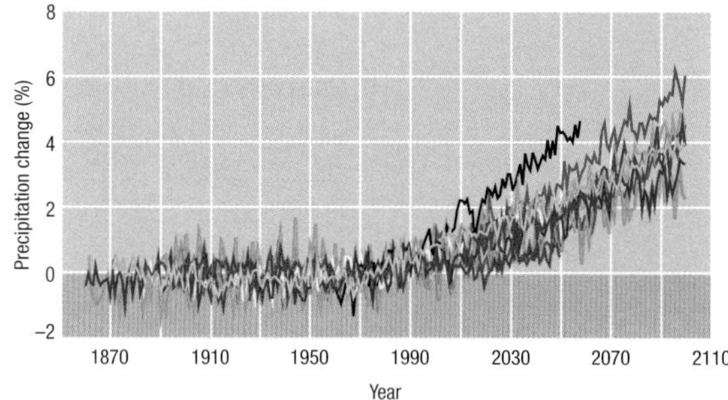

Various climate models predict that by the end of the 21st century the global mean precipitation will rise by a few per cent. The multiple model ensemble map for the end of the 21st century predicts that some regions, including the polar regions, will be wetter but others will be drier by some tens of per cent, relative to the 1961–90 Normals.

Thermohaline circulation

Most models show a weakening of the ocean thermohaline circulation and a reduction of ocean heat transport into high latitudes of the Northern Hemisphere. Even in models where the reduction in heat transport is considerable, however, there is still warming over Europe due to the effect of increased greenhouse gases. None of the models exhibits a complete shutdown of the thermohaline circulation by 2100.

Extreme events

The disruptive ecological and socio-economic impacts of extreme events means that they are likely to command disproportionate attention in consideration of the effects of future climate change. Current understanding suggests that we can expect an increased frequency of some extreme events, such as heavy rainfalls and thunderstorms, in a warmer climate. In addition, an increase in the number of very hot days, the frequency and strength of extreme winds, exceptional precipitation events, and the length and severity of droughts are all of concern.

The incidence of various extremes may alter substantially with a relatively small change in average temperature or climatic variability (see 1.5), so reliable measurements are vital. This requires high-quality daily data, which is a challenge for the current global meteorological observing networks to provide on a reliable basis. Although there is some evidence that the occurrence of extremes has changed in certain areas, a global pattern is not yet apparent. It is a matter of urgency that we find out more about regional shifts in the occurrence of extreme events as any changes are bound to have a profound influence on international action to reduce the impact of human activities on the climate.

A good example of the challenges involved is to identify how human-induced global warming might influence the incidence of specific weather regimes. Thus, developing models that can predict how patterns of natural variability may be affected by human activities remains a pressing research question.

Planning to meet the threat of climate change requires detailed predictions of the challenges to be faced in different parts of the world. This will involve using regional climatic models to explore various scenarios that identify the most likely range of future conditions. Robust features of these scenarios provide a useful guide on where to concentrate our efforts.

Whatever the scale of global warming in the next 100 years, our lives will continue to be influenced by variability and disrupted by extremes of the climate. Planning for these fluctuations from year to year and over the decades will be an essential part of the function of future climate services. How these fluctuations change in the future must be an integral part of the predictions of future climate change. Policy makers must then consider using predictions of how a warming climate will alter conditions in the regions where they have responsibilities, and what this means for social and economic services. This will involve analysing not only how average conditions change but also how the incidence of extremes will shift and whether the weather will become more variable.

Using scenarios

The principle of using scenarios is to map out a range of possible future events and to identify policy options that might best be able to meet the challenges posed. The decision to explore particular scenarios does not reflect our belief that these particular futures will occur, but that they represent a reasonable basis for testing policy options. While we are confident that without drastic action the level of greenhouse gases in the atmosphere will rise appreciably in the coming decades, and that this will lead to further global warming, we are less certain about the magnitude and impact of these rises and what they will mean in terms of regional climate changes. We therefore need to consider several scenarios to estimate the potential impact of global warming on the natural variability of the climate.

As has already become clear, there is a wide range of possible outcomes in the climatic conditions resulting from the build-up of greenhouse gases in the atmosphere. It is not realistic to consider all the permutations of different emission scenarios with all the varying climatic sensitivities that might result from rising concentrations of greenhouse gases. We have to make some judicious choices about combinations of conditions with which to generate a range of scenarios of future climate.

Once we have decided which scenarios provide the best basis for planning, we then must estimate how these conditions will alter future economic and social conditions, including the change in climatic risk factors affecting life and property. These will range from direct consequences, such as the rise in sea level, through the most obviously weather-sensitive activities (e.g. agriculture, forestry and fisheries), to those areas where long-term shifts in the climate will affect major investment decisions (e.g. transport infrastructure and buildings and construction). At all stages in this process we have to recognize that uncertainties in the climatic scenarios will be compounded with the uncertainties of trying to predict how our social structures will respond to the challenges of climate change.

Faced with such tremendous uncertainty, it would be all too easy to say it is just too difficult and we should not bother. This is not acceptable. We have to do the best we can to minimize life's hazards by developing robust and resilient community structures, while recognizing the limitations because of the uncertainties. What we must not do is blindly assume that we have chosen the right scenarios and act accordingly. As time goes by, the quality of any predictions will improve and we will have to refine our thinking. The essential feature of any planning process is that it must not be too rigid and should be capable of adapting to better information.

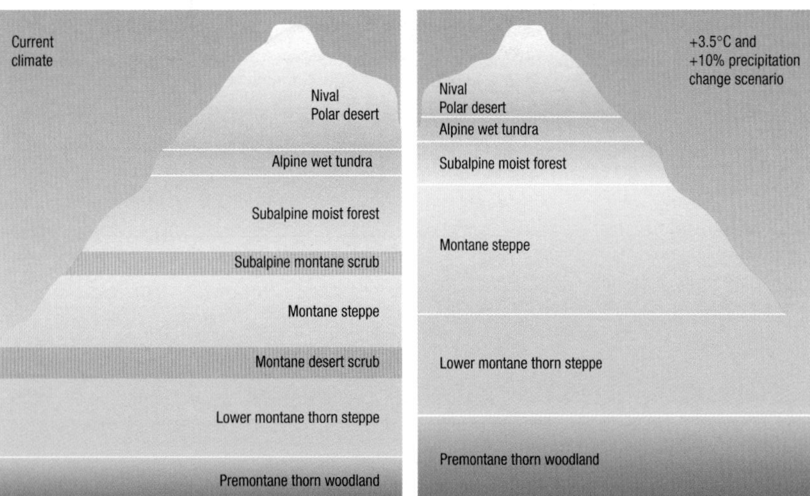

In a warmer world climate zones will move to higher altitudes in mountainous regions with flora and fauna in the coldest zones being squeezed into smaller refuges.

Regional climate models (RCMs)

Currently, we cannot use coupled global climate models to explore detailed regional policy options. Even in their most advanced form, coupled global models lack the resolution to consider regional impacts. Furthermore, they already require massive amounts of computer time to generate a limited number of scenarios. While scientific and technological advances may enable us eventually to use global models, for the present we can employ regional models embedded within global models. Regional climate models can provide more spatial detail of simulated climate compared with coupled global models. This is especially true in regard to the surface hydrological budget.

The increased resolution of RCMs also allows the simulation of a broader spectrum of weather events, in particular the higher-order climate statistics such as daily precipitation intensity distributions. In many parts of the world a serious limitation to evaluating these models is a general lack of climate data. In these parts it is difficult to verify the performance of models as a guide to their predictions. Nevertheless, where there are data there have been encouraging results, with some RCMs showing better agreement with observations than obtained from global models. Nevertheless, a consistent set of RCM simulations of climate change for different regions, which can be used for work on scenarios, is not yet available. By extending our studies to more regions, and obtaining ensembles of simulations with different models, we should be able to explore enough scenarios to provide useful information for planning.

Every year the lives and livelihoods of hundreds of thousands of people around the world are devastated by climate-related natural disasters. Effective strategies to help vulnerable areas become more resilient to such disasters will require reliable climate scenarios and an understanding of the economic and social conditions of the communities.

Examples of scenarios

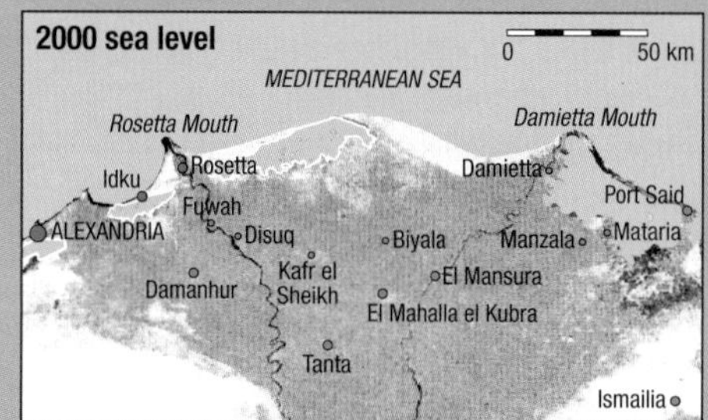

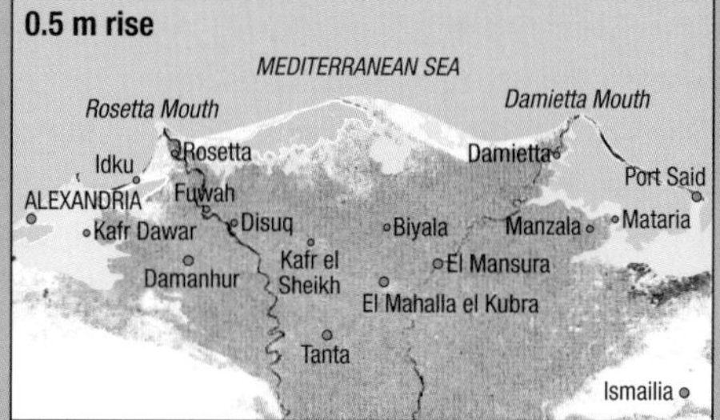

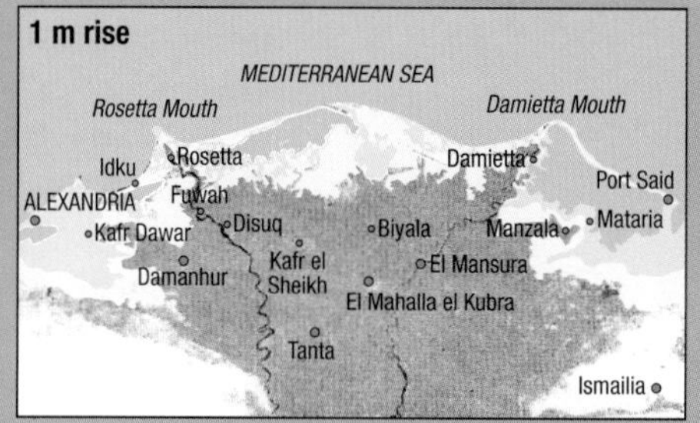

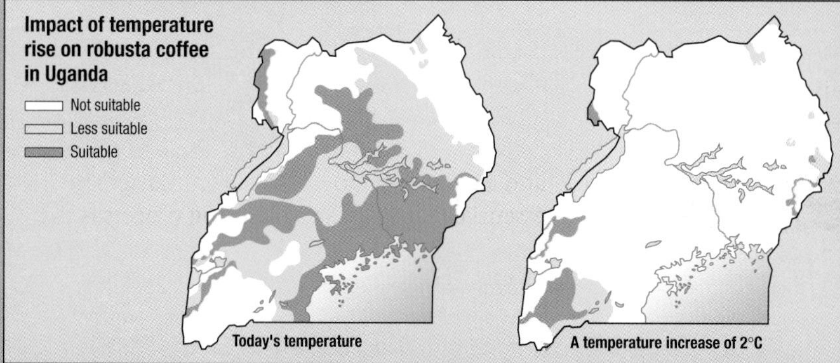

Some of the simplest forms of scenarios examine the direct impact on regional economies that might occur as a result of predicted climate change. For instance, in Uganda, the total area suitable for growing robusta coffee under a temperature increase of 2°C would be dramatically reduced. Only higher areas would remain productive, while the other areas would become too hot for the coffee variety to be economically viable. This is a clear example of the vulnerability of a country whose economy relies principally on one or two agricultural products.

The impact of rising sea level can be predicted in the same way. In the Nile Delta, a rise of 0.5 m could destroy weak parts of the sand belt, which is essential for the protection of lagoons and the low-lying reclaimed lands. One third of Egypt's fish catches are made in the lagoons. Sea-level rise therefore could affect nearly 4 million people. It would change the water quality and affect most freshwater fish. Some 1800 km^2 of valuable agricultural land could also be lost. Vital, low-lying installations in Alexandria and Port Said would be threatened. Beach facilities for recreation and tourism would be endangered, and essential groundwater could be salinated. Construction of dykes and other protective measurements could probably prevent the worst flooding for a rise of up to 0.5 m, but there could still be serious groundwater salination and damage from increasing wave action. A rise of one metre would cause much greater disruption, endangering 4500 km^2 of cropland, and affecting more than 6 million people.

5.4 The uncertainty factor

There are many uncertainties in our estimates of the future behaviour of the climate. These range from uncertainties in our basic understanding of the properties of clouds, through how the frequency of extreme events might affect regional climates, to how the oceans and the major ice sheets could respond to global warming.

The range of outcomes from different computer models highlights the fact that many aspects of the physics of the climate system have to be simplified to make them manageable and, further, how the different underlying assumptions in the simplifications contribute to the spread of results. When looking at what the models say about future variability, we have to recognize how much the results depend on the simplifying assumptions, especially those with respect to clouds.

Then, there is the possibility of what could be called 'major surprises'. Such events, which might involve a sudden and substantial shift in some aspect of the global climate, and external forces such as volcanism and solar activity, are a natural consequence of the potentially chaotic nature of the climate, and as such, are likely to be highly unpredictable. The fact that there is limited evidence of such events occurring in the climate record of the last 10 000 years or so does not mean that they could not happen in the future.

More long-term variability?

In recent years the performance of global coupled atmosphere–ocean models has led to improved estimates of natural variability of the climate on long timescales. There are, however, some inconsistencies:

- Several model-based studies show that in the future El Niño events will be more frequent. Other studies show little change in this phenomenon.
- Models generally predict enhanced year-to-year variability in the Asian monsoon but the amount of change varies with each model.
- There is no consensus on circulation statistics (e.g. paths of storms, frequency of middle latitude blocking).
- The models cannot yet handle changes in the characteristics of low-frequency variability (e.g. the North Atlantic Oscillation).

Possible irreversible changes in the Atlantic Ocean

The possibility of a collapse in the oceanic conveyor belt circulation is of particular relevance to European climate. If global warming provoked a reduction in the thermohaline circulation of the North Atlantic, this might be irreversible. However, we do not know whether the present climatic state is close to the temperature and salinity threshold for a sudden collapse of the circulation. While the models tend to show a reduction of the thermohaline circulation under a global warming scenario, none of the current projections exhibits a complete shutdown during the next 60 years. However, since natural variability in the climate system is not fully predictable, it follows that there are inherent limitations to predicting transitions and thresholds. The possibility of an irreversible

Radiative forcing

Human activities cause changes in the atmospheric composition and the reflectance of the land surface, and each changes the radiative forcing of the climate system. Natural phenomena, such as changes in solar output and volcanic activity, also contribute to change the radiative forcing. For the build-up of greenhouse gases in the atmosphere our confidence (as shown by the vertical bars) in the magnitude of the radiative forcing is reasonably high. The estimates for the photochemical destruction of ozone in the stratosphere and its creation in the lower atmosphere are moderately satisfactory. When it comes to the other impacts, however, such as those associated with the formation of aerosols, aircraft and the effects of changing land use, our confidence is much lower. These uncertainties, and how they are incorporated in climate models, are bound to make us cautious about our estimates of future climate change.

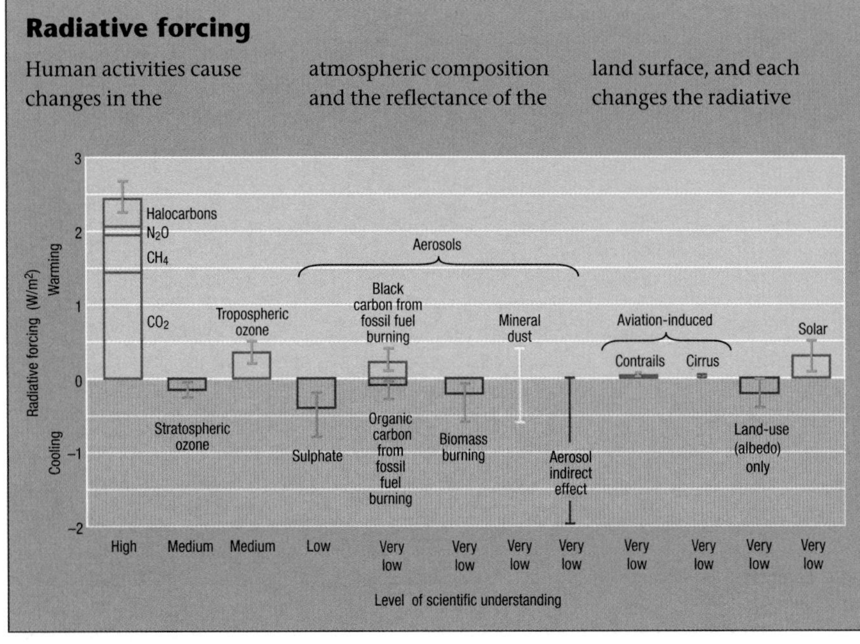

change of the thermohaline circulation remains, therefore, an unanswered question.

The collapse of the Greenland and Antarctic ice sheets?

The most important contribution to projected sea-level rise in the long term would be changes in the size of the Greenland and Antarctic ice sheets. At present there is no agreement from models and observations on whether either ice sheet is currently expanding or contracting. The Greenland ice sheet looks more vulnerable to losses during the 21st century. Conversely, the Antarctic ice sheet is likely to grow because higher temperatures would lead to an increase in regional snowfall. Overall, it is likely that increased losses from Greenland will be largely counteracted by gains in Antarctica for most of the 21st century.

The more dramatic scenarios – the complete melting or collapse of the Greenland or Antarctic ice sheets – appear to be more distant. Several modelling studies suggest that continued warming into the 22nd century could have dramatic consequences for the Greenland ice sheet. If the level of greenhouse gases rose beyond the equivalent carbon dioxide concentrations of twice the present value, the ice sheet might melt irreversibly. Although this level is not expected to occur until well into the 22nd century, and could be averted if appropriate action were taken in good time, once set in motion, such melting would have dramatic consequences. While the whole ice sheet would take at least 1000 years to melt, the rate of sea-level rise could be anything between 10 and 70 cm per century over the period.

Similar questions surround the stability of the West Antarctic ice sheet, which rests on a rock bed well below sea level. Depending on the physical assumptions made about this situation, models produce equivocal predictions about whether this ice sheet could collapse catastrophically with global warming beyond some threshold value. Catastrophic collapse of the ice sheet could lead to a sudden rise in sea level of several metres – far greater than anything predicted as a result of other consequences of global warming. This scenario has been the subject of numerous studies. At temperatures expected over the 21st century the West Antarctic ice sheet is considered to be stable. With snowfall over Antarctica increasing, the ice sheet is expected to make a negative contribution to sea level until global temperature has risen at least 8°C, something that is not predicted to happen until 2200 at the very earliest.

As for the stability of the remainder of the Antarctic ice sheet, its threshold for disintegration is estimated to involve warming in excess of 20°C, a situation that has not occurred for at least the last 15 million years. The ice sheet would still take a period of several tens of thousands of years to decay fully.

There is great uncertainty about how quickly the Antarctic and Greenland ice sheets will melt and possibly collapse if the global temperature rises appreciably in the next several hundred years.

Clouds

"I've looked at clouds from both sides now From up and down and still somehow It's clouds illusions I recall I really don't know clouds at all." Joni Mitchell (a Canadian singer)

Any detailed examination of climate model results highlights how little we know about the part clouds play. We do know they play a dominant role in the Earth's radiation balance, but cannot say whether climatic change will lead to coherent changes in cloudiness that reinforce any change (a positive feedback mechanism) or damp down the change (a negative feedback mechanism). As clouds exert such a strong influence on both incoming solar radiation and outgoing terrestrial radiation, the treatment of cloud type and distribution in the models has a major impact on whether the feedback is positive or negative.

There is nowhere near enough computer power to handle all the physics of cloud formation in the models. We therefore make informed judgements about their average behaviour on the scale of the model grid sizes. Even the most advanced models cannot say with confidence how much cloudiness will change with global warming, or how this might change the rate of warming. Although there was evidence of a rise in cloudiness around the world during the 20th century, we cannot yet be certain that a warmer world would be a cloudier world, and how any change in cloudiness would affect future warming.

Another question is whether warmer oceans pumping more water vapour into the atmosphere will alter the precipitation processes in clouds. If more moisture in the lower atmosphere leads to heavier rainfall from clouds, the impact will be limited. If, however, the result is an increase in humidity throughout the troposphere, it will lead to a positive feedback producing greater warming, because water vapour is the most important greenhouse gas. Finally, we must find out more about how much sunlight the sulphate particulates or aerosols absorb and reflect, and whether they have an indirect impact on cloud formation, their extent and duration, and the rain that falls from them.

The most immediate climatic threats to humankind relate to increased variability in storm and rainfall patterns, more heatwaves in major urban areas, and the impact of rising sea level on low-lying coastal regions. How we face up to these challenges may define our ability to mitigate longer-term changes.

Tarawa Atoll, capital of the archipelago nation of Kiribati, is a good example of the many small islands in the Pacific and Indian Oceans that are particularly vulnerable to a rise in sea level due to global warming.

Even though we do not know precisely how the climate will change in the future we can take prudent precautions. It seems wise to assume that the climate will remain at least as variable as in recent decades, sea level will continue to rise at the current rate, and our economic exposure to extreme events will increase. All of these changes pose urgent and growing challenges to how we manage many aspects of our lives.

Future patterns of drought and flooding

As the weather extremes that cause the greatest harm to society are droughts and flooding, we need improved predictions of how rainfall patterns will change with global warming. Existing models are consistent with each other in predicting increased precipitation intensity. This suggests that flooding will become even more prevalent. Furthermore, in terms of mean annual rainfall there is a consistent picture of increased rainfall in the tropics, parts of Northern Africa and South Asia. In addition, precipitation is likely to increase in middle and high latitudes, and generally to decrease in the subtropical belts.

Many of the communities most vulnerable to flooding are in the tropics, where rainfall is likely to increase. Not only will the risk of flooding be greater, but also the impact of disasters could be magnified by the spread of mosquito-borne and water-borne diseases (e.g. malaria, dysentery and cholera). In these parts of the world, plans for effective flood-control strategies are a major economic priority. The possible decline in rainfall in subtropical regions is likely to make the problems of drought and land degradation an even greater challenge in sub-Saharan Africa, Central Asia and parts of Australia. In these areas, long-term plans to alter agricultural practice and develop water conservation strategies should be a high priority.

The threat to the cities

The predicted rise in global temperatures during the 21st century means heatwaves in some locations will probably become more intense. As excess mortality rises sharply when daytime temperatures rise above 35°C and nights stay warm, any increase in the incidence of these heatwaves will have a disproportionate impact on public health. The

Public health

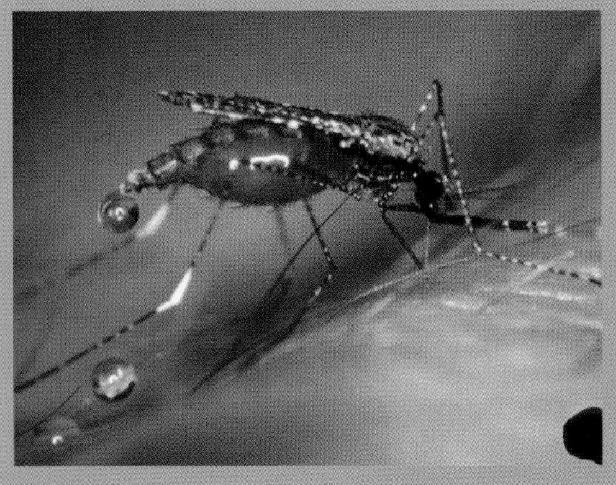

A warmer climate plus changes in precipitation patterns could expand the area of incidence of many tropical diseases, especially mosquito-borne varieties (e.g. malaria, filariasis and dengue fever). As a consequence, public health services will come under increasing pressure. Malaria has been the subject of particular analysis. At present over 2 billion people in the tropics and subtropics are at risk and some 270 million are infected each year. Most deaths are among the very young. Major epidemics are rare, but, on the fringes of its normal domain, where it is present for only a few months each year, malaria can do much greater harm. However, where it extends into new areas it may lead to increased death rates as high as 20 per cent.

urban heat island effect in the big cities will compound this impact. If average summer temperatures rise by 2–4°C, the excess mortality each summer in cities like Chicago or New York could rise by several hundred. In countries where the use of air conditioning is not widespread these figures are likely to be even greater.

Cost-effective solutions to specific problems such as urban heat islands would not only reduce the effects of heat mortality, but also would have a beneficial impact on amenity and pollution levels. The formation of photochemical smog is highly sensitive to temperatures. Cooling a city by about 3°C would have a marked impact on smog concentration. Low-cost solutions, such as painting dark surfaces white to reduce the amount of sunlight absorbed, can have a significant effect on temperatures. Longer-term solutions include the design of buildings to improve the airflow in central areas. Planting trees and bushes can also help to reduce urban temperatures as well as make cities greener and more pleasant places to live.

The threat of sea-level rise to vulnerable countries

No one wins with sea-level rise. Its challenge will have to be addressed at every stage, from sea defences for local communities, to massive international aid programmes to help the most endangered nations should the threat eventuate. During the last 100 years sea level has risen between 10 and 25 cm. Given that the range of predictions for the 21st century is for a further rise of between 9 and 88 cm, long-term planning for coastal communities must assume that sea level will continue to rise at least as fast as it has in the recent past. Moreover, because of the thermal inertia of the oceans, the global mean temperature will probably increase beyond 2100 and sea level will continue to rise at a similar rate in future centuries, even if greenhouse gas concentrations were stabilized by then.

This inexorable rise will involve hard-nosed economic decisions about whether or not to provide government money to protect shoreline communities. An alternative is to require them to rely on the insurance market. However, the desperate situation facing the heavily populated low-lying areas of South Asia and the small island states of the Indian and Pacific Oceans will require different solutions, possibly involving natural remedies such as mangrove expansion, coral reef growth and dune movements to work to the advantage of communities.

It is estimated that a sea-level rise of one metre could displace some 70 million people in both Bangladesh and China. In terms of land loss, low-lying atoll nations in the South Pacific, including the Marshall Islands, Tuvalu, Nauru, Kiribati and Tokelau, are extremely vulnerable to even small rises in sea level. A one-metre rise could result in the loss of 80 per cent of Majuro atoll in the Marshall Islands, home to half of the nation's population, and 12.5 per cent of the land mass of Kiribati. Erosion will further reduce land area on low-lying Pacific Islands, as well as increase the swampiness and salinity of land that remains above sea level.

Rising sea temperatures in the tropics will cause more bleaching of corals, with widespread damaging consequences for the ecology of reefs.

These threats are compounded by the fact that the economic well-being of most Pacific Islanders is primarily derived from activities carried out in low-lying coastal zones. Mangrove forests are a critical ecosystem in many places, important for commercially valuable fish, crustacean species and rare fauna. They also provide a source of construction materials, firewood, tannin and herbal medicines. Mangroves also act as a buffer zone, helping to ameliorate the impacts of storms. Mangrove ecosystems naturally cope with sea-level rise where sedimentation rates match or exceed local sea-level rise.

The fate of coral reefs is also vital to the survival of these islands. They protect coastal areas from erosion and storms. The maximum sustained vertical accretion rate of these reefs is estimated to be 10 mm per year, which would be adequate to keep up with a sea-level rise of one metre or less during the 21st century. The resilience of the reef ecosystems, however, could be greatly reduced through human impacts such as pollution. Coral reefs are also under severe threat from bleaching, as the temperatures of the tropical oceans rise with global warming.

Mangrove trees that form protective barriers along many tropical seashores are being harvested for timber or cleared for development, thus exposing the coastlines to erosion.

5.6 Future technologies

New technologies will further expand our understanding of the climate system and help us face up to the immediate challenges of climate change. Advances in measurements, data handling and computation will involve the further development of existing technologies. Some technologies may also provide means of reducing the scale of climate change.

The latest generation of geostationary satellites provides vast amounts of information about the world's weather.

The increasing pace of technological change will exert a profound influence on how we handle climate issues in the future. At the simplest level, the huge amount of data that will be automatically generated from satellites and *in situ* sensors, will be managed using ever-more powerful computers and digital storage databases. Furthermore, advances in analysis techniques will be necessary to process all the data into useful information for research and the widening array of services needed to address future climate challenges. Technological change itself, however, raises challenges of data compatibility. As new techniques and instruments are introduced, it is essential that temporal data homogeneity is maintained.

More powerful satellites

Throughout the closing decades of the 20th century, new satellite systems increased the type and number of observations about the climate system and this trend is expected to continue. The second generation of the European Geostationary Meteosat will produce 20 times as much data as the former system. These satellites will not only produce more accurate and frequent observations of more parameters, but also will be more accessible to weather services across Africa. These data and the forecasts they support will help develop better and more timely information for natural disaster early warning, improved food security, better health management, more efficient water use and safer transport.

Improved remote sensing

Given the challenges of maintaining and extending traditional observation networks, improved monitoring of global environmental change will depend critically on more sensitive remote-sensing systems. Many of these will be flown on satellites. They will include new efficient laser systems to measure wind speeds in the middle and upper atmosphere. Radar systems will also be used more widely to monitor specific events such as flood extent and changes in floodwater levels.

Improved weather forecasts

Industry estimates suggest that computer-processing speed will have increased by well over a million times by 2050 – a development central to improved numerical weather forecasts. An increasing proportion of the effort in preparing forecasts will be devoted to collecting and assimilating all the available observations especially those from space. The most accurate possible estimate of the initial conditions largely underpins the accuracy of computer predictions. Improving this estimate is essential to improving the accuracy and extending the temporal range of weather forecasts. Whatever improvements are made, however, attempts to extend day-to-day explicit weather forecasts much beyond about 10 days may still run up against fundamental limits in predictability due to the chaotic behaviour of the atmosphere.

Greater benefits will inevitably flow from improved shorter-term forecasts. The increasing economic cost of weather disasters provides huge incentives to produce more effective early warnings. Better measurements and more powerful computers will help to produce improved local forecasts. Regional models, with a grid scale of about 5 km or less and including better representation of hydrological processes, streamflows and soil moisture, could make predictions with much greater accuracy. This will offer substantial improvements in predicting flooding. In addition, reliable predictions more than a week ahead will assist decision-making in the agricultural sector.

Better climate models

The ever-expanding power of computers will also be used in producing better climate models for seasonal to interannual predictions, as well as better assessments of climate change. With additional computer power, climate modellers will perform many simulations with different starting conditions and better distinguish real climate signals from the background noise. What is clear is that future global climate models will need this computing muscle if they are to tackle some of the more difficult climate issues. Although the models do a reasonable job on the broad feature of the climate, they have considerable difficulties with the finer detail.

One answer is to calculate the climate in much greater detail. Changing the horizontal resolution by a factor of two (from, say 100-km to 50-km grid-point interval) increases the number of grid points by four. Because of computational constraints it is also necessary that the calculations in the model are carried out twice as often throughout the simulation. Overall, then, doubling of the grid resolution requires an eightfold increase in computation effort. Furthermore, with greater resolution, it may be necessary to include much greater refinement of the physical processes, and hence even more computation. The technique of combining global models of relatively low resolution with regional models, which have much more detail for specific processes and local topography, is likely to continue to make rapid progress.

Technological fixes to prevent global warming?

The inevitability of further increases in greenhouse gases in the atmosphere, and of carbon dioxide in particular, means that potential solutions to the threat of global warming are to increase the use of non-fossil fuels and to increase the rate of removal of the gas from the atmosphere. There are many technologies that could contribute to achieving the objective. The real issue is whether they are both economically viable and environmentally acceptable. On the one hand, it is technically feasible to remove carbon dioxide from the emissions of fossil-fuel-fired power stations but this is prohibitively expensive in current circumstances. On the other hand, nuclear power can generate electricity without emitting any greenhouse gases, but is, in many countries, regarded as unacceptable on environmental grounds.

In between, scientists and engineers are promoting a variety of technologies. These include exploiting the fact that the biological productivity of much of the oceans is limited by the availability of nutrients. Seeding the oceans with iron in a form that can be used by marine life may lead to substantial increases in the amount of biological growth and carbon dioxide uptake in the oceans. In principle, this process could be used to lock large amounts of carbon in marine organisms. When they die, this carbon would be deposited on the ocean floor.

On land, increased growth of plants and trees could also be used to lock up more carbon. This process of sequestration of carbon could involve anything from more widespread use of fertilizers and compost, to developing fast-growing genetically modified crops, which can store far more carbon than existing plants. Clearly, these options range from the environmentally attractive to the highly controversial. Difficult choices will have to be made in deciding whether the prevention of environmental damage due to climate change justifies running the risk of other forms of potential damage that may occur with, for example, nuclear energy or the use of genetically modified plants.

Wind-driven electricity-generating turbines provide one of the most economically competitive forms of renewable energy generation and are increasingly being used in many parts of the world.

Renewable energy

Renewable sources of energy are the most promising way to reduce carbon dioxide emissions. These sources, including hydroelectric power, already meet 15 to 20 per cent of world energy demand, and could well meet up to half of the total demand by the middle of the 21st century. There is already growing evidence that wind, solar, geothermal and biomass can compete economically with fossil fuel sources, and tidal and wave-power schemes have the potential to generate significant amounts of energy. The growing pressure to minimize the environmental impact of fossil-fuel energy sources is likely to increase the competitive advantage of renewables.

Robotic aircraft

After making important contributions to early upper atmosphere meteorology, using aircraft as research platforms has been largely restricted to large-scale specific projects which can afford the high cost of aircraft operations. The recent development of small robotic aircraft has changed this situation, as they provide a platform for small instruments to make special observations of weather systems. These aircraft, weighing less than 15 kg with a wingspan of no more than 3 m can stay aloft for up to 30 hours and travel 3000 km. Aircraft positions can be located precisely using the Global Positioning Satellite (GPS) navigation system and their observations transmitted via its low Earth-orbiting communications satellites.

5.7 Adapting to future change

Successful adaptation to climate change will depend on the validity of predictions and on sound political and economic policies implemented at national and international levels.

Predictions based on computer models provide some indications of potential future climatic shifts, but it is our ability to respond to the challenges of climate change that will be paramount. At present, many features of our lives are designed to manage for the most part the ups and downs of our local climate. A gradual warming or a drift to either wetter or drier conditions could probably be absorbed as part of the continual process of investment in the fabric of society. A sudden increase in a particularly damaging form of extreme (e.g. drought in parts of the world or tropical storms in others) would cause far greater difficulties, while more extreme shifts might stretch some communities, and even governments, to breaking point.

The barrage on the River Thames has provided protection for London, England, from storm surges and rising sea levels since 1983 and is expected to meet future challenges for several decades more.

Thriving on change

An unknown sage observed, "An optimist is someone who thinks the future is uncertain". The human race has always had to deal with the unexpected, and the inevitable changes are often deeply distressing. The fact that we may face greater climatic variability in the future is only another form of the challenges that have been faced throughout human history. We have adapted to a wide range of climates around the world, so accommodating climate change could be a viable option.

How we adapt will vary, depending on where we live, our current lifestyles and the resources we have at hand. It will also reflect the degree of uncertainty about future changes. Some solutions will be part of general developments. Agriculture will develop new temperature- and pest-resistant crop varieties, and new technologies to increase crop yields and improve irrigation efficiency. Energy industries will develop more efficient forms of supply and, as communities, we will seek to use energy more efficiently.

In other cases there will be many activities that involve 'no regret' actions, in that they cost little to implement and can yield multiple benefits. For example, in the case of reducing the impact of sea-level rise on tropical islands, the prohibition of extraction of reef and sand, and prevention of the removal of mangrove swamps, are low-cost options that conserve and best use natural resources. By comparison, building sea walls involves major investment with consequent environmental impact. The economic value of the foreshore developments that are being protected needs to be carefully assessed in this context.

Practical 'win-win' options

There are many areas where adopting the right policy mix can be both environmentally friendly and beneficial to all concerned. The most obvious example is making more efficient use of energy at every level from generation to final use. There are many instances where adopting new technologies and reducing waste offer major cost savings for relatively small investments. Similarly, more fuel-efficient vehicles will become increasingly attractive as oil supplies decline and fuel prices rise.

The harvesting of biomass for energy is another example of a 'win-win' philosophy that has both climatic and environmental advantages. Also, developing a mix of crops, and especially native species, to maximize the retention of soil nutrients in a locality and minimize the build-up of pests, is one route to sustainable agriculture. Harvesting practices should encourage regeneration and replanting and the maintenance of biodiversity. In addition, if crops are managed to increase soil organic matter, especially of carbon in the roots and other woody material left on site, the climatic benefits are maximized.

Enhancing our adaptive capacity

Adaptability is an essential ingredient in 'win-win' options. Many of the options for adaptation will involve changes in peoples' lives, and so their willingness and ability to embrace change will be crucial. The swings of the El Niño/Southern

Oscillation and the ups and downs of the North Atlantic Oscillation, for example, are part of the natural variability of the climate from year to year and decade to decade. They have, at least for several millennia, produced patterns of extreme weather around the globe with serious consequences for many communities. While the nature of the outcomes may change in the future, much of what we will face in the 21st century will repeat what was seen in previous centuries.

Becoming more adaptable will also involve the development of improved climate services. Better planning and design of all aspects of community living will reduce the deleterious impacts of climatic variability and enable us to capitalize on the benign swings. A future where greater variability and more dangerous extremes are the norm will place even greater emphasis on adaptation, preparedness and early warning.

Shoppers in Dhaka go about the business of choosing fresh produce from an attractive street stall, even though up to their knees in monsoon flood waters. Societies adapt to local weather conditions, and thrive in them, but can be caught unawares by sudden changes or extreme events.

Rising food production

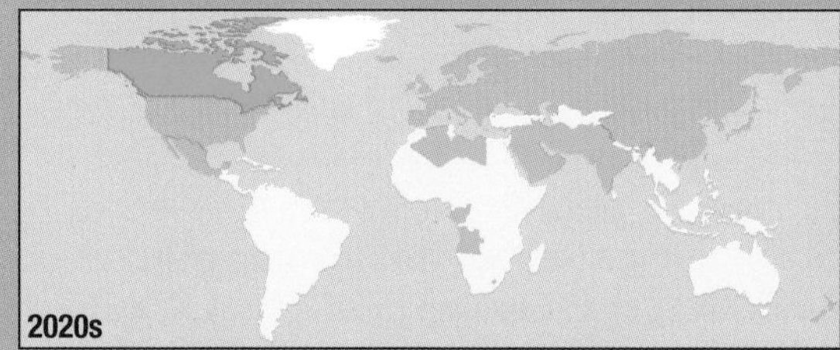

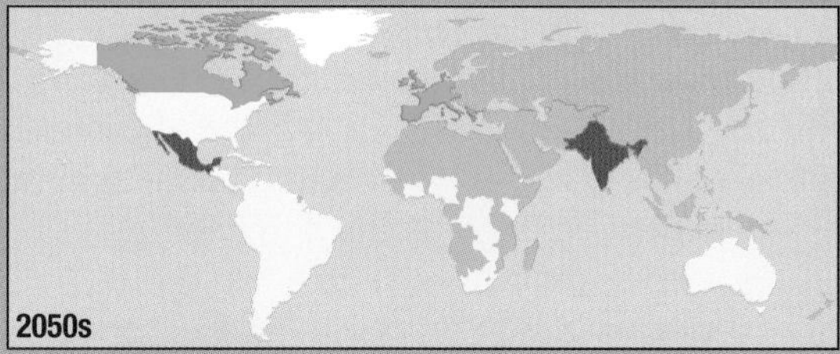

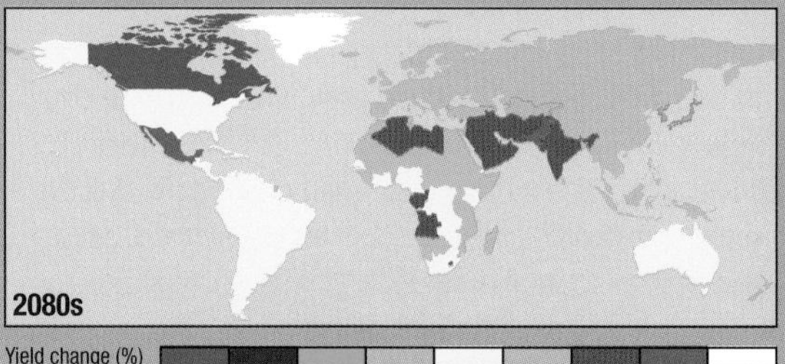

Rises in both temperature and carbon dioxide levels will lead to changes in crop yields, affecting food prices and world food trade patterns. Various studies have sought to predict how the world will adapt to these changes. Some studies have combined local knowledge with analysis of how crop yields might respond to biochemical and genetic advances in the coming decades. Computer models have then examined how global warming together with the direct physiological effects of increased carbon dioxide might alter crop yields.

The broad conclusion of these studies is that the global production of grain will, in the absence of climate change and rising carbon dioxide levels, nearly double by the second half of the 21st century. Climate change and rising carbon dioxide levels will alter this figure. How agricultural practices adapt to these changes also matters. Farmers will make personal choices about crops, planting dates and irrigation levels. Where climate change has a major impact, there will be political pressure for greater investment by governments in infrastructure and agricultural management.

The predicted changes in global grain production due to climate change and carbon dioxide fertilization are relatively modest. While climate change alone could reduce production by between 10 and 20 per cent, the inclusion of the direct effects of carbon dioxide fertilization and various adaptation strategies reduce falls in predicted output to less than 5 per cent. More important are the predicted latitudinal responses, which highlight the potential pluses and minuses of climate change. Near the high-latitude boundaries of production, rising temperatures would lengthen growing seasons, and the greater warmth in summer would increase yields. Together with the benefits from increased carbon dioxide fertilization, the gains are predicted to be considerable, notably in Canada throughout the 21st century.

In middle latitudes, where yields are already high, the additional warmth is likely to shorten crop development periods and reduce yields. Increased heat and water stress levels are not expected to be significant and overall any adverse climatic effects are outweighed by the benefits of increased carbon dioxide, notably in Europe, China and other parts of eastern Asia. At lower latitudes, however, the combination of rapid crop development plus heat and water stress levels may lead to significant and increasing reductions in output, especially in much of Africa, the Middle East, India and Central America by the end of the 21st century. If these predictions are correct, they suggest a disparity in the agricultural vulnerability to climatic change between low-latitude and high-latitude countries, and would have significant implications for national economies and world food policies.

5.8 Protecting the planet and its people

International cooperation is the only way to produce a coherent response to global climate issues. How nations participate in international activities, and how they design their own national programmes to fit in with these efforts, will determine how successfully we face the challenges.

International negotiations are the key to successful global action to mitigate the impact of human activities on the climate. The Secretary-General of the World Meteorological Organization, Professor G.O.P. Obasi (second from the right), addressed the 1992 Earth Summit on the important roles of weather, climate and water in sustainable development.

Human survival depends on clean air and water, and nutritious food. All of these are related to the 'health' of the climate system. The United Nations Universal Declaration of Human Rights in 1948 stated that: "Everyone has the right to a standard of living adequate for the health and well-being of himself and of his family, including food, housing and medical care ..." The 1974 Report of the World Food Conference identified the need to provide the human rights framework to deal with climate variability and climate shifts.

To produce and maintain a clean atmosphere, we must recognize that air flows unhindered, and carries any load of acquired pollutants, across political borders. The atmosphere, therefore, must be managed collectively for the common good of all countries and individuals. The United Nations Conference on the Human Environment held in Stockholm identified this principle in 1972, where Article 21 of its Declaration makes states responsible "... to ensure that activities within their jurisdiction or control do not cause damage to the environment of other States or of areas beyond the limits of national jurisdiction".

Combustion of fossil fuels, such as coal, petroleum and natural gas, by power stations, industrial enterprises and in our cars and homes, releases carbon dioxide into the Earth's atmosphere and contributes to global warming.

Water stress: the need for international action

The growing demand for freshwater, and the possible intensification of the hydrological cycle as the climate warms, will exacerbate existing national and regional shortages. It could also lead to conflict between nations. This scope for conflict could be compounded by the conversion of forests to agricultural land, where the more rapid runoff would result in more flooding of downstream communities.

Binding international agreements about the allocation of resources and their efficient use will become more urgent, particularly in the arid and semi-arid regions of the subtropics. These agreements will have to be built around both an understanding of the climatic conditions governing a region and technologies to use water more efficiently.

Stabilizing carbon dioxide concentrations

Emissions due to burning fossil fuels will be the dominant factor in the build-up of carbon dioxide concentrations in the atmosphere during the 21st century. Reducing carbon dioxide emissions is the most important factor for lowering the global impact of human activities on the climate system.

Immediate action is required to stabilize atmospheric carbon dioxide concentrations. To stop the concentration rising above 450 parts per million (ppm) compared with the current level of around 370 ppm would require global anthropogenic carbon dioxide emissions to drop below 1990 levels within a few decades. If such reductions were not achieved until around 2100 the carbon dioxide concentrations would rise to about 650 ppm, and if stabilization of emissions were delayed another century, then concentrations would rise to 1000 ppm before stabilizing.

Can international agreements have the desired effect?

The international understandings that emerged from major United Nations sponsored conferences like those held in Rio de Janeiro in 1992, Berlin in 1995 and in particular the emissions targets agreed in Kyoto in 1997, suggested that in terms of curbing the build-up of greenhouse gases in the atmosphere, there was a measure of agreement on

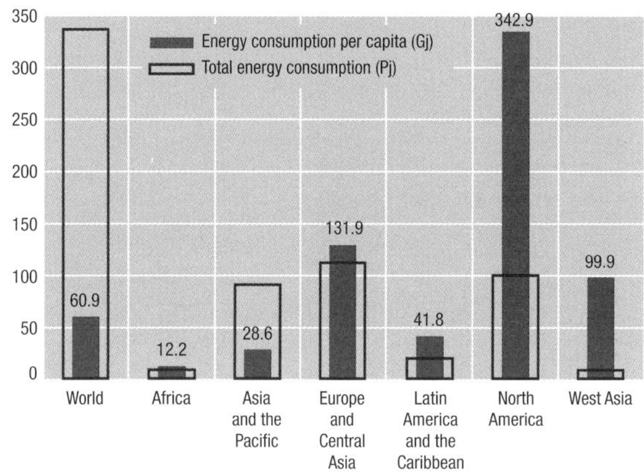

As seen in these 1995 data, enormous differences in per capita energy consumption between developed and developing countries poses a major point of contention for achieving successful global strategies to reduce the build-up of greenhouse gases in the atmosphere.

what action should be taken. The subsequent difficult negotiations at international meetings held since Kyoto have revealed the challenges that the international community faces in turning broad principles into practical implementation strategies.

Leaving aside the basic uncertainty about how rapidly carbon dioxide levels will rise during the 21st century, there is the tricky political issue of apportioning the reduction of emissions between the industrialized and industrializing nations. The former use much more energy per person than the latter, and there is a strong case for developing nations being allowed leeway to maintain higher rates of economic growth, which may enable them to narrow the gap with the developed world. Options, such as trading emissions between countries, are one way forward but will require ongoing delicate negotiations to convert agreed principles into workable solutions. Any progress will be coloured by how the climate actually behaves, and shifting perceptions about whether other human activities are making important contributions to the observed changes. Whether deforestation should be counted as a national contribution to global greenhouse gas emissions, whereas afforestation can be offset against emissions, are hotly debated issues. Similarly, some countries may argue that changes in agricultural practice, which sequester carbon in the soil, should count in their favour.

The negotiations will get tougher as the economically attractive and politically acceptable options ('no regrets') are used up. Industrialized nations may be willing to reduce carbon dioxide emissions by sensible energy conservation measures, improved public transport in major urban areas, and closing down ageing coal-fired power stations. Should it come to the imposition of unpopular taxes on energy consumption or the restriction of the use of private vehicles, the resolve to meet demanding targets may weaken.

Curbing the impact of motor vehicles on the local environment, and reducing their emissions of greenhouse gases, is the most pressing environmental challenge for many national governments.

It is possible to bring about real change. This is demonstrated by the fact that the concentrations of a number of ozone-depleting substances have stopped increasing in the atmosphere, and some have even decreased. The Vienna Convention to Protect the Ozone Layer, and its associated Montreal Protocol, provide a precedent of successful international action. This in itself is an encouraging sign that the challenges identified by the 2002 Johannesburg World Summit on Sustainable Development can be met if political will can be turned around and brought to bear.

Setting the agenda at every level

Although governments will play a pivotal role in setting policies, their implementation will require action at every level of both the public and private sectors. For example, effective decisions on sea defences would integrate local knowledge with predictions of future sea-level rises. Public investment may be needed where the greatest losses are likely to occur. Elsewhere, the answer could lie in changing zoning laws or using private insurance. Zoning restrictions can reduce the number of people living in vulnerable areas, whereas insurance allows people to make decisions for themselves about the risk of continuing to live in such areas.

An analysis of the vulnerability of 1300 km of sea defences around the shores of England and Wales to various aspects of severe storms, and the flooding they could cause, concluded that the needs of the communities could be met by the insurance market, albeit at a price. A similar set of conclusions was reached in an analysis of the economic costs of predicted greenhouse-induced sea-level rises for developed property in the USA. Depending on whether or not people exercise foresight, the potential is there in some locations for market-based adaptation in anticipation of the threat of rising seas.

The global mean tropospheric abundance (ppt) of CFC-11 rose from near zero in the 1950s to a peak in the early 1990s. The success of international agreements to reverse the build-up (as seen in the recent decline) is a measure of the effectiveness of global environmental action.

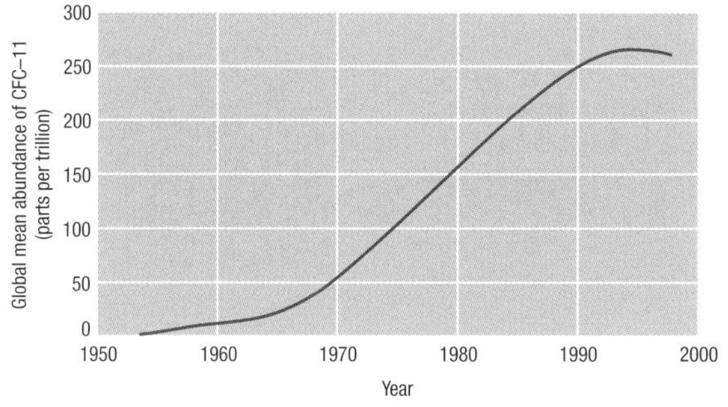

5.9 Future surprises?

Many aspects of future climate will surprise us. The changes, however, will only be part of a variety of unexpected developments concerning the environment and, more generally, wider economic, political and social developments. We will need to keep a sense of perspective about how the climate fits into this array of challenges.

As we have seen throughout this book, climate is an integral part of our lives. Any view of the future must consider how changes in the climate will combine with other developments. This perspective is essential, as many of the important challenges facing humankind in the 21st century will not be related to the climate.

Rapidly expanding urban populations, with their basic requirements for clean air, adequate supplies of food and freshwater, and for safe housing, will be one of the greatest challenges for humankind in the future. Extreme weather events will continue to disrupt and threaten people's lives.

Health

Although climate change may pose significant threats to human health, far greater challenges already confront us. The spread of human immunodeficiency virus (HIV) is overwhelming some countries and is threatening other parts of the world. Similarly, although the provision of adequate clean water supplies is linked with climatic developments, in many parts of the developing world basic public health investment in safe water supplies would do far more to reduce infant mortality, notably from diarrhoea, than any other intervention. Furthermore, new drug-resistant forms of old diseases (e.g. tuberculosis and plague) could be a much more virulent threat than the spread of malaria as the climate warms.

The problems of providing clean drinking water to all the world's population are compounded by weather- and climate-related disasters such as the floods in Mozambique in 2000.

Environmental change

The climatic factors in such environmental developments as deforestation, desertification, overfishing and the introduction of alien species into new habitats have been explored in various places in this book. In many cases, however, the climatic aspects of these environmental threats are only a part of the story. How we respond to different threats will depend on our perceptions of the potential scale of any catastrophe and our capacity to respond effectively. For instance, the disastrous shrinkage of the Aral Sea is clear for all to see, but finding a way to reverse the decline may prove impossible. Even if a solution is found it will be many decades before the damage can be undone.

What is clear is that our ability to confront potential threats depends on the collective will to embrace policies that curb certain human activities. In the future, governments may be galvanized into action by unexpected and dramatic shifts in environmental conditions. Precipitous drops in fish stocks or rapid expansion of deserts could do more to produce moves on cutting the emissions of greenhouse gases than the more measured pronouncements of scientific assessments of climate change. However great the threat, political reluctance will remain a major factor militating against early and effective action.

The introduction of alien species, such as the water hyacinth in Lake Victoria, represents an environmental challenge that requires international action.

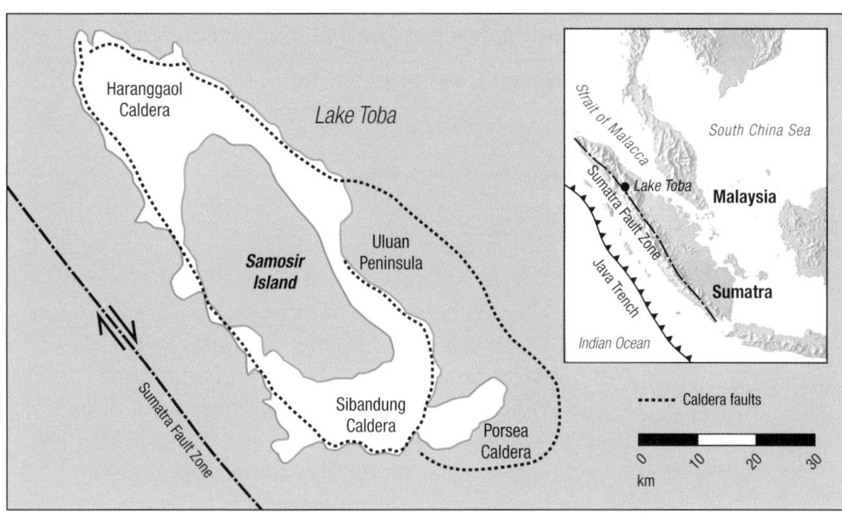

The caldera formed by the massive eruption of the supervolcano Toba, Indonesia some 74 000 years ago, measures 30 by 100 km and provides a graphic illustration of the scale of this event.

Genetically modified crops

The development of genetically modified crops, including those more tolerant to climatic extremes, is among the most contentious issues in modern science. Depending on one's point of view, this technology could provide solutions for feeding the world, and even provide a way of locking away carbon emissions, or they could produce immensely damaging species that could dislocate agriculture and the communities that depend on it. The directions the current political debates take around the world, together with the outcomes of field trials, will decide how fast this technology is developed and the contribution, both positive and negative, that it can make to meeting the challenges we face in the 21st century.

Natural catastrophes

While it could be difficult to come to terms with some of the consequences of rapid global warming, these pale in significance when compared with the most extreme natural catastrophes. The best-researched examples of such disasters are an asteroid impact, the eruption of a supervolcano or even a significant increase in volcanic activity globally. These types of event have occurred relatively frequently during the Earth's history and have had disastrous consequences, both for the climate and for many other aspects of life on Earth. They are cited as the cause of a number of mass biological extinctions throughout geological history, including the event at the end of the Cretaceous Era, 65 million years ago, which is thought to have brought an end to the 'Age of the Dinosaurs'.

What is clear is that if one of these massive events were to occur there is little we could do about it. In the case of a large asteroid impact, it is conceivable that, if we were able to detect its approach in good time, there might be some possibility of taking action to divert it from its collision course. As for a supervolcano, there is no prospect of our being able to do anything about it, and it is exceedingly unlikely that we would have much warning of its eruption. The last known supervolcano was Toba, in Sumatra, 74 000 years ago. It blasted vast clouds of ash across the world. The resultant caldera formed Lake Toba, 100 km long and 30 km wide. The huge dust veil may have caused the global temperature to drop 5°C, triggering a volcanic winter. A similar event today would have catastrophic consequences for humankind.

The possibility of such immense natural disasters should not be exaggerated, but it should give us a sense of perspective when we consider other climatic and environmental challenges. All that we do carries some degree of risk. What we must do is concentrate on those areas where the risks are unreasonably high and where we can do something practical to reduce them.

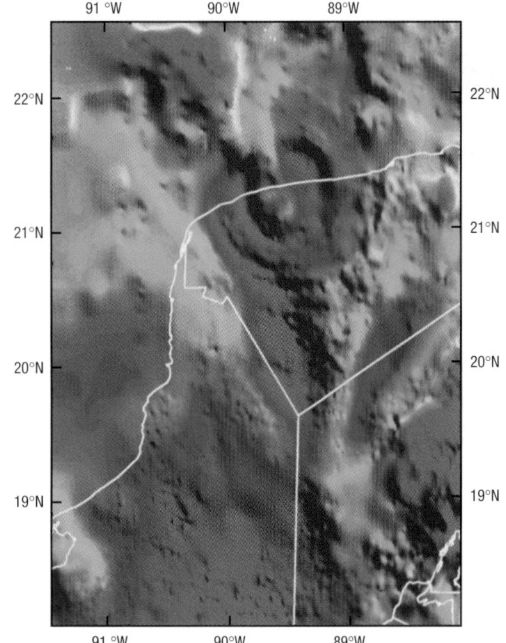

An asteroid approximately 10 km in diameter struck the Earth some 65 million years ago, in Yucatan, Mexico. The crater is not exposed to the surface, but variations in the magnitude of the gravity field in the vicinity show a horseshoe shaped impact crater some 70 km across.

5.10 From the past to the future

The 20th century witnessed enormous growth in understanding and monitoring of the climate system. Beneficial outcomes have been early warning of dangerous weather and climate events, but tempered by the knowledge that humanity may be contributing to changes in global climate.

We live at the bottom of an extraordinarily thin layer of air that surrounds planet Earth. There is no doubt that we now have the capacity to radically alter the physical character of this fragile envelope. To ignore the warning signs may prove costly for future generations.

We have seen how, by building on the knowledge of the previous centuries, scientists and technologists of the 20th century developed a global infrastructure for monitoring the climate system and a capability to predict some of its ever-changing patterns. The systems support early warning of dangerous phenomena, including severe storms, tornadoes, tropical cyclones, middle-latitude storms and seasonal droughts. Nevertheless, many challenges still remain.

At the beginning of the 20th century there was some understanding of many of the basic physical processes driving the climate. Even the potential threat from uncontrolled build-up of carbon dioxide had been identified. A critical piece of foresight from which we are now benefiting, was the decision to build up archives of data as the basis for local climate studies and comparing climates around the world.

A century of progress

European and North American research into weather systems during the first quarter of the 20th century provided the key to understanding many of the processes associated with middle-latitude weather systems. It was also a period when the processes of the Indian monsoon and its variability were becoming understood. Flying machines and their safety provided new challenges. Some scientists found the study of clouds and their violent motions of special interest.

The aviation industry went through a period of rapid development during the second quarter of the 20th century and made new demands on weather services. Much of the expansion of observing sites to all inhabited continents and the use of balloons to sound the upper atmosphere winds were spurred by the need to increase the safety of aircraft operations. Later, World War II gave further impetus to the expansion of meteorological services.

From the 1950s onwards, high-quality research and technological achievement were the spur to success. Starting with the International Geophysical Year of 1957–58, there was a period of rapid advancement as balloons began to unravel the mysteries of the stratosphere. Regular observations commenced over Antarctica. Radar, computers and satellites all entered the meteorologist's arsenal. The World Meteorological Organization's World Weather Watch harnessed an international observing network with global telecommunications and data-processing centres to put weather forecasting on a sound footing.

Multidisciplinary studies, that integrate knowledge from different Earth science disciplines, have advanced our knowledge of the climate in recent decades. By the late 1970s, evidence of a period of prolonged cooling and the accumulated evidence from previous Ice Ages were raising concerns about the future climate. Ironically, at the same time, the observations of increasing carbon dioxide concentrations from fossil-fuel burning were alerting us to the potential for global warming. Developments in numerical weather prediction extended the limits of useful weather forecasting to near a week, and coupling of these

The vision for the 21 century

J.W. Zillman, Director of Meteorology, Australia and President, World Meteorological Organization (WMO)

The sister sciences of meteorology (including climatology), hydrology and oceanography advanced dramatically and contributed greatly to the welfare of humanity and the protection of the global environment through the 20th century. With the dawning of the 21st century they stand as the major component disciplines of Earth system science whose sound development and wise application will hold the key to many aspects of human progress and planetary survival through the third millennium.

The WMO vision for the 21st century is for a world which will benefit even more greatly from the cooperation built up during the 20th century and the scientific and technological advances that have been the product of that cooperation. It commits the WMO Member States and Territories to doing their utmost to build a world in which, *inter alia*:

- Nations will continue to work together within the framework of cooperation which underpins the WMO Convention;
- The principle of free and unrestricted international exchange of basic meteorological data and products will be maintained and strengthened;
- The potential of meteorological and hydrological warning systems to reduce the loss of life and property in severe weather, flood and drought situations will be fully realized and implemented early in the century;
- Collaboration among the geosciences will lead to the development of skilful systems for seasonal forecasts and climate predictions such as on El Niño and global warming; and operational systems will be put in place to foster effective use of this information for the benefit of all nations;
- An integrated global environmental monitoring and service system will serve the needs of national and international users for weather, climate and related environmental services.

The last decade of the 20th century saw increasingly strong linkages develop across the geosciences and between the research and operational communities. The science and technology are developing rapidly, and the scientists and scientific institutions are strongly committed to realizing the benefits of international collaboration for the benefit of society in the 21st century.

models with ocean forcing identified the basis of seasonal anomalies, especially from El Niño events. Fully coupled climate models that include bio-hydrological feedback clearly demonstrated the direction of climate change as greenhouse gas concentrations increase.

The challenges ahead

Our ability to meet the challenges of the climate system has grown immensely. As recurring natural disasters in different parts of the world continue to show, however, our adaptation strategies are not always successful. We must improve planning, warning, and information and prediction services to make food and water resources more secure, and where we live safer. Predictions of climate anomalies on seasonal and longer timescales offer real benefits, but to realize their full potential requires more research into climate processes, more complete observations and better climate models. This research will also aid policy-making on the impacts of future climate change.

Perhaps the biggest challenge in the field of climate science is to ensure that the knowledge and services that benefit society are available to all. As a global community it will be essential to narrow the divide of opportunity so that all people can benefit from early warnings for their own safety and for the security of their essentials of life, including food and water reserves, their homes and the infrastructure supporting their livelihoods. Some countries still have deficient, or even non-existent, observing networks. Not only are the peoples of these countries deprived of adequate services, but also the global infrastructure on which everyone depends is less effective. Technology transfer and intergovernmental programmes of assistance will continue to be an integral part of international cooperation in weather and climate.

Appendices

Adaptation
An adjustment in a natural or human system in response to actual or expected climatic stimuli or their effects, which moderates harm or exploits beneficial opportunities.

Aerosols
A collection of airborne solid or liquid particles, with a typical size between 10 nm and 10 000 nm, residing in the atmosphere for at least several hours. Aerosols may be of either natural or anthropogenic origin.

Afforestation
Planting of new forests on lands that historically have not contained forests.

Air mass
An extensive body of the atmosphere whose physical properties, particularly temperature and humidity, are relatively constant in the horizontal. It may extend over an area of several million square kilometres and over a depth of several kilometres.

Air temperature
The temperature indicated by a thermometer exposed to the air in a place sheltered from direct solar radiation.

Albedo
The fraction of solar radiation reflected by a surface or object, often expressed as a percentage.

Antarctic circumpolar current
A Southern Ocean current that flows around the entire globe, driven by the circumpolar westerly winds.

Anthropogenic
Resulting from or produced by the activities of humans.

Anticyclone
A region of the atmosphere where the mean sea level pressures are high relative to those in the surrounding region, and from which winds are rotating anticyclonically outwards (clockwise in the Northern Hemisphere, anticlockwise in the Southern Hemisphere). Also known as an 'area of high pressure' or a 'high'.

Aquifer
A permeable water-bearing geological formation capable of yielding exploitable quantities of water.

Arid regions
Ecosystems receiving less than 250 mm of precipitation per year.

Atmosphere
The gaseous envelope surrounding the Earth. The dry atmosphere consists almost entirely of nitrogen and oxygen. It also contains small quantities of argon, helium, carbon dioxide, ozone, methane and many other trace gases. In addition, the atmosphere also contains water vapour, clouds and aerosols. See also *Greenhouse gases*.

Atmospheric circulation
Motions of the Earth's atmosphere. See also *General circulation*.

Atmospheric pollution
Contaminants present in the atmosphere, such as dust, gases, fumes, mist, odour, smoke or vapour in such quantities and with such characteristics and durations as to be injurious to human, plant and animal life or to property.

Atmospheric pressure
The force per unit area exerted by the atmosphere on any surface by virtue of its weight.

Avalanche
Mass of snow and ice falling suddenly down a mountain slope, often taking with it earth, rocks and rubble.

Biodiversity
The number and relative abundance of different genes (genetic diversity), species and ecosystems (communities) in a particular area.

Biodiversity hot spot
Area with high concentrations of endemic species facing extraordinary habitat destruction.

Biomass
The total mass of living organisms in a given area or volume; recently dead plant material is often included as dead biomass.

Biosphere (terrestrial and marine)
The part of the Earth system comprising all ecosystems and living organisms, in the atmosphere, on land (terrestrial biosphere) or in the oceans (marine biosphere), including derived dead organic matter, such as litter, soil organic matter and oceanic detritus.

Blizzard
Violent winter storm, lasting at least three hours, which combines below-freezing temperatures and very strong wind laden with blowing snow that reduces visibility to less than 1 km.

Carbon cycle
The term used to describe the flow of carbon (in various forms, e.g. as carbon dioxide) through the atmosphere, ocean, terrestrial biosphere and lithosphere.

Carbon dioxide (CO_2)
A naturally-occurring gas, also a by-product of burning fossil fuels and biomass, as well as from land-use changes and other industrial processes.

Catchment
An area that collects and drains rainwater.

Celsius scale
The temperature scale that designates 0° as the freezing point and 100° as the boiling point of water (at sea level and standard atmospheric pressure).

Chlorofluorocarbons (CFCs)
Synthetic compounds used as refrigerants, propellants in aerosol products, solvents and as intermediates in the synthesis of other fluorine compounds.

Climate
Climate, in a narrow sense, is usually defined as the 'average weather'. In a wider sense, climate is the state of the climate system, including a statistical description in terms of the mean and variability of relevant quantities over a period of time ranging from months to thousands or millions of years.

Climate change
A statistically significant variation in either the mean state of the climate or in its variability, persisting for an extended period (typically decades or longer). Climate change may be due to natural internal processes or external *forcings*, or to persistent anthropogenic changes in the composition of the atmosphere or in land use.

Climate feedback
An interaction mechanism between processes in the climate system is called a climate feedback when the result of an initial process triggers changes in a second process that in turn influences the initial one. A positive feedback intensifies the original process, and a negative feedback reduces it.

Climate model
A computer-based numerical representation of the climate system incorporating the physical, chemical and biological properties of its components, their interactions and feedback processes, and accounting for all or some of its known properties. Generally, climate models are general circulation models (GCMs) of the ocean and atmosphere coupled at the Earth–atmosphere interface by exchanges of heat, moisture and momentum.

Climate prediction (or forecast)
A description of the expected evolution of the climate system at a future time (e.g. seasonal, interannual or longer timescales).

Climate scenario
A plausible and often simplified representation of the future climate, based on an internally consistent set of climatological relationships.

Climate sensitivity
In the context of global climate change, climate sensitivity refers to the change in global average near-surface air temperature (°C) following a unit change in radiative forcing (W/m^2). More generally, climate sensitivity is the magnitude of a climatic response to a perturbing influence.

Climate system
The system, encompassing the atmosphere, hydrosphere, cryosphere, land surface and biosphere, which determines the Earth's climate as a result of mutual physical, chemical and biological interactions, and responses to external influences. The climate system evolves in time under the influence of its own internal dynamics and because of external forcing such as volcanic eruptions and solar variations, and human-induced forcing such as the changing composition of the atmosphere and land-use change.

Climate variability
Variations in the mean state and other statistics (such as standard deviations, the occurrence of extremes, etc.) of the climate on all temporal and spatial scales beyond that of individual weather events. The term is often used to denote deviations of climatic statistics over a given period of time (e.g. a month, season or year) from the long-term statistics relating to the corresponding calendar period. In this sense, climate variability is measured by those deviations, which are usually termed anomalies. See also *Patterns of climate variability*.

Climatic anomaly
Departure of the value of a climatic element from its normal. See *Climate variability*.

Climatic classification
Division of the Earth's climates into a worldwide system of regions, each of which is defined by the relative homogeneity of its climatic elements. Examples are Köppen's, Trewartha's and Thornthwaite's climate classifications.

Climatic oscillation
A fluctuation in which the variable tends to move gradually and smoothly between successive maxima and minima.

Climatic trend
A climatic change characterized by a regular increase or decrease of the average value of a climate element over the period of record.

Climatological station network
Group of observing stations which together are sufficient to define regimes and

patterns on a scale appropriate to climatological studies.

Cloud condensation nuclei (CCN)
Hygroscopic airborne particles that serve as initial sites on which water vapour can condense.

Condensation
The transition from the gaseous to the liquid state; the physical process by which water vapour is transformed into dew, fog or cloud droplets.

Convection
Unstable overturning within a layer of air leading to the vertical transport of mass, heat, moisture and momentum. Moist convection is associated with condensation and cloud formation.

Coral bleaching
The paling in colour of corals resulting from a loss of symbiotic algae. Bleaching occurs in response to physiological shock such as abrupt change in temperature, salinity or turbidity.

Cryosphere
The component of the climate system consisting of all snow, ice and permafrost on and beneath the surface of the Earth and ocean.

Cumulonimbus
Heavy and dense cloud, with a considerable vertical extent, resembling a mountain or huge tower, associated with moist convection.

Cyclone
An area of low atmospheric pressure with inwardly rotating winds (anticlockwise in the Northern Hemisphere, clockwise in the Southern Hemisphere). A cyclone of middle and high latitudes is called a depression. See also *Depression*.

Deepwater (or bottomwater) formation
See *Thermohaline circulation*.

Deforestation
Conversion of a forest to non-forest land use.

Depression
Region of the atmosphere in which the mean sea level pressures are lower than those of the surrounding region, often the early stage of cyclone formation. In middle to high latitudes a cyclone is sometimes referred to as a depression.

Desert
An ecosystem with less than 100 mm of precipitation per year.

Desertification
Land degradation in arid, semi-arid and dry sub-humid areas resulting from various factors, including climatic variations and human activities.

Diurnal temperature range
The difference between the maximum and minimum temperature during a day.

Doppler radar
Radar that makes use of the Doppler shift of an echo received from a moving target to measure its velocity.

Downburst
Violent and damaging downdraught reaching the Earth's surface, associated with a severe thunderstorm.

Dropsonde
Radiosonde, launched from an aircraft, that measures atmospheric properties and transmits the observations during its descent.

Drought
The phenomenon that exists when precipitation has been significantly below normal recorded levels, causing serious hydrological imbalances that adversely affect land resource production systems and human settlements. Drought is a relative term. A shortage of precipitation during the growing season that results in crop damage is deemed an agricultural drought. A shortage of precipitation that results in reduction in water supplies is called a hydrological drought.

Ecosystem
A system of interacting living organisms together with their physical environment.

El Niño
In its original sense, El Niño is a warm water current which periodically flows down the coast of Ecuador and Peru, disrupting the local fishery. This oceanic event is now associated with an extensive warming of sea surface temperatures across the central and eastern equatorial Pacific Ocean lasting from several months to more than a year.

El Niño/Southern Oscillation (ENSO)
The coupled atmosphere–ocean phenomenon that links the fluctuating temperatures of the surface layers of the central and eastern equatorial Pacific Ocean (the El Niño and La Niña) with the changing atmospheric circulation and cloud and rainfall patterns across the tropical Indo-Pacific region (the Southern Oscillation). See *El Niño* and *La Niña*.

Electromagnetic radiation
Energy emitted by substances by virtue of their temperature and propagated through space or material media by wave disturbances in electric and magnetic fields.

Endemic
Restricted or peculiar to a locality or region.

Energy balance
Averaged over the globe and over long time periods, the energy budget of the climate system is assumed to be in balance. As the climate system derives all its energy from the Sun, this balance implies that, globally and at the top of the atmosphere, the amount of incoming solar radiation is, on average, equal to the sum of the outgoing reflected solar radiation and the outgoing infrared radiation emitted by the climate system.

Epidemic
Occurring suddenly in numbers clearly in excess of normal expectancy, said especially of infectious diseases but applied also to any disease, injury or other health-related event occurring in such outbreaks.

Erosion
The process of removal and transport of soil and rock by weathering, mass wasting and the action of streams, glaciers, waves, winds and underground water.

Evaporation
The process by which a liquid becomes a gas; the process by which liquid water is transformed into vapour.

Evapotranspiration
The combined process of evaporation from the Earth's surface and transpiration through vegetation.

Extreme
In climatology, the highest or lowest value of a climatic element observed during a given time interval. The greatest extreme for the full period of record during which observations are available is the absolute extreme.

Extreme weather event
An event that is rare within its statistical reference distribution at a particular place. Definitions of 'rare' vary, but an extreme weather event would normally be as rare as or rarer than the 10th or 90th percentile. By definition, the characteristics of what is called extreme weather may vary from place to place.

Eye of tropical cyclone
The relatively clear and calm area inside the circular, rotating wall of convective clouds, the geometric centre of which is the centre of the tropical cyclone.

Faculae
Bright patches on the Sun. The area covered by faculae is greater during periods of high solar activity.

Fahrenheit scale
A temperature scale that assigns values of 32° to the freezing point of water and 212° to the boiling point of water (at sea level and standard atmospheric pressure).

Flash flood
A flood that rises very rapidly, usually as the result of an intense rainfall over a small area, or possibly because of the collapse of an ice jam or a dam failure.

Flood plain
That part of a valley floor subject to flooding when streamflow exceeds the carrying capacity of a river channel.

Fog
Suspension of very small, usually microscopic, water droplets in the air; generally reducing the horizontal visibility at the Earth's surface to less than 1 km.

Forest
A landscape type dominated by trees.

Freshwater lens
A lenticular, fresh groundwater body, usually underlain by saline water. Particularly relevant to coral atolls.

Front
The interface or transition zone between air masses of different densities (temperatures, humidities).

Frost
Frost occurs at or near the ground when the air temperature is equal to or less than 0°C. A frost period that is severe enough to end the growing season is called a killing frost. Frost is also the term used to describe the layer of ice crystals that forms on a freezing object by direct deposition of water vapour to solid ice.

Fujita scale
Also known as the F-scale, this is an index that relates the intensity of a tornado to the damage it causes to landscapes and structures. The windspeeds associated with the six-point scale are:
F0 (light damage): 18–32 m/s
F1 (moderate damage): 33–49 m/s
F2 (considerable damage): 50–69 m/s
F3 (severe damage): 70–92 m/s
F4 (devastating damage): 93–116 m/s
F5 (incredible damage): 117–142 m/s.

General circulation
The large-scale motions of the atmosphere and the ocean as a consequence of differential heating on a rotating Earth, aiming to restore the energy balance of the system through transport of mass, heat, moisture and momentum.

General circulation models (GCMs)
See *Climate model.*

Geostationary meteorological satellite
A meteorological satellite orbiting the Earth at an altitude of approximately 36 000 km with the same angular velocity as the Earth and within the equatorial plane, thus remaining above a fixed point on the surface. It can provide nearly continuous information in an area within a range of about 50° from a fixed sub-satellite point at the equator.

Glacial epoch
An interval of geological time which was marked by a major equatorial advance of ice.

Glacier
A mass of land ice flowing downhill (by internal deformation and sliding at the base) and constrained by the surrounding topography e.g. the sides of a valley or surrounding peaks.

Global Data-processing System (GDPS)
The coordinated global system of meteorological centres and arrangements for the collection, processing, storage and retrieval of meteorological information within the framework of the World Weather Watch.

Global Observing System (GOS)
The coordinated system of facilities deployed on land, at sea, in the air and in near-outer-space for making observations on a worldwide scale within the framework of the World Weather Watch.

Global near-surface temperature
The area-weighted global average of (i) the sea surface temperature over the oceans and (ii) the surface-air temperature over land.

Global Telecommunication System (GTS)
The coordinated global system of telecommunication facilities and arrangements for the rapid collection, exchange and distribution of observational data, processed information and related products within the framework of the World Weather Watch.

Global warming
An overall increase in the Earth's average near-surface temperature.

Greenhouse effect
Retention on the Earth of heat that would otherwise be radiated back to the universe, by radiatively active trace gases such as water vapour, carbon dioxide, etc. These gases in the atmosphere absorb infrared radiation emitted by the Earth's surface, by the atmosphere and by clouds and re-emit it in all directions, including downward to the Earth's surface. By trapping heat within the surface–troposphere system they maintain the Earth's surface at a higher temperature than would be produced by the absorption of direct solar radiation alone. This is called the natural greenhouse effect.

Greenhouse gases
Those radiatively active gaseous constituents of the atmosphere, both natural and anthropogenic, that absorb and emit radiation at wavelengths within the spectrum of infrared radiation emitted by the Earth and atmosphere. The primary greenhouse gases in the Earth's atmosphere are water vapour (H_2O), carbon dioxide (CO_2), nitrous oxide (N_2O), methane (CH_4) and ozone (O_3), all naturally occurring.

Hail
Precipitation in the form of balls or lumps of ice. An individual unit of hail is called a hailstone.

Halocarbons
Compounds containing halogen (chlorine, bromine or fluorine) and carbon atoms. Such compounds can act as powerful greenhouse gases in the atmosphere. Halocarbons containing chlorine and bromine are also involved in the depletion of the ozone layer.

Heat island
An area, within an urban zone, characterized by higher ambient temperatures than those of the surrounding region, resulting from the absorption of solar energy by materials such as concrete and asphalt.

Heatwave
A period of unusually (and uncomfortably) hot weather, often humid, which lasts generally from a few days to a few weeks.

Heinrich events
Massive discharges of icebergs into the North Atlantic Ocean, six of which have been confirmed to have occurred over the last 75 000 years. These 'armadas' of icebergs carried a lot of crushed rock which, when the icebergs melted, was laid down in identifiable layers (Heinrich layers) on the ocean floor.

Holocene
The relatively warm epoch which started around 10 000 years ago following the last Ice Age and runs up to present time. It is marked by several short-lived particularly warm periods, the most significant of which, some 6000 years ago, is called the Holocene optimum.

Homogeneous series
Data consisting of a sequence of values of a variable which have been observed under the same (or similar) conditions and with the same (or similar) equipment.

Hook echo
Characteristic curved or hooked shape radar echo from the lower portion of a severe thunderstorm, often associated with a tornado outbreak.

Humidity
The water vapour content of the air.

Hurricane
Name given to a tropical cyclone with maximum surface wind of 118 km/h (64 kts) or greater (hurricane-force wind) occurring over the North Atlantic, the Caribbean, the Gulf of Mexico or in the Eastern North Pacific Ocean. See also *Tropical cyclone.*

Hydrological cycle
The succession of stages through which water passes from the atmosphere to the Earth and returns to the atmosphere: evaporation from the land or sea or inland waters; condensation to form clouds; precipitation; accumulation in the soil or in bodies of water; and re-evaporation.

Hydrosphere
That part of the Earth covered by water and ice.

Ice Age
Particular period of a geological era during which extensive ice sheets (continental glaciers) covered many parts of the world.

Ice sheet
A mass of land ice which is sufficiently deep to cover most of the underlying bedrock topography, so that its shape is mainly determined by its internal dynamics (the flow of the ice as it deforms internally and slides at its base). There are only two large ice sheets in the modern world, on Greenland and Antarctica.

Ice shelf
A floating ice sheet of considerable thickness attached to a coast; often a seaward extension of ice sheets.

Impacts (climate)
Consequences of climate on natural and human systems.

Infrared radiation
Radiation with wavelength longer than the visible wavelengths. Part of the Sun's radiation is in the infrared, but it forms a much higher proportion of the radiation emitted by the Earth's surface, the atmosphere and the clouds, which is also known as terrestrial or long-wave radiation.

Infrastructure
The basic equipment, utilities, productive enterprises, installations and services essential for the development, operation and growth of a system, organization, city or nation.

Instrument shelter
Structure to protect meteorological observing instruments from exposure to radiation and weather while at the same time ensuring sufficient ventilation. Thermometers are usually housed in the standard shelter, the Stevenson screen.

Intertropical convergence zone (ITCZ)
A zone in the equatorial region marked by a band of convective clouds where the Trade winds of the two hemispheres meet.

Jet streams
Flat tubular, quasi-horizontal current of air, generally near the tropopause, whose axis is along a line of maximum wind speed and which is characterized by great speeds and strong vertical and horizontal wind shears.

La Niña
An extensive cooling of the sea surface temperatures of the equatorial Pacific Ocean lasting from seasons to more than a year. See also *El Niño/ Southern Oscillation.*

Land use
The total of arrangements, activities and inputs undertaken in a certain land cover type (a set of human actions).

Land use change
A change in the use, or management, of land by humans, which may lead to a change in land cover.

Landslide
A mass of material that has slipped downhill by gravity, often triggered by heavy rainfall that saturates the soil material.

Latent heat
Amount of energy released or absorbed during a change of state of a substance. In meteorology, the important changes are those in water, energy being released to the surrounding air in the changes from vapour to liquid to solid and absorbed from the air during changes in the opposite direction.

Latent heat of fusion
Also called 'latent heat of melting'. Energy released when a substance freezes or consumed when a substance melts. For water, the latent heat of fusion is about 330 J/g.

Latent heat of vaporization
Also called 'latent heat of condensation'. Energy consumed when a substance evaporates, released when a

substance condenses. For water, the latent heat vaporization is about 2 500 J/g.

GODAE
A pilot ptoject of GOOS to develop a global system of observations, modelling and assimilation to deliver regular comprehensive information on the current state of the oceans

Lightning
Luminous manifestation accompanying a sudden electrical discharge which takes place from or inside a cloud or, less often, from high structures on the ground or from mountains.

Lithosphere
The upper layer of the solid Earth, both continental and oceanic.

Long-wave radiation
Radiation with wavelengths greater than 4000 nm. See also *Infrared radiation* and *Terrestrial radiation.*

Mean sea level (MSL)
The arithmetic mean of hourly sea-level heights observed over some specified period.

Mean sea level (MSL) pressure
The estimated atmospheric pressure at MSL obtained by correcting an observed pressure for its height of measurement above MSL and temperature.

Metadata
Information that provides enhanced knowledge pertinent to the content of the data. In meteorology and climatology, for example, it is the set of information describing the meteorological records, including details about the observation site, the methods and instruments of observation and their management over the period of operation of the observing site.

Microburst
Downburst with a small lateral extent, about 1 to 4 km, and lasting only a short time.

Microclimate
The physical state of the atmosphere close to a very small area of the Earth's surface. Microclimates are strongly influenced by local surroundings.

Milankovitch cycles
Large variations in climate which are due to variations in the Sun's radiation reaching the Earth's surface caused by the varying eccentricity of the Earth's orbit; the varying tilt of the Earth's rotational axis; and the precession of the solstices and equinoxes.

Mitigation
A human intervention to reduce the predicted impact of a climate hazard, including potential climate change.

Monsoon
A persistent seasonal wind and typically with a pronounced change in direction from one season to the other.

Non-linearity
A process is called 'non-linear' when there is no simple proportional relation between cause and effect. The climate system contains many such non-linear processes, resulting in a system with potentially very complex behaviour.

Normals
Averages of climate elements calculated over a uniform and relatively long period covering at least three consecutive ten-year periods. The current standard Normal period is 1961–90, as defined by the World Meteorological Organization (WMO). The quantities are most often surface variables such as temperature, precipitation and wind. See also *Climate.*

North Atlantic Oscillation (NAO)
The North Atlantic Oscillation consists of opposing variations of barometric pressure near Iceland and near the Azores. The pressure difference between Iceland and the Azores fluctuates on timescales of days to decades, and can be reversed at times.

Numerical weather prediction
Forecast of the evolution of the atmospheric circulation by numerical solution of the hydrodynamical equations (subject to specified initial conditions).

Ocean conveyor belt
A term for the thermohaline driven global current. It results from two counteracting density forces: (1) high-latitude cooling and low-latitude heating, and (2) haline forcing (net high-latitude freshwater gain and low-latitude evaporation, which acts in the opposite direction. In the North Atlantic, the thermal forcing dominates, thus the conveyor's upper current flows from south to north.

Ozone
Ozone, the triatomic form of oxygen (O_3), is a gaseous atmospheric constituent. In the troposphere it is created both naturally and by photochemical reactions involving gases resulting from human activities ('photochemical smog'), and is considered a pollutant. It is corrosive and in high concentrations it can be harmful to a wide range of living organisms. In the stratosphere, ozone is created naturally by the interaction between solar ultraviolet radiation and molecular oxygen (O_2). Depletion of stratospheric ozone by increased concentrations of *halocarbons* results in an increased ground-level flux of harmful ultraviolet (UV)-B radiation.

Ozone hole
See *Ozone layer.*

Ozone layer
The region of the stratosphere that contains large concentrations of naturally occurring ozone. The layer extends from about 12 to 40 km, and ozone concentration reaches a maximum between about 20 and 25 km. Every year, during the Southern Hemisphere spring, a very strong depletion of the ozone layer takes place over the Antarctic region, accelerated by human-made *halocarbons* in combination with the specific meteorological conditions of that region. The region of depletion is called the ozone hole.

Paleoclimate
Climate of periods prior to the development of measuring instruments, including historic and geologic time, for which only proxy records are available.

Paleoclimatology
The study of paleoclimates and the causes of their variations.

Parameterization
Simplified representation, in climate models, of processes or physical effects taking place that have dimensions less than the model grid scale, rather than requiring such effects to be consequences of the dynamics of the system.

Particulates
Very small solid particles, especially those emitted during the combustion of fossil and biomass fuels. Particulates may consist of a wide variety of substances. Of greatest concern for health are particulates of less than or equal to 10 nm in diameter, usually designated as PM_{10}.

Patterns of climate variability
On seasonal and longer timescales, natural variability of the climate system occurs predominantly in preferred spatial patterns, through the dynamical non-linear characteristics of the atmospheric circulation and through interactions with land and ocean surfaces. See also *North Atlantic Oscillation* (NAO) and *El Niño/Southern Oscillation* (ENSO).

Permafrost
Perennially frozen ground that occurs wherever the temperature has been continuously below 0°C from a few to several thousands of years.

Phenology
The study of natural phenomena that recur periodically (e.g. blooming, migrating) and their relation to climate and seasonal changes.

Photochemical smog
A mix of photochemical oxidant air pollutants produced by the reaction of sunlight with primary air pollutants, especially hydrocarbons.

Photosynthesis
The process by which plants take CO_2 from the air (or bicarbonate in water) to build carbohydrates, releasing O_2 in the process.

Plankton
Minute floating marine organisms. Phytoplankton are the dominant plants in the sea; the basis of the entire marine food web. Zooplankton, classified as animals, are also an important step in the global food chain.

Pleistocene climate
The climate of the Pleistocene division of geological time (from about 2.5 million to 10 000 years ago).

Polar-orbiting satellite
A satellite whose orbital phase passes above the North and South Poles.

Polynyas
Areas of open water within the ice pack or sea ice.

Precipitation
An ensemble of liquid- or solid-phase aqueous particles, falling or suspended in the atmosphere. The forms of precipitation are: rain, drizzle, snow, snow grains, snow pellets, diamond dust, hail and ice pellets.

Probable maximum precipitation
The theoretically greatest depth of precipitation for a specific duration which is physically possible over a particular drainage area at a certain time of the year.

Proxy climate indicator
A local record that is interpreted, using physical and biophysical principles, to represent some combination of climate-related variations at a certain time. Climate-related data derived in this way are referred to as proxy data. Examples of proxies are: tree ring records, characteristics of corals, and various data derived from ice cores.

Quasi-biennial oscillation (QBO)
Alternation of easterly and westerly wind regimes in the stratosphere, within about 12° of the equator, with a period varying from about 24 to 30 months.

Radar
Radio method of determining at a single station the direction and distance of an object. The distance is determined by the time taken by signals emitted by the station to reach a distant object and return. The term radar is derived from 'RAdio Detection And Ranging'.

Radiative forcing
The change in the net vertical irradiance (expressed in watts per square metre: W/m^2) at the top of the atmosphere (usually taken as the tropopause) due to an internal change or a change in the external forcing of the climate system, such as, for example, a change in the concentration of carbon dioxide or the output of the Sun. See also *Energy balance*.

Radiatively active gases
See *Greenhouse gases*.

Radiometer
Instrument designed to measure radiation (in meteorology, particularly solar and terrestrial radiation).

Radiosonde
Instrument intended to be carried by a balloon up through the atmosphere, equipped with devices to measure one or several meteorological variables (pressure, temperature, humidity, etc.), and provided with a radio transmitter for sending this information to the observing station.

Remote sensing
The collection and recording of data from a distant point, e.g. radar and satellite-based observations of the atmosphere, as opposed to on-site (*in situ*) sensing.

Ridge
A region of the atmosphere with an elongated axis of high pressure.

Rossby waves (atmospheric)
Waves (sometimes referred to as long waves or planetary waves) in the atmospheric circulation, in one of the principal zones of the westerly winds, characterized by a great length and significant amplitude.

Salinization
The accumulation of salts in soils.

Sea level pressure
See *Mean sea level pressure*.

Sea level rise
A long-term increase in the mean level of the ocean.

Sea surface temperature (SST)
Temperature of the surface layer (generally the upper few metres) of the sea as measured by *in situ* instruments. Satellites also estimate SST from the emission of long-wave radiation. This is sometimes referred to as sea surface skin temperature.

Semi-arid regions
Ecosystems that have greater than 250 mm of precipitation per year, but are not highly productive; usually classified as rangelands.

Semi-permanent anticyclone
A region where anticyclones largely predominate during a major portion of the year and for which a region of high pressure appears on the mean monthly pressure charts (e.g. Azores, Bermuda, North Atlantic, Siberia, South Atlantic High).

Semi-permanent depression
A region where cyclones largely predominate during a major portion of the year and for which a region of low pressure appears on the mean monthly pressure charts (e.g. Aleutian, Icelandic Low).

Severe weather
Any atmospheric condition potentially destructive or hazardous for human beings and socio-economic systems. It is often associated with extreme convective weather (tropical cyclones, tornadoes, severe thunderstorms, squalls, etc.) and with storms of freezing precipitation or blizzard conditions.

Short-wave radiation
Radiation with wavelengths from 300 nm to about 4000 nm, including visible wavelengths. See also *Solar radiation*.

Smog
See *Photochemical smog*.

Snow gauge
Instrument to measure the water equivalent of a snowfall, either by weighing the snow or by melting it and measuring its volume.

Soil moisture
Water stored in or at the land surface and available for evaporation.

Solar ('11 year') cycle
A quasi-regular modulation of solar activity with varying amplitude and a period of between 9 and 13 years.

Solar radiation
Radiation emitted by the Sun. It is also referred to as short-wave radiation. Solar radiation has a distinctive range of wavelengths (spectrum) determined by the temperature of the Sun. See also *Infrared radiation*.

Solstice
The times each year that mark the northern or southern limits of the latitude of overhead sun. On the June solstice (approximately 21 June) Northern Hemisphere latitudes have their longest day of the year and Southern Hemisphere latitudes have their shortest day.

Southern Oscillation (SO)
A large-scale atmospheric fluctuation centred in the equatorial Pacific Ocean, exhibiting a pressure anomaly, alternately high over Australasia and high over the South Pacific. Its period is variable, averaging 2.3 years. The variation in pressure is accompanied by variations in wind strengths, ocean currents, sea surface temperatures, and precipitation in the surrounding areas.

Southern Oscillation Index (SOI)
A measure of the state of the Southern Oscillation. Commonly, this index is the sea-level pressure anomaly at Tahiti minus the sea-level pressure anomaly at Darwin, Australia, divided by the standard deviation of that quantity.

Spectrum (wavelength spectrum)
The distribution of the energy emitted by a radiating body as a function of the wavelength.

Standard deviation
The square root of the variance. See also *Variance*.

Stevenson screen
Standard, wooden louvered instrument shelter used mainly to house temperature and humidity instruments.

Storm surge
The temporary increase, at a particular locality, in the height of the sea or lake due to extreme meteorological conditions (low atmospheric pressure and/or strong winds).

Stratosphere
The highly stratified region of the atmosphere above the troposphere extending from about 10 km (ranging from 9 km in high latitudes to 16 km in the tropics on average) to about 50 km.

Sublimation
The transition of a substance from the solid phase directly to the vapour phase, or vice versa, without passing through the liquid phase.

Sunspots
Small dark areas on the Sun. The number of sunspots is higher during periods of high solar activity, and varies with the solar cycle.

Supercell
Persistent, single, intense updraught and downdraught coexisting in a thunderstorm in a quasi-steady state rather than in the more usual state of an assemblage of cloud cells each of which has a relatively short life.

Synoptic
Relating to or displaying characteristic weather conditions as they exist simultaneously over a broad area.

Synoptic scale
The scale of the high- and low-pressure systems of the lower atmosphere whose typical dimensions range approximately from 1000 to 2500 km.

Teleconnection
The weather and climate around the world in one place is generally strongly linked to that in other places through atmospheric linkages. These teleconnections link neighbouring regions mainly through large-scale, quasi-stationary Rossby waves. Some arise from natural preferred modes of the atmosphere associated with the mean climate state and the land–sea distribution. Several are linked to SST changes.

Temperature
A physical quantity that characterizes the mean random motion of molecules (or heat energy) in a physical body.

Temperature inversion
Vertical temperature distribution such that temperature increases with height.

Terrestrial radiation
Long-wave radiation emitted by the Earth, including its atmosphere. See also *Infrared radiation*.

Thermal expansion (of the ocean)
The increase in volume (and decrease in density) that results from warming of the ocean. The expansion from an overall warming of the ocean leads to an increase in sea level.

Thermistor
An electronic device whose electrical resistance changes with temperature, which can be adapted as a thermometer.

Thermocline
The gradient in the world's ocean, typically at a depth of a few hundreds of metres (in low and middle latitudes), where temperature decreases rapidly with depth and which marks the boundary between the surface mixed layer and the cold deep ocean waters.

Thermohaline circulation
The formation of cold bottomwater in restricted sinking regions of the North Atlantic Ocean and the Antarctic Ocean brought about by cooling of highly saline surface water of tropical origin (the North Atlantic Ocean) or release of sea salt under sea ice (the Antarctic).

Thermometer
An instrument which measures temperature.

Thunder
A sharp or rumbling sound which accompanies lightning. It is emitted by rapidly expanding gases along the channel of a lightning discharge.

Thunderstorm
Sudden electrical discharges manifested by a flash of light (lightning) and a sharp rumbling sound (thunder). Thunderstorms are associated with convective clouds (cumulonimbus) and are, more often, accompanied by precipitation, generally in the form of rainshowers or hail.

Tornado
A violently rotating storm of small diameter; the most violent weather phenomenon (sometimes known as a twister). It is produced in a severe thunderstorm and appears as a funnel cloud extending from the base of a cumulonimbus to the ground.

Trade winds
Persistent winds, mainly in the lower atmosphere, which blow over vast regions from a subtropical anticyclone to the equatorial regions. Their predominant directions are NE in the Northern Hemisphere and SE in the Southern Hemisphere.

Transmittance
The ratio of the light energy falling on a body to that transmitted through it.

Transpiration
The emission of water vapour from the surfaces of leaves or other plant parts.

Tropical cyclone
The generic term for a non-frontal synoptic scale cyclone originating over tropical or subtropical waters with organized convection and definite cyclonic surface wind circulation.
Note:
Tropical disturbance – light surface winds with indications of cyclonic circulation;
Tropical depression – wind speeds up to 33 kts;
Tropical storm – maximum wind speed of 34–47 kts;
Severe tropical storm – maximum wind speed of 48–63 kts;
Hurricane – maximum wind speed of 64 kts or more;
Typhoon – maximum wind speed of 64 kts or more;
Tropical cyclone (southwest Indian Ocean) – maximum wind speed of 64–90 kts;
Tropical cyclone (Bay of Bengal, Arabian Sea, southeast Indian Ocean, South Pacific) – maximum wind speed of 34 kts or more.

Tropopause
The boundary between the troposphere and the stratosphere.

Troposphere
The lowest part of the atmosphere, up to about 10 km in altitude in middle latitudes (ranging from 9 km in high latitudes to 16 km in the tropics on average) where clouds and weather phenomena occur. In the troposphere temperatures generally decrease with height.

Trough
A region of the atmosphere with an elongated axis of low atmospheric pressure.

Turbidity
Reduced transparency of the atmosphere to radiation (especially visible) caused by absorption and scattering by solid or liquid particles other than clouds.

Typhoon
Name given to a tropical cyclone with maximum sustained winds of 64 kts or more near the centre in the western North Pacific. See *Tropical cyclone*.

Ultraviolet (UV)-B radiation
Solar radiation within a wavelength range of 280–320 nm, the greater part of which is absorbed by stratospheric ozone. Enhanced UV-B radiation suppresses the immune system and can have other adverse effects on living organisms.

Upper-air (or upper atmosphere)
Term without precise definition; used mainly in synoptic meteorology to signify the region above the lower troposphere to the upper limit of routine balloon sounding in the lower stratosphere (about 30 km).

Upwelling (water)
Transport of deeper cold water to the surface, usually caused by horizontal movements of surface water.

Urban climate
Climate of cities, which differs from that of the surrounding areas because of the influence of the urban settlement.

Urban heat island
See *Heat island*.

Urbanization
The conversion of land from a natural state or managed natural state (such as agriculture) to cities.

Variability
See *Climate variability*.

Variance
A statistical measure of the variability within a sample, which is the mean of the sum of the squared deviations of a set of observations from the corresponding sample mean.

Vulnerability
The degree to which a system is susceptible to, or unable to cope with, adverse effects of climate, including extremes associated with climate variability and long-term climate change. Vulnerability is a function of the character, magnitude, and rate of climate variation to which a system is exposed, its sensitivity and its adaptive capacity.

Walker circulation
Direct zonal tropical circulation, thermally driven, in which air rises over the warm western Pacific Ocean and East Asia, and sinks over the cold eastern Pacific Ocean.

Water vapour
Water in the gaseous phase.

Wavelength
The distance between two successive peaks or troughs on a harmonic (sinusoidal) wave.

Wave number
In an electromagnetic wave, the reciprocal of the wavelength (i.e. the number of waves in a unit distance expressed as number/cm^1).

Weather
State of the atmosphere at a particular time, as defined by the various meteorological elements. See also *Extreme weather event* and *Severe weather*.

Weather modification
Intentional or unintentional change in the weather caused by human activities.

Westerlies (or westerly winds)
Zone, situated approximately between latitudes 35° and 65° in each hemisphere, where the air motion is mainly from west to east, especially in the high troposphere and low stratosphere. Near the Earth's surface, the zone is particularly well marked in the Southern Hemisphere.

Wind
Air motion (generally horizontal) relative to the Earth's surface.

World Weather Watch (WWW)
The worldwide, coordinated system of meteorological facilities and services provided by WMO Members for the purpose of ensuring that all Members obtain the meteorological information required both for operational work and for research. The essential elements of the WWW are the Global Observing System (GOS); the Global Data-processing System (GDPS); and the Global Telecommunication System (GTS).

Sources: IPCC TAR WG I and WG II Reports, the WMO Glossary, the AMS Glossary of Meteorology, the Meteorological Glossary of the UK Meteorological Office, and the Author.

Acronyms and abbreviations

AOC Western and Central Africa HYCOS

API Annual Parasitic Incidence Index

ARCHISS Archival Climate History Survey project

AWS Automatic Weather Station

BaPMON Background Air Pollution Monitoring Network

CET Central England Temperature

CFC Chlorofluorocarbon

CLICOM Climate computing (WMO)

CLIPS Climate Information and Prediction Services

DARE Data rescue

ECMWF European Centre for Medium Range Weather Forecasts

ENSO El Niño/Southern Oscillation

EUMETSAT European Organization for the Exploitation of Meteorological Satellites

EVI Enhanced Vegetation Index

FLOPS Floating Point Operations per Second

FY-1, -2 Polar-orbiting Satellite (China)

GAW Global Atmosphere Watch

GCM General Circulation Model

GCOS Global Climate Observing System

GOES Geostationary Operational Environmental Satellite (USA)

GMS Geostationary Meteorological Satellite (Japan)

GODAE Global Ocean Data Assimilation Experiment

GOMS Geostationary Operational Meteorological Satellite (Russian Federation)

GOOS Global Ocean Observing System

GPS Global Positioning System

GSN GCOS Surface Network

GUAN GCOS Upper Air Network

HDD Heating degree days

HYCOS Hydrological Cycle Observing System

IGAD Eastern Africa HYCOS

IKONOS High spatial resolution satellite (USA)

IMO International Meteorological Organization (predecessor of WMO)

IPCC Intergovernmental Panel on Climate Change

IPO Interdecadal Pacific Oscillation

ITCZ Intertropical convergence zone

JMA Japan Meteorological Agency (Japan)

MED Mediterranean HYCOS

METEOR Polar-orbiting Satellite (Russian Federation)

METEOSAT EUMETSAT series of meteorological geostationary satellites

MJO Madden-Julian Oscillation

MODIS Moderate Resoluting Imaging Spectroradiometer

MSG METEOSAT Second Generation Satellite

MSL Mean sea level

NAO North Atlantic Oscillation

NASA National Aeronautics and Space Administration (USA)

NASDA National Space Development Agency (Japan)

NCAR National Center for Atmospheric Research (USA)

NCEP National Centers for Environmental Prediction (NOAA, USA)

NH Northern Hemisphere

NOAA National Oceanic and Atmospheric Administration (USA)

ODAS Ocean Data Acquisition System

QBO Quasi-Biennial Oscillation

RCM Regional Climate Model

SADC Southern Africa HYCOS

SAR Second Assessment Report (IPCC)

SOI Southern Oscillation Index

SST Sea Surface Temperature

START System for Analysis, Research and Training

TAO Tropical Atmosphere-Ocean

TAR Third Assessment Report (IPCC)

TIROS Television Infrared Observation Satellite

TOGA Tropical Ocean Global Atmosphere project (WCRP)

TOMS Total Ozone Mapping Spectrometer

TOPEX/ POSEIDON US/French Ocean Topography Satellite Altimeter Experiment

TRMM Tropical Rainfall Measuring Mission

UNEP United Nations Environment Programme

UNESCO United Nations Educational, Scientific and Cultural Organization

UNFCCC United Nations Framework Convention on Climate Change

USSR Union of Soviet Socialist Republics

UV Ultraviolet radiation

VORTEX Verification of the Origins of Rotation in Tornadoes Experiment

WCP World Climate Programme

WCRP World Climate Research Programme

WHYCOS World Hydrological Cycle Observing System

WMO World Meteorological Organization

WWW World Weather Watch (WMO)

January average MSLP (hPa) for 1961–1990

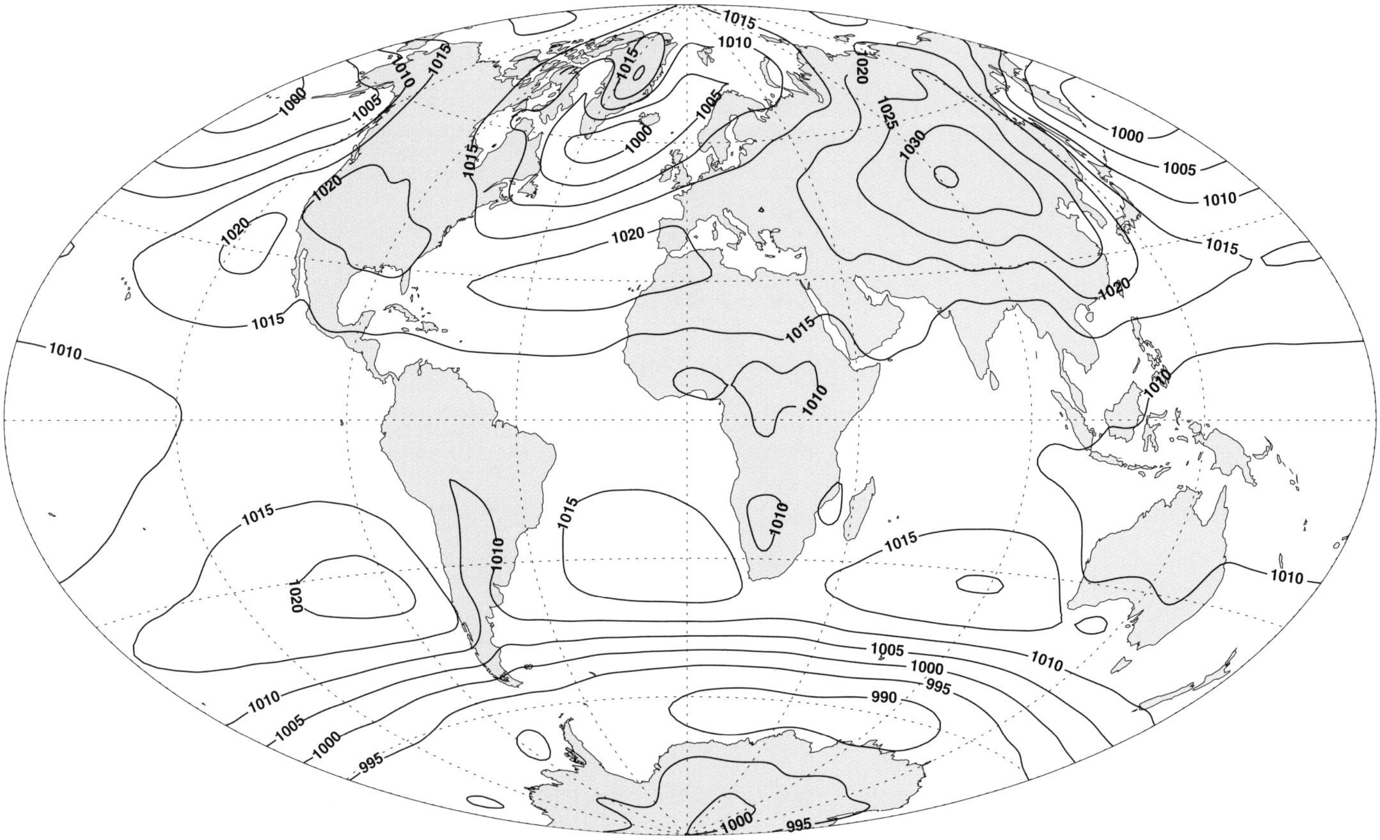

January average 500 hpa heights (gpm) for 1961–1990

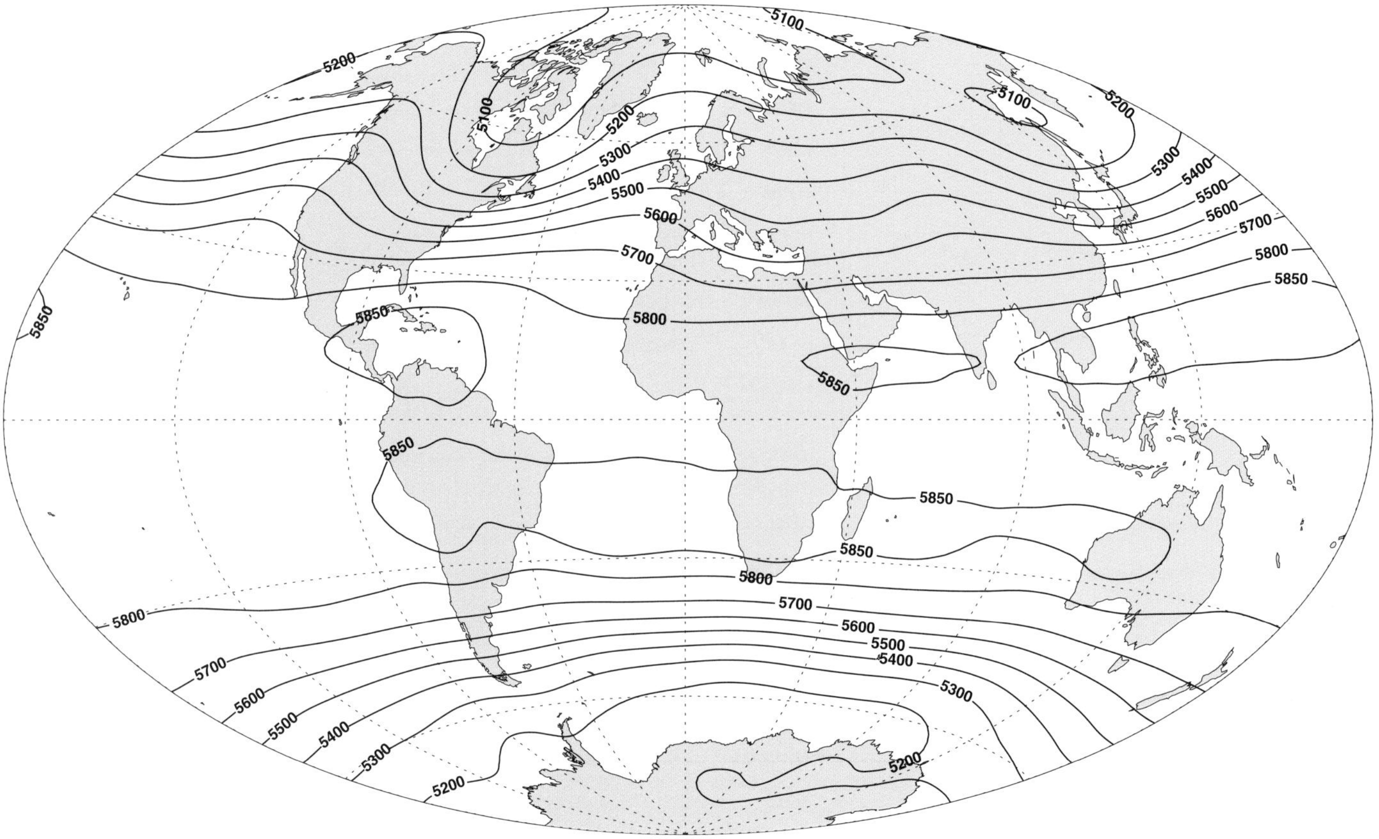

July average MSLP (hPa) for 1961–1990

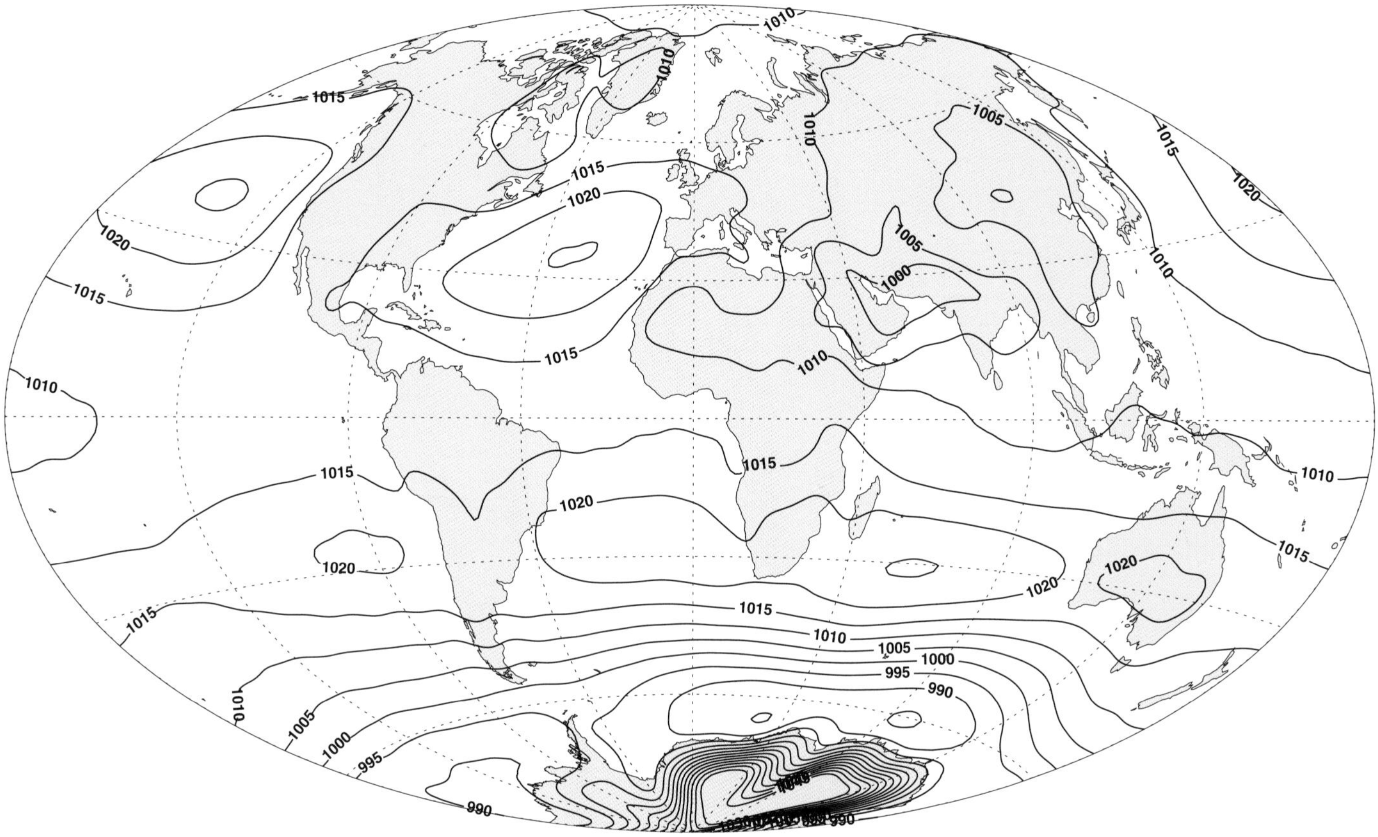

January surface temperature (degrees Celsius) for 1961–1990

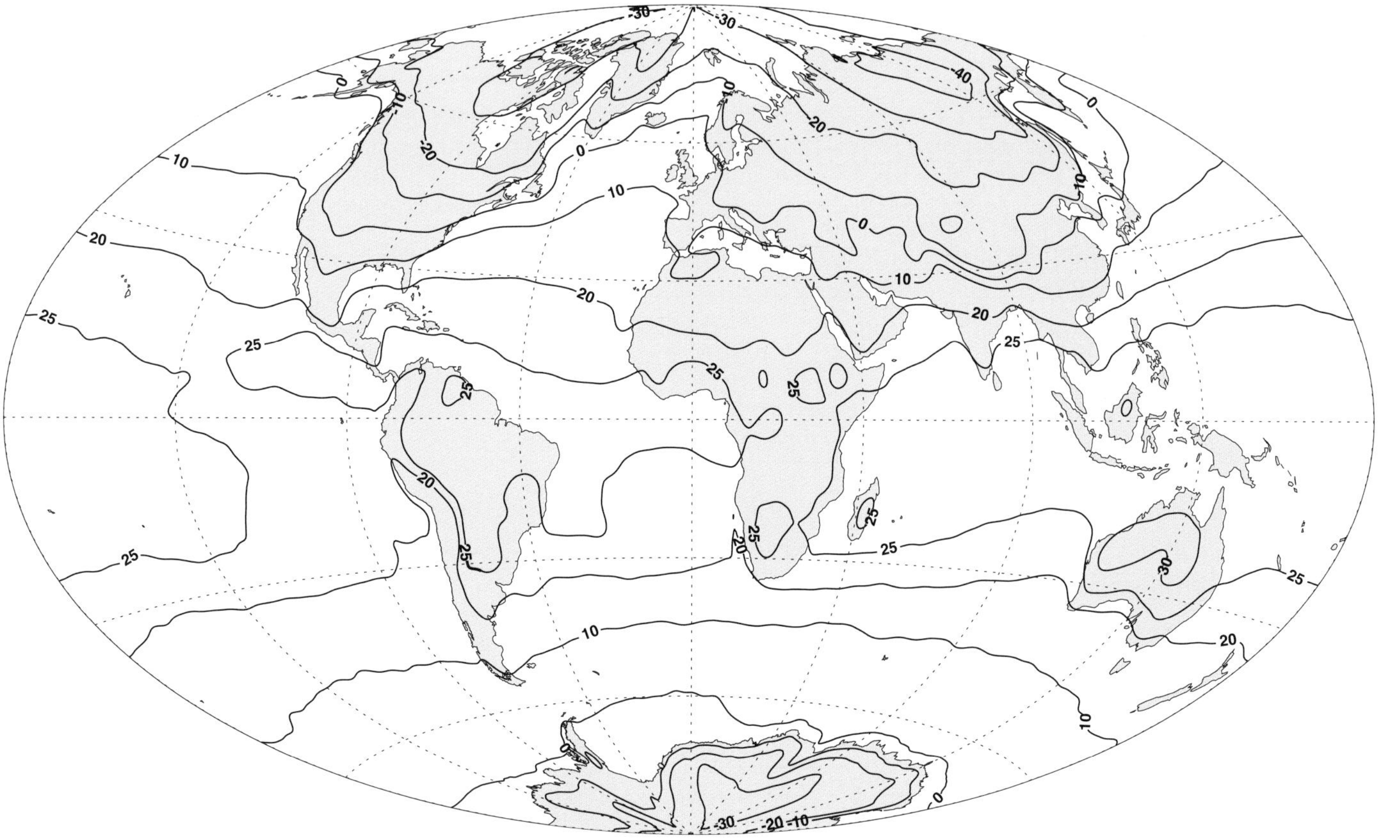

Units

SI (*Système Internationale*) units:

Physical quantity	Name of unit	Symbol
length	metre	m
mass	kilogram	kg
time	second	s
thermodynamic temperature	kelvin	K
amount of substance	mole	mol

Special names and symbols for certain SI-derived units:

Physical quantity	Name of SI unit	Symbol for SI unit	Definition of unit
force	newton	N	$kg\ m\ s^{-2}$
pressure	pascal	Pa	$kg\ m^{-1}\ s^{-2}$ (= $N\ m^{-2}$)
energy	joule	J	$kg\ m^2\ s^{-2}$
power	watt	W	$kg\ m^2\ s^{-3}$ (= $J\ s^{-1}$)
frequency	hertz	Hz	s^{-1} (cycles per second)

Decimal fractions and multiples of SI units having special names:

Physical quantity	Name of unit	Symbol for unit	Definition of unit
length	Ångstrom	Å	10^{-10} m = 10^{-8} cm
length	micron	mm	10^{-6} m
area	hectare	ha	$10^4\ m^2$
force	dyne	dyn	10^{-5} N
pressure	bar	bar	$10^5\ N\ m^{-2}$ = 105 Pa
pressure	millibar	mb	$10^2\ N\ m^{-2}$ = 1 hPa
mass	tonne	t	10^3 kg
mass	gram	g	10^{-3} kg
column density	Dobson units	DU	2.687×10^{16} molecules cm^{-2}
streamfunction	Sverdrup	Sv	106 $m^3\ s^{-1}$

Atmospheric and climate services use a number of convenient derived units including:

Practical combination units:

GtC	gigatonnes of carbon (1 GtC = 3.7 Gt carbon dioxide)
PgC	petagrams of carbon (1 PgC = 1 GtC)
MtN	megatonnes of nitrogen
TgC	teragrams of carbon (1 TgC = 1 MtC)
Tg(CH_4)	teragrams of methane
TgN	teragrams of nitrogen
TgS	teragrams of sulphur

Non-SI units:

°C	degree Celsius (0 °C = 273 K approximately) [Temperature differences also given in °C (= K) rather than the more correct form of "Celsius degrees"]
bp	before present
h	hour
yr	year
ky	thousands of years
cal	4 184 joules
t	optical depth
TWh	terawatt hours
ppm(v)	parts per million (10^6) (volume)
ppbv	parts per billion (10^9) by volume
pptv	parts per trillion (10^{12}) by volume

Fractional prefixes used with SI and related units

Fraction	Prefix	Symbol	Multiple	Prefix	Symbol
10^{-1}	deci	d	10	deca	da
10^{-2}	centi	c	10^2	hecto	h
10^{-3}	milli	m	10^3	kilo	k
10^{-6}	micro	μm	10^6	mega	M
10^{-9}	nano	n	10^9	giga	G
10^{-12}	pico	p	10^{12}	tera	T
10^{-15}	femto	f	10^{15}	peta	P

Chemical symbols

C	carbon (there are three isotopes: ^{12}C, ^{13}C, ^{14}C)
CFC	chlorofluorocarbon
CFC-11	$CFCl_3$ (trichlorofluoromethane)
CFC-12	CF_2Cl_2 (dichlorodifluoromethane)
CH_4	methane
CO	carbon monoxide
CO_2	carbon dioxide
HCFC	hydrochlorofluorocarbon
H_2O	water vapour
H_2SO_4	sulphuric acid
NO	nitric oxide
NO_x	nitrogen oxides (the sum of NO and NO_2)
N_2O	nitrous oxide
O_3	ozone
SO_2	sulphur dioxide

Conversion factors

1 kilometre	≈ 0.62 miles	≈ 3280.84 feet
1 km/h	≈ 0.28 m/s	
	= 0.62 m/h	
	≈ 0.54 knots (nautical miles per hour)	
Temperature (°F)	= $\frac{9}{5}$ [Temperature (ΥC)] + 32	
Temperature (°C)	= $\frac{9}{5}$ [Temperature (ΥF – 32)]	
Degrees Kelvin (°K)	= [Temperature (ΥC)] + 273.16	
Temperature (°C)	= [Temperature (ΥK)] – 273.16	

Acknowledgements

Climate: Into the 21st Century was developed with the advice and guidance of scientists from around the world, representing numerous climate-related disciplines. WMO would like to extend its appreciation to all those involved in preparing this book although they are too numerous to be acknowledged individually. WMO is particularly grateful for the contributions and commitment of the lead author, Dr William Burroughs; the members of the project task team: Brad Garanganga (Drought Monitoring Centre, Zimbabwe); Phil Jones (Climatic Research Unit, University of East Anglia (CRU, UEA), UK); Dave Phillips (Meteorological Service of Canada (MSC), Canada); Chet Ropelewski (International Research Institute for Climate Prediction (IRI), USA); and the team lead, Mary Voice (Bureau of Meteorology (BOM), Australia); the focal points for major sections of the book: Neville Nicholls (Bureau of Meteorology Research Centre (BMRC), Australia); Ann Henderson-Sellers (Australian Nuclear Science and Technology Organisation (ANSTO), Australia); Michael Glantz (Environmental and Societal Impacts Group, National Center for Atmospheric Research (ESIG NCAR), USA); Reid Basher (International Research Institute for Climate Prediction (IRI), USA); Hiroki Kondo (Meteorological Research Institute (MRI), Japan Meteorological Agency (JMA), Japan); Stephen Schneider (Stanford University, USA); and Richard Moss (Office of the US Global Change Research Program, USA); and those who reviewed the various drafts of the manuscript: M. Crowe (National Oceanic and Atmospheric Administration National Climatic Data Center (NOAA NCDC), USA); C. Folland (the Met Office (UKMO), United Kingdom); H. Grassl (Max-Planck-Institut, Germany); G. Gruza (Institute for Global Climate and Ecology (IGCE), Russian Federation); G. McBean (University of Western Ontario, Canada); M. Nicholls; and C.C. Wallén. The book's artwork and design were created by bounford.com.

This book would not have been possible without the financial support of sponsoring agencies. WMO sincerely thanks its Member countries Australia, Canada, France, Germany, Japan, Norway, Sweden, Switzerland, the UK and the USA for their generous support.

* IPCC Third Assessment Report images are reproduced or adapted from: *IPCC, Climate Change 2001: The Scientific Basis*; J.T. Houghton *et al.* ed., Cambridge University Press, or *Climate Change 2001: Synthesis Report*; Robert T. Watson *et al.* ed., Cambridge University Press

Credits: graphics and artwork

Abbreviations: B=bottom; C=center; L=left; R=right; T=top

Section 1

12 WMO
14 Adapted from © IPCC 2001 with permission
15 W.J. Burroughs (*set of seven*)
16 (*pair*) P.D. Jones, CRU, UEA
18 (*pair*) Météo-France, Réunion
19 G. Trewartha, U. Wisconsin, Madison
20 © IPCC 2001. Reproduced with permission (*set of three*)
21 (*pair*) P.J. Wynne and *Scientific American*
24 WRI
25 UNDP/Disaster Management Unit Viet Nam
28 BOM
31 M.J. McPhaden, NOAA PMEL
32 WRI
36 Adapted from © IPCC 2001 with permission (*set of four*)
37T Modified from T. Crowley
37B K. Alverson/Past Global Changes (PAGES)
38T Reprinted by permission from *Nature* 403: 'Temperature trends over the past five centuries reconstructed from borehole temperatures', 756-758, © 2000 Macmillan Magazines Ltd.
38B P.D. Jones, CRU, UEA
39 P.D. Jones, CRU, UEA
40T P.D. Jones, CRU, UEA
40B P.D. Jones, CRU, UEA (*set of three*)
41 P.D. Jones, CRU, UEA

Section 2

44 Adapted from © IPCC 2001 with permission
45T WMO
45C WMO
46T WMO
46B NASA JPL/Caltech
47B WMO
48 BOM
50 P.D. Jones, CRU, UEA
51 T. Sparks, CEH
52 © IPCC 2001. Reproduced with permission
53 J. Lazier, Department of Fisheries and Oceans (DFO), Canada)
55 BOM
56 A. Couper and RMS
57 NOAA Geophysical Fluid Dynamics Laboratory (GFDL)
59L Météo-France
59R MétéoSuisse
60 D.A. Robinson, Rutgers U. (*set of three*)
61T JMA
61B W.J. Burroughs, data from Met Office
62 E. Cook, LDEO (*set of three*)
64 WMO
65 T.T. Fujita and S. Fujita
66 © IPCC 2001. Reproduced with permission
69 © IPCC 2001. Reproduced with permission
70L (*pair*) WMO
70R (*pair*) WMO
72 M.J. McPhaden, NOAA PMEL and NCAR/ESIG (*set of three*)
73T NASA GSFC
73C P.D. Jones, CRU, UEA
74 P. Nobre, J. Shukla © 1996 AMS
76 (*pair*) NOAA NCEP
77T NOAA NCEP
77B J.S. Watson, NOAA, National Weather Service (NWS) and S. Colucci, Cornell U.
78T P.D. Jones, CRU, UEA
78B (*pair*) M. Visbeck, LDEO
79 W.B. White, reprinted by permission of *Nature* 380: 699-710, © 1996, Macmillan Magazines Ltd.
81T Météo-France
81C A. Ben Mohamed, Institut des Radio-Isotopes, Université Abdou Moumouni
84 R. Spencer, Global Hydrology and Climate Center, NASA Marshall Space Flight Center (MSFC)
85 © IPCC 2001. Reproduced with permission
86 D.A. Robinson, Rutgers University
87 the Met Office's Hadley Centre for Climate Prediction and Research
89 M.P. McCormick, Hampton U.; NASA Stratospheric Aerosol and Gas Experiment II (SAGE II) (*set of four*)
91T J.L. Lean, Naval Research Laboratory (NRL), USA and C. Fröhlich, Physikalisch-Meteorologisches Observatorium Davos/World Radiation Center (PMOD/WRC), Switzerland)
91B Solar Influences Data Analysis Centre (SIDC), Observatoire Royal de Belgique, Belgium
92 E. Jáuregui, National U., Mexico
95 P. Tans, NOAA CMDL Carbon Cycle Greenhouse Gases (CCGG)
100 NOAA CMDL
101T Institute for Atmospheric Science, Swiss Federal Institute of Technology
101B WMO Global Atmosphere Watch (GAW)
102 © IPCC 2001. Reproduced with permission

Section 3

107 Munich Re
109 W.J. Burroughs, data from IFRC (*set of three*)
113C Conservation International
113B E. Beaubien, Devonian Botanic Garden, U. Alberta; from Plantwatch Program: www.devonian.ualberta.ca/pwatch
116 Institute for Global Climate and Ecology, Moscow
118L W.J. Burroughs
118R NOAA NCEP
120 Royal Society
121 U.S. De, IMD
123 C. Landsea, NOAA AOML
124 K.C. Sinha Ray, IMD
125 U.S. De, IMD (*set of three*)
126 M. Hulme, Tyndall Centre, UEA
131 P. Zhai, National Climate Centre, China Meteorological Administration (CMA)
135 E. Cook, LDEO
136 (*pair*) NOAA CPC NCEP
138 (*pair*) NOAA CPC NCEP
139 BOM
140 BOM
141C WMO
141B WMO, data from P.D. Jones, CRU, UEA

143 H. Diaz, NOAA Climate Diagnostics Center (CDC) and G. Poveda, Cooperative Institute for Reaserch In Environmental Sciences (CIRES), U. Colorado
144T L.S. Kalkstein, U. Delaware
144B G. Jendritzky, DWD
145 U.S. De, IMD
146T www.metoffice.com
146B South Coast Air Quality Management District, CA
147 Population Division of the Department of Economic and Social Affairs of the United Nations Secretariat (2001). *World Urbanization Prospects: The 1999 Revision, Key Findings.*
152 WRI
153 State Institute of Oceanography, Moscow

Section 4
156 WMO
157 WMO
160 C. Folland, © British Crown Copyright, Met Office
161 *(pair)* WMO
164 NASA GSFC
167 WMO
169 U. Berne, HADES
170 *(pair)* JMA
172 M.J. McPhaden, NOAA PMEL
173L European Centre for Medium-Range Weather Forecasts (ECMWF) and Climate Variability and Predictability (CLIVAR)
173R International Research Institute for Climate Prediction (IRI), USA
174 B. Naujokat, U. Berlin
175L W.J. Burroughs
175R the Met Office's Hadley Centre for Climate Prediction and Research
186 *National Atlas of Sweden, Climate Lakes and Rivers*, Eds. Leif Wastenson, Birgitta Raab, Haldo Vedin
188 WRI
192 NOAA NCEP Environmental Modeling Centre *(set of three)*
193C *(pair)* BOM
193B *(pair)* W.J. Burroughs
195T Max-Planck-Institut für Meteorologie *(set of four)*
195B Max-Planck-Institut für Meteorologie
196 © IPCC 2001. Reproduced with permission *(set of three)*
197 © IPCC 2001. Reproduced with permission
198 National Geographic, W.H. Bond
199 © IPCC 2001. Reproduced with permission *(set of three)*

Section 5
202L © IPCC 2001. Reproduced with permission
202R *(pair)* P. Rekacewicz, GRID-Arendal, Norway, 2000
203 © IPCC 2001. Reproduced with permission
204 *(pair)* © IPCC 2001. Reproduced with permission
205 *(pair)* © IPCC 2001. Reproduced with permission
206 P. Rekacewicz, GRID-Arendal, Norway, 2000 *(all)*
207 P. Rekacewicz, GRID-Arendal, Norway, 2000
208 © IPCC 2001. Reproduced with permission
215 M. Parry, Jackson Environment Institute (JEI), UEA, UK *(set of three)*
217T United Nations Environment Programme (UNEP) GEO 2000
217B © IPCC 2001. Reproduced with permission
219 Modified from Knight, Ellwood *et al.*, with permission

Appendices
228 *(pair)* Jones and Ropelewski
229 *(pair)* Jones and Ropelewski

Credits: photographs and satellite pictures

Section 1
13T M. Golay
13B *(pair)* National Snow and Ice Data Center (NSIDC), USA
15TL NOAA Photo Library
15TR WMO/B. Genier
16 NOAA Photo Library
17T Canadian Coast Guard
17B NOAA Photo Library
18T Bryan and Cherry Alexander Photography
18C WMO/J. Stickings
19 DWD
21 MSC
22T M. Golay
22B J. Roberts, CEH
23T J. Richardson
23B IFRC
25T IFRC
25B IFRC
26T Photo Klopfenstein AG
26B KNMI
27T E. Gorre-Dale
27B D. Malin, Anglo-Australian Observatory (AAO)
28T National Geographic Society Image Collection/G. Ludwig
29T NCAR/University Corporation for Atmospheric Research (UCAR), USA
29B © the Nobel Foundation
30T D. Hardy, University of Massachusetts
30B NOAA/NSSL
31 © British Crown Copyright, Met Office
32 AFP/J. Uzon
33T AFP/P. Kittiwongsakul
33B AFP/D. Chowdhury
34T Munich Re/W. Kron
34B IFRC/L. de Toledo
35 IFRC
36T NOAA slide set
36B Deutsches Museum
39 U. Schotterer

Section 2
47T R. Sterner, JHUAPL
49T WMO/E.H. Al-Majed
49C Det Norske Meteorologiske Institutt (DNMI), Norway
49B JMA
51T T. Holden via British Trust for Ornithology (BTO) (UK)
51B NOAA slide sets *(set of four)*
52 NOAA Photo Library
53 R. Munns, courtesy of C. Wunsch
54 Mary Evans Picture Library
57 M.J. McPhaden, NOAA PMEL
58 *Sevenoaks Chronicle*
59 NOAA Photo Library
61 M. Golay
62 NOAA Photo Library
63T BOM
63B NOAA slide sets
64 NOAA NCDC
66 NOAA AOML
67T Texas State Library
67B *(pair)* NOAA Photo Library
68T M.L. Altinger de Schwarzkopf
68B U. Chicago
69 D. Zaras, NOAA NSSL
71T IFRC
71B AFP/Saiful Islam – STR
73 G. Argent
75T Météo-France
75B JMA
77 JMA
79 BOM AAD/G. Dixon
80T European Organization for the Exploitation of Meteorological Satellites (EUMETSAT)
80B United Nations/Department of Public Information (UN/DPI)
81 SeaWiFS project, NASA GSFC and Orbimage
82T NASA GSFC
82B V. Gudkov
83 NASA TERRA MODIS
84 K.L Thalmann
85C M Schäffer (courtesy Whyte Archives of the Canadian Rockies)
85B B. Luckman, U. Western Ontario
86 Canadian Space Agency/Agence spatiale canadienne
87 *(pair)* NSIDC
88C USGS David A. Johnston Cascades Volcano Observatory (CVO) (USA)/J.N. Marso
88B P.D. Jones, CRU, UEA
89 J. Zillman, BOM
90 The Solar and Heliospheric Observatory (SOHO), (ESA and NASA)
92 E. Jáuregui, National U., Mexico
93T PhotoDisc/M. Downey
93C PhotoDisc/D. Buffington
93B F. Landsberg
94T Still Pictures/B. Yoshida
94B NOAA Mauna Loa Observatory (MLO)
95 *(pair)* SeaWiFS project, NASA GSFC and Orbimage
96T Still Pictures/M. Edwards
96B Still Pictures/K. Lohua
97T NASA GSFC Ozone Processing Team
97B NASA GSFC Ozone Processing Team
98 NASA Earth Observatory
99TL FAO/I. Balderi
99TR FAO/P. Cenini
99B Still Pictures/M. Edwards
100 NASA GSFC Ozone Processing Team
101 Coherent
103T © National Gallery, London
103B AFP/F. Pages – STR

Section 3
106 Still Pictures/N. Dickinson
107 AFP/C. Bouroncle – STF
108TL AFP/C. Allegri
108TR National Geographic Society Image Collection/S. Raymer
109 AFP/M. Alvarez – STR
110 American Society of Civil Engineers (ASCE), reproduced with permission from 'The Yellow River Problem', O.J. Todd and S. Eliassen in *Transactions ASCE*, Vol. 105 (1940)
111T USGS/G. Pflafker
111B J. Dubief
112C Centre for Environment, Fisheries and Aquaculture Science (CEFAS), UK
112B Still Pictures/A. Newman
113T BOM AAD/D. Monselesan
113B H. Flygare: Courtesy of E. Beaubien, Devonian Botanic Garden, U. Alberta
114T Imperial War Museum, London
114B © Musée de l'Armée, Paris

115 United States Coast Guard (USCG) Historian's Office
116 AFP – STF
117 National Geographic Society Image Collection/D. Conger
118 KNMI
119 Munich Re Archives
120T KNMI
120B het Keringhuis
121 D. Pitchford
122 BOM
123 R. Sterner, JHUAPL
125T Still Pictures/J. Schytte
125B F.W. King; © 1996 Florida Museum of Natural History
126 AFP/J. Robine – STF
127T NASA GSFC
127C NASA GSFC Scientific Visualization Studio (SVS)
127B FAO/F. McDougall
128 WMO/B. Genier
129T NASA Earth Observatory
129B FAO/P. Cenini
130T DRIK Picture Library Ltd./N. Mahmood
130 DRIK Picture Library Ltd./S. Alam
131T Still Pictures/Y. Feng/UNDP
131B *(pair)* T. Venzin
132L CatPress Agenzia di Stampa
132R RMS
133T AFP/K. Swiderski – STF
133B W. Balfer
134 *(pair)* NOAA NCDC
135 Federal Emergency Management Agency (FEMA) USA
137T NASA JPL Caltech
137B AFP/Ganis – STR
139T NASA JPL Caltech
139C AFP/J. Ngwenya – STF
140 AFP/La Industria de Trujillo
141 R.M. Bourke, Australian National U.
142T AFP/G. Johnson – STR
142B *(pair)* G.M. Wellington, U. Houston
143 NOAA Photo Library
144 AFP/A. Datta – STR
145 BOM
146 NREL/D. Parsons
147 Still Pictures/A. Doto
148C United States Department of Agrigulture (USDA)
148B AFP/R. Laberge – STR
149T NOAA Photo Library
149B International Coffee Organization
150 Centro Internacional de Mejoramiento de Maíz y Trigo (CIMMYT), Mexico
151T IFRC
151B BOM
153T FAO/M. Marzot
153B *(pair)* Landsat images, USGS

Section 4

156 MSC
157 *(pair)* NOAA NCDC
158C WMO
158B MSC
159T Météo-France
159B JMA
160 Finnish Meteorological Institute (FMI), Finland
162T BOM
162B © D.G. Barton
163T MSC
163C NOAA Storm Prediction Center (SPC)
163B NOAA Photo Library
164 NASA TERRA MODIS
165 NASA JPL Caltech
166T WMO
166B WMO
167 South African Airways
168 WMO/H. Kontongomde
169 Diário de Pernambuco/Jaquiline Maia, courtesy of Instituto Nacional de Meteorologia (INMET), Brazil
171 FAO/G. Tortoli
175 P. Mitchell
176T WMO
176B BOM
177C Météo-France/D. Stuber
177B WMO/M.V.K. Sivakumar
178C Still Pictures/L. Olesen
178B J.G. Goldammer, The Global Fire Monitoring Center (GFMC), Max Planck Institute for Chemistry
179L Consultative Group on International Agricultural Research (CGIAR)
179R MSC
180L BOM AAD/ J. Hosel
180R MSC
181TL John Laing plc
181TR M. Ryan, *The Gulf News*, Port aux Basques, Nfld, Canada
181B Hong Kong Airport Authority
182T FAO/R. Faidutti
182B AFP/A. Datta – STF
183 AFP/Pictor
184L M. Burgess, GSC, NRCAN
184R US Army Corps of Engineers
185T NREL/W. Gretz
185B Brutsch et Brutsch
186 NOAA Photo Library
187T Still Pictures/H. Schwarzbach
187C FAO/A. Wolstad
187B Shell International
188 G. Brooks, GSC, NRCAN, reproduced with the permission of the Minister of Public Works and Government Services Canada, 2001
189T R.S. Schemenauer
189B WMO
190L National Geographic Society Image Collection/T. Tomaszewski
190R Still Pictures/D. Blell
191T B. Jones, Winds of Hope Foundation © Breitling
191C National Geographic Society Image Collection/D. Bremner
191B V. Lauwers
192T © Fr. Schuiten 2002
192B U. Pennsylvania
194 A. Pearce, Commonwealth Scientific and Industrial Research Organization (CSIRO), Australia
199 © IPCC 2001. Reproduced with permission. Photo credit: Science Photo Library

Section 5

203 NOAA Photo Library
207 IFRC
209T NOAA Photo Library
209B NOAA Photo Library
210T AFP/T. Blackwood – STF
210B World Health Organization (WHO)/Special Programme for Research and Training in Tropical Diseases (TDR)
211T Still Pictures/D. Fleetham
211B Still Pictures/D. Halleux
212 European Space Agency (ESA)
213T NREL/C. Babcock
213B BOM
214 Environment Agency, UK
215 AFP/M. Munir – STF
216T WMO
216B Still Pictures/K. Andrews
217 Still Pictures/J-L. Dugast
218T IFRC
218C Still Pictures/R. Giling
219T UNDP/R. Chalasani
219B V.L. Sharpton, Lunar and Planetary Institute (LPI), USA, from the slideset "Terrestrial Impact Craters, 2nd edition"
220 NASA
221 BOM

Acronyms

AAD Australian Antarctic Division (Australia)
AFP Agence France-Presse
AMS American Meteorological Society
AOML Atlantic Oceanographic and Meteorological Laboratory (NOAA)
BOM Bureau of Meteorology (Australia) [All photos © Commonwealth of Australia]
CEH Centre for Ecology and Hydrology (UK)
CMDL Climate Monitoring and Diagnostics Laboratory (NOAA)
CPC Climate Prediction Center (NOAA NCEP)
CRU Climatic Research Unit (UEA, UK)
DWD Deutscher Wetterdienst (Germany)
ESA European Space Agency
ESIG Environmental and Societal Impacts Group (NCAR)
FAO Food and Agriculture Organization of the United Nations
GRID Global Resource Information Database centre (Arendal, Norway)
GSC Geological Survey of Canada
GSFC Goddard Space Flight Center (NASA)
IFRC International Federation of Red Cross and Red Crescent Societies
IMD India Meteorological Department
IPCC Intergovernmental Panel on Climate Change
JHUAPL Johns Hopkins University Applied Physics Laboratory (USA)
JMA Japan Meteorological Agency
JPL Jet Propulsion Laboratory, California Institute of Technology (NASA)
KNMI Koninklijk Nederlands Meteorologisch Instituut (Netherlands)
LDEO Lamont-Doherty Earth Observatory (USA)
MSC Meteorological Service of Canada
NASA National Aeronautics and Space Administration (USA)
NCAR National Center for Atmospheric Research (USA)
NCDC National Climatic Data Center (NOAA)
NCEP National Centers for Environmental Prediction (NOAA)
NOAA National Oceanic and Atmospheric Administration (USA)
NRCAN Natural Resources Canada
NREL National Renewable Energy Laboratory (USA)
NSSL National Severe Storm Laboratory (NOAA)
PMEL Pacific Marine Environmental Laboratory (NOAA)
RMS Royal Meteorological Society (UK)
UEA University of East Anglia (UK)
UNDP United Nations Development Programme
USGS United States Geological Survey
WMO World Meteorological Organization
WRI World Resources Institute (USA)

Index

Figures and tables in *Italic*, material in boxes shown by B after page number